SUDOKU
PUZZLE BOOK

VOL 3

This Book
Belongs To:

_____

_____

_____

FROM VERY EASY TO HARD

FOR ALL LEVEL

RULES FOR SUDOKU GRID SIZE 9X9:

There are three rules to follow

1- The numbers 1 through 9 only appear once in each **3X3 SQUARE**

		6	1	3	7	4	2	8
		6				9		7
4		1	8		2		6	5
3		5	2	7	6	1		4
			3	8	9	6	5	2
6		2	4	1	5	7	8	3
8		9		4	3	2		1
7			9					6
5	2	3	7	6	1	8	4	9

2 - The numbers 1 through 9 only appear once in each **ROW**

		6	1	3	7	4	2	8
		6		4	9			7
4		1	8		2		6	5
3	8	5	2	7	6	1	9	4
			3					2
6		2		1	5	7		3
8		9		4				1
			9					6
5	2	3	7	6	1	8	4	9

3 - The numbers 1 through 9 only appear once in each **COLUMN**

		6	1	3	7	4	2	8
		6			4	9		7
4		1	8		2		6	5
3	8	5	2	7	6	1	9	4
			3	8	9	6	5	2
6		2	4	1	5	7	8	3
8		9		4	3	2	7	1
7			9		8	5	3	6
5	2	3	7	6	1	8	4	9

Puzzle 1 - Very Easy

		9	8		4	7	6	5
5			7	3	6	1		2
2		7	1	5	9	4	3	
4	9	2			5	6		
8	5	3	6	7	1	9	2	4
1	7		4	9	2	5	8	
6		8	5	1		2	7	
7	2					3	5	6
9	3	5	2	6	7			1

Puzzle 2 - Very Easy

7			4					8
	8	6	7			4	5	9
4	3		8		9	2	7	6
	7			4		6	8	1
	1	4	6	9	8	5	2	7
	6	8	1	5		3	9	4
9	2	1		8	6	7	4	3
6	5	3	9	7			1	2
8	4	7	2	3				5

Puzzle 3 - Very Easy

1	3	9	4	2	6	8	7	
8	4	7	1	5	9	6	3	2
2	6	5		3	7	9	4	1
4	8	2	6		3	1	5	7
7		6	2	8	1	3	9	
9		3	5	7	4	2		
	9	4	7		2		6	
	7		9			4	2	
6		8	3	4		7	1	9

Puzzle 4 - Very Easy

	1	3	7	6		4		8
	5	9	4	3	8	6	1	2
4	6		5	2	1	7		3
	3	7	6	4			2	
6	2	4	8		3	1	7	5
8				1	7	3		6
5	8	1	3	7	2		6	4
9	4	2	1	8	6	5		7
3	7	6	9		4	2		1

Puzzle 5 - Very Easy

1			2	7	3	9	8	5
	3	9		2	4	1	6	7
7	4	5	1	8	6		3	9
	7	6	3		1		8	2
			4	9	7		1	6
9	1	3	2	6		7	4	5
	8		9		3	5	2	
3		7	6		2	4	9	8
		1	8			6		

Puzzle 6 - Very Easy

5	9	8	4	6	7	1	3	
2	3		9	5	8	4	6	7
	7	4	1	3	2	5	8	9
3			7		5			4
	6	5	3		9	8	7	1
	8		6			9	5	
	4	3	2		6	7	1	5
1				7		3	9	8
		7	8	1			4	6

Puzzle 7 - Very Easy

8	1	9	6			7	2	4
5		2	8	1	4	9	3	6
4	6		2	9		5	1	
7	9	1	5		3	8	4	
6	2		4	7	9	1		
3	4	5	1	2		6	7	9
	5	6	9	4		3	8	7
9	3		7	8		2	6	
2	8		3	5		4	9	

Puzzle 8 - Very Easy

6	3	7		8	4	1	2	
8	4	2	1	7	5	9	3	6
1	9		6	3		8	7	4
9	1	6		4	7	5	8	3
	8		3		6			7
2		3	8	5	1			9
3	2	9	5			7	4	1
4			7		3		9	2
7	6	1	4	2				

Puzzle 9 - Very Easy

2	5	3		6	9	7	8	
7	8	9	3	5	1	2		6
6	4		8	7	2		9	
		2	7	8	6	1	3	4
1	3	4	2		5	8	6	7
8		6		4		9		5
4	6	7	5		8	3	1	9
9	1	8	6	3			5	2
3	2	5		1		6	7	

Puzzle 10 - Very Easy

2		3	7	8	4		9	5
9	1	8	3		6			2
5	4	7		2	1	6	3	8
	5			7		3	8	4
		9	5	1	8	2		7
8	7	6	4		2	5		9
	8	4	1		5	9	2	3
3	2	1	8		9	7	5	6
7	9	5	2	6		8	4	1

Puzzle 11 - Very Easy

6	2	5		8	4	7	3	
	4	1	7			8		
9	7	8		2			4	5
8		9	4	6	3		2	7
1		4	8	7		5		
7	6	2	1	9	5	4	8	3
5	9	7		4		3		6
	1	6	5	3	9		7	8
2	8	3	6	1		9		4

Puzzle 12 - Very Easy

1			2	3	7	5		9
5	3	7	4	1		6	8	2
	2	9	8	6			3	7
7	5	3			1	4	9	8
2	8		7	9	4			5
	9		3	5	8		7	1
3			9	8	2	7	1	6
8	7	2		4			5	3
	1		5	7	3	8		4

Puzzle 13 - Very Easy

6	3			1			9	4
8		9	2		6			7
				9		6	3	2
2		4	6		1	9		3
5	6	3	8	7		4	2	1
	1	7	3	4	2	5	8	6
4	2			8	3	7		9
1	7	6	9	2	5	3		8
3	9	8	7	6	4		1	5

Puzzle 14 - Very Easy

5	7	6		1	4	8		2
3	4	9	2				5	6
2	1	8					3	4
	2	4	1	9	5	3	8	7
	5	3	4	7			6	9
	9	7	6	2	3	4	1	5
9	3	2		4		6		
		5	8	3		9		
4	8	1			9	5	2	3

Puzzle 15 - Very Easy

6	3	9	2	8		5	1	
7	4	8	3	1	5	2	6	9
	5	1	6	9	4	3	8	
	6	7	1	2	9	8	4	3
		4	5	7		1		
1		2	8		3		5	6
4	7		9	5				
	2	5	7	6	1			8
8		6	4	3	2	9		5

Puzzle 16 - Very Easy

9	2		8	4	3	1	5	6
8	3	6	2	1		4	7	9
4	5		6					2
2	6	5		8	4	7	9	
7	8		9	6	1	3		5
1			7			6	4	8
6			5		9	2	1	
5	1	2	4				6	3
	7	9	1	2	6		8	4

Puzzle 17 - Very Easy

8	3			7	9	5	2	
5	1	7	8				3	4
6	2	9	5	3	4	1	8	7
	5	3	4	8	1	2	6	
9		2	3	6	7	4	5	1
1		6	2			3	7	8
	9	8	6	4		7	1	
3	7	1	9		8		4	
4		5	7		2	8	9	3

Puzzle 18 - Very Easy

1		7	5	8	4	2	9	6
		8	1	3	6		7	
	5		9	7	2	8	3	1
	1	5	4	2		3		7
2	8		3	1		6	4	
7			8	6	5		2	
5	6	4	2		8	7	1	3
3	7	2	6	4	1	9	5	
	9	1		5	3	4		2

Puzzle 19 - Very Easy

2	3	7	6	8	5	4	9	
6	1		2	4	7	8		5
5		8		9	1		7	
	9		1		4		6	3
	5	6	9	2	3	7		
		3	5	6			1	9
9		5	4	1	6	3	2	
3	6		7	5	2	9	4	8
7			8	3		1	5	6

Puzzle 20 - Very Easy

9	1		3	8		2		5
4	5		6	2	7		9	1
			5	9		8	6	4
8	9	4	1	7	3		2	6
6		5				1		9
	3	2	9	5	6	7	4	8
2	6	9		1		4	8	3
	4	3	2	6		9		7
7		1		3	9	6		2

Puzzle 21 - Very Easy

8		3	5	2	9	1	7	4
		7	3	1	8	2		9
	1	2	4	7	6	8	5	3
	8		7	5	1	3		
1					3	7	9	
	3	5	9	4			8	1
	7			9	4	5	1	6
5	4	6	1		7	9	2	
2	9		8	6	5	4		7

Puzzle 22 - Very Easy

	9	8	1	5	7	3	6	4
4		7					5	9
	3		4	6	9		2	7
7		5	9	2	4	6		
9					6		1	5
6	8	4	3		5	7	9	2
1	4	2	6	9	8	5	7	3
3		9	5	4	1	2	8	6
8				3	2	9	4	1

Puzzle 23 - Very Easy

2	8	3	4	9	5			7
5	6	1	7	2	3	9	4	8
7	4		8		6	5	3	
4	2	6	5	8	1	7		3
9	7	8	6	3	4	2		1
3			9	7	2	4	8	6
1			3			6	2	
6	3		1	4	9			5
	5		2	6	7	3		9

Puzzle 24 - Very Easy

3	1		7		4		6	
7		9	5	3	6		2	
5	4	6		8	1	9	3	7
8	5	1	3	2	7	6	4	9
2	9		6		5	8	1	3
4	6	3	9		8		5	2
9	3	5	1	7	2	4	8	6
1	2			6	9		7	5
6	7		4		3		9	

Puzzle 25 - Very Easy

6			8		5	1	2	7
1	3	8	2	6	7	9	4	5
	5	7	1	4				
8	1		5			3	9	4
5		4	3		8	7		6
7	6			1	4	8		2
4	2	1	7	8	3	5		
3	8	5	6	9	2	4	7	1
9	7	6			1	2	8	3

Puzzle 26 - Very Easy

2	3	7	9		1		4	8
	6	5	3	4	2	7	9	1
	9	4	8	5	7	6	2	3
3	8	2	7	9		1	5	
6			1	4		3	8	
4	5		2	8	3	9	7	
5	2	3	1	7	8		6	9
9		8	6		5			7
7			4		9	8		

Puzzle 27 - Very Easy

6	3	5		1	7	9		2
1			6	9		7	8	3
7		9	3	4	2	1	6	
3	1	6	9	7	5	8	2	4
8			6	3	4	5	1	9
5		4		2	8	6	3	
2	7	8		5	1	3		
4	6	1	7	9		2	5	8
9		3	2		6	4	7	

Puzzle 28 - Very Easy

5	2	6	4	3			8	9
8	4	7		6	2	5		3
9	3		8	5		4		
2	5	8	3		4	9		6
1	9		5	7	6	2	4	8
6	7	4	2	9	8	3	5	1
	1	2		8	5	6	9	
4	6	5	1	2	9	8	3	7
7						1	2	

Puzzle 29 - Very Easy

1		8		4	9		2	3
	4	6	3	8	7		1	5
	7	9	5	1	2		4	6
5	8		2	9	6	3	7	1
7				3		5	6	2
6	3			7		4	8	9
9				2	4		5	7
8	1			6	3			4
4	2	7	9	5	1	6	3	8

Puzzle 30 - Very Easy

1		8	5	3	9	2	7	4
	4				1	5	3	6
	3	5	4	6		1	9	
			9		5		1	2
6	1	2				7	5	9
7	5	9	2	1	6	8	4	3
8	7	3	1	9	2		6	5
4	9	1	6	5	8	3		7
5		6	7	4	3	9		1

Puzzle 31 - Very Easy

8	3		7	6	5			9
6	5	7	1	9	4	8	3	2
4	1	9	8	2	3	5	6	7
7		8	2	1	9		5	
	9					2	8	4
2	6	3		5		7		
1		6		4	2	9	7	5
	2	5	9	7	6		4	
9	7		5	8			2	6

Puzzle 32 - Very Easy

		8	1	5	9	3	7	6
9	5	7	4	3				2
3	6	1	2	7				4
	9	3	6	4	7	2	8	
5	4	2	9		3		6	
8		6	5	1	2	4		9
6	8	4	7			5	9	
2	1	9	3		5	7	4	8
7	3		8	9	4	6	2	

Puzzle 33 - Very Easy

9	3	1	2				6	5
2	7	5		1	8	9	3	4
4		8	9		5		1	2
	9	2	3			1	4	8
				2	1	5		6
5	1	6		8	9	3		7
3	8	4	1	6	7		5	9
1	5			4	2			3
6	2		5	9	3	4	8	

Puzzle 34 - Very Easy

7		5	8	9	1	4	2	6
4	2	8	7	6	5	9		1
		9	2	3	4		7	8
8	9	6		1				
1		2	6	7	9	8		3
5	7		4	8	2	6	1	9
	8	1	9	4	7	3	6	
3	6	7		5	8	2	9	4
9			3	2	6		8	7

Puzzle 35 - Very Easy

4	1	5	2		8	7		6
	8		7		9	1	2	5
2	9	7		1		8		4
7	5			2	6	4	1	9
6		1	9	5	3		7	8
8	2	9	1	7		5	6	3
9	7			6	2	3		1
5					1		4	7
1	3	8	4	9	7		5	

Puzzle 36 - Very Easy

7	4	9	5	6	1	8		3
5	2	6	9	8		4	1	7
1	3		7	2	4	9	5	6
	5	4	2	7				8
3		7		4	6			5
	9	2	1	3	5	6	7	4
	7	5		1	8	3		2
4	8	1	3	5		7	6	
2		3		9	7	5	8	

Puzzle 37 - Very Easy

8	6	4	3	7	5		9	
1	2	5	9	8	6	3	4	
	9	7	4	1		5		8
4			6	5	9	7	8	1
5	1		7		8		2	3
	7	8			1	9		
7	5	3		6	4		1	9
9	4	6		2	3	8	7	5
2	8	1	5	9	7	4	3	6

Puzzle 38 - Very Easy

	2		6		5	3	4	1
7	1	6		9	3		8	
3	4		1	2	8		6	
5	9	4	8	1	7			3
1	7	8	3	6	2	4	5	9
6	3	2	9	5		7	1	
		7	5			1	3	6
4	5		7	3	6	8	9	2
9			2	8	1	5	7	4

Puzzle 39 - Very Easy

2	9	8	5	1	3	6	4	
	4	3		6	7		9	
7	6		2		9	8	1	
3	2	4	7		5	1	8	
6	7	1		8	2		5	9
8	5	9		3	6	7	2	
	8	7			4		6	1
5	3	6		2	1		7	8
4	1	2	6		8	9	3	5

Puzzle 40 - Very Easy

7	3			1	6	2		8
	5	1	8	3		6	7	
8	6	2	7	9		1	5	3
6			4	7	3	9	2	
2	4	7	1	6		3	8	5
9		3		5	8	7	6	4
5	9	6	3		7	4	1	2
3		8		4	1		9	
1	7		9	2	5	8	3	

Puzzle 41 - Very Easy

		4	6	7	2	3	9	5
9			4		5	1	8	
5	6	3		8		2	7	4
2		1	7	9	8	6	4	3
7	9	6			1	5		8
3	4	8	2	5	6	9	1	
	2		8	1	3		6	9
6		9		2	7	4	3	1
	3		9	6	4		5	2

Puzzle 42 - Very Easy

9			7	4	3	5	2	6
	6	7	8	9			3	1
3	4				1	7	9	8
7		1	9	5	4	8		
4	2	5			6	3	7	9
6		9	3	2		1	5	4
	5		2	3		6	8	7
8	9	6		7	5	2	1	
	7		6	1		9		5

Puzzle 43 - Very Easy

8				3	2	5	4	7
2	3	7	4	9	5	6		1
	1		6	7	8	3		9
1	2	8	9	4	6	7	5	3
	5	9	8	1	3	2		4
	6	4			7	1	9	8
9	7	2	3	5	4	8		6
6	4			8	1		7	5
5		1		6	9	4		

Puzzle 44 - Very Easy

2	8		7	6	9		1	3
1	6	3		5	2	8		9
4	9			3	8			6
		4	5	8		1	9	7
8		9	3	7	4	2	6	5
7	5	6			1	3		4
	4	2	9	1	5		3	8
5	7		6	2	3			1
9		1	8	4	7	6	5	2

Puzzle 45 - Very Easy

2		7	6	8			1	5
1	9	5	4		7		3	6
6	8		1	5	9	2	4	7
	6	2	9		4	5		1
5	7		2	1	8			3
	1	8	3	6	5	7	2	9
8			5			3	7	4
7			8		1	6	9	2
9	3		7		2	1	5	8

Puzzle 46 - Very Easy

			4	3	7	9	5	6
3	4	5	8	6		2		7
6		7	5	1	2	4	8	3
1	8	9	3	5		7	2	
4		2		8		6		
5		6	7	2				
7	6	1	2	9	5	3	4	8
8	2	4		7	3	5	6	9
	5	3		4	8		7	2

Puzzle 47 - Very Easy

6			8	5	9			7
7	3	5			4	2	9	
8	1	9		2	3	6	4	5
9			4	3	6		8	1
4	5	8	9	7	1	3	2	
1		3	5	8	2	9	7	4
5	8		3		7		1	
2	9				8		5	3
3	4	7	2	1	5		6	9

Puzzle 48 - Very Easy

5	6	9		3	1	2	8	
4		1	6	5	8		3	9
	3		2	9	4			
2	5		4	7		8	9	1
8	4	3			2	5	7	6
	1	7			6	4	2	3
3			1		5	9	6	7
	9	5	8	6	7		4	2
6	7	2	3	4	9	1	5	8

Puzzle 49 - Very Easy

3		7	6	8	1	2	4	5
2	4	1	5	3		8		
6	8	5	4	7	2	1	9	3
	6	9	1		7		2	
7	2	8		9	6		5	1
1		4			5	9		6
4	1		7	6	8	5		
	7	3		5		6	1	
9	5	6	2	1	3			4

Puzzle 50 - Very Easy

	6	9		2	3	5	7	8
7	1	3	9	8	5	6	4	2
2	5	8	6	4	7			1
	7	4	2	5		1	6	3
	2	1	4			8	5	7
3	8	5		1	6	9	2	
1		6	3	9		7	8	
8	9	2			1	4	3	6
5			8	6	4	2	1	

Puzzle 51 - Very Easy

1		3	6	8	9	4	7	2
8		4	7	5	2	9	3	
7	2	9	4	3		6	8	5
	7	6	9	1		2	4	
9	8		5		4	3	6	7
3		2		7			1	
6		7		4		8		
2	1			6	8	7	9	4
4	3	8		9	7			6

Puzzle 52 - Very Easy

		2	9	1	7	4	6	
		1		8		2		9
9	4	3	6	2	5	8	1	7
3	1	9	4	6	8	7	2	5
5	2		1	7		3		6
6		4	2	5		9	8	1
			4	1			9	8
	8	6	7			5	3	4
4	9	5	8	3	6	1	7	2

Puzzle 53 - Very Easy

		5			4	7	1	3
7	4	6	8	1			2	
2	3	1	7	5		6	4	
6	2	8	4	7	5	9	3	
	5	7	1	9	8	4	6	2
1	9		2	3	6		7	5
5	6	9	3		2	1	8	7
8	1	2		6	7		5	
	7		5		1	2		6

Puzzle 54 - Very Easy

	9	2	1	7	6	8	5	4
	6	4	2	3			9	7
1	7		5		4	2	6	3
6	5	7	8			4		
8	4	1	9	5	7	6	3	
9	2		6	4			7	
2	8	5	7		9		4	1
7		6		1	2	9	8	5
4		9	3	8	5	7	2	6

Puzzle 55 - Very Easy

2		1	9	8	4		6	7
		8			1	5	2	9
6		7			5			8
4	3	9		5	7	2	1	6
7		5	2	1	6	9	3	4
1	6	2	4					
9	2	3	6	4	8	7	5	1
5	7	4	1	9	3		8	
		6	5	7	2	4		3

Puzzle 56 - Very Easy

	2		9	5		4	6	
3	5	9	4	6	8	7	2	
4	6	8	7		2			9
9	8	4	6					
7	3	5	1	8	4	2		6
	1	2	3	7	9		4	
5			3	6		9	7	2
2			5	4	7	1	8	3
8		3			1	6	5	4

Puzzle 57 - Very Easy

	4					9	1	
3		2	4	9		7	5	8
7	9		8		2	3	6	
4	2	9	1	8	5	6	3	7
6			9		7			
1	8		6	2			9	5
5		6	2			8	7	9
9	7	4	5	3	8	1	2	6
2		8	7	6	9	5	4	3

Puzzle 58 - Very Easy

	5	7	4		9	6		
	2		5			4		9
9	4		2	8	6	5	3	
4		6	8	2	5	1	7	3
3	8	2			7	9	5	
7		5	9	4	3	8	6	2
2	6		3	1		7	9	5
1	7	9	6		2	3	4	8
5		8	7				1	6

Puzzle 59 - Very Easy

		8	1		6	3	5	
1	5	4	3	9	2	7		8
6	2	3			8	9		
9	1		2	8	7	4	3	6
4	6	7	5	1	3	8	2	
	3			6	4	1	7	
2	8		6		1	5		3
5	7		4	3	9	2	8	1
	4	1	8	2	5	6	9	7

Puzzle 60 - Very Easy

3	5	9		8	7		4	
	7	4	9			8	2	3
8	6	2	3			7	9	5
9		3				5	1	4
5			4	1	8	3	6	9
6	4	1	5	9		2	7	8
4		6		2	5	1	3	7
7	1			3			8	
2	3	8	1		4	9	5	6

Puzzle 61 - Very Easy

2		4		1	6	5	3	9
7			8	9		1		4
9	1			3	4	7	8	6
3	5	2	4	7	8	9	6	
8	7	9	3	6	1	4	5	2
1	4	6	9	5		8	7	3
4	3			8			1	5
6	9		5	2	7		4	
5	2			4	3		9	7

Puzzle 62 - Very Easy

	9	2		3	7	1	8	6
6	8		2		1		3	4
	4	1	5			9	7	2
9	7	3		4	2	6		8
2	6	8	9			4	1	3
4	1			6			2	9
7	3	9	6	1	8	2		5
8	2	4	7		9	3		1
1	5			2				7

Puzzle 63 - Very Easy

3	5	2	8		1	9	4	7
4	6			3	9		8	2
9	1	8			4	5		
5	8	4			3	6	2	
7	9	1		2	8		3	5
2	3	6	1	4		8	7	
		9	3	5	2	7	1	8
8	7	3	9			2	5	4
			4	8	7		9	6

Puzzle 64 - Very Easy

	8	2		7	6			1
6	1	5	4	9		7		8
		3	1	8	2	5		9
7		1	8	6	4	2	3	5
	4	8	3	1	5	6	9	7
	3	6	9	2	7	1		
	6	4	2	5		8	7	3
3	5	7	6		8		1	2
8	2	9	7	3	1	4	5	

Puzzle 65 - Very Easy

9	1	8	7	5		2		3
	5	4	1		9	7	6	8
3	6		8	4	2	5	9	1
7		3		1	4	8	5	
8		1	3				7	2
6	9	5		8	7	3		4
1	7	6	4	2	3		8	5
			9		1	4	3	
		9	5		8	1	2	7

Puzzle 66 - Very Easy

3				8			7	1
2	1	5	9	3	7	6		
6		7	1	2		5	9	3
9	5	4		7	3	1	8	
1			8	5	9	7	2	4
8	7	2	6		1	3		9
4		8		1	5	9	3	
7			3	9	8	4	6	5
5	9	3	4	6			1	7

Puzzle 67 - Very Easy

	4	7	2	9	3	5	1	
	2	1	6	8		9	4	7
			4	7	1	2	6	3
		3		4		6	5	1
1	8				6	3	7	2
7	6	5	1	3	2	8	9	4
8		6		1	4			9
5	1	9	7		8		3	6
	7	2		6			8	5

Puzzle 68 - Very Easy

7	6	3					4	8
4	1	2	7	6	8	9	5	3
9	5	8	2	4	3	7	6	1
2	3	9	1	8			7	4
8	4	5	9	7	6		1	2
6	7	1			4		8	9
			6	1	7	8		
1			8	5	9	4	3	6
5	8			3	2	1		7

Puzzle 69 - Very Easy

			1		3	9	2	
3		2	7	9	4	8	5	6
	4	9		8	5	3	7	1
4	3	6	8	1	2			5
8	7	1		5	9	2	6	3
2		5	3			4	1	
	8	3	6	2	7	5	4	9
5		4	9		1	6	8	7
9	6			4	8	1	3	2

Puzzle 70 - Very Easy

9	1	6	5	7		4	2	
5		7	1	2	4	8		
8			3	9	6	5	1	
1	2	3	9	5		6	8	4
4		5	6	8	3	2		
		8		4		3	9	5
6	5		8	3	9			
2	7		4	6	5	1	3	
3	8	4	7	1	2	9		6

Puzzle 71 - Very Easy

	8	6			1		4	5
7	1	9	5	4	6	3	8	2
5	4	3	7	2	8	6		9
		5	8		3	2	6	
	2	8	6	1	9	4	5	7
1	6	7		5		8		3
8				6		9		
6	7	2	4	9		1	3	
9	3	4	1	8		5	2	6

Puzzle 72 - Very Easy

		9			5	6		2
	6		1	9	2	4	5	3
5	4	2	6	3	8	9	7	
		4	3	1	7		9	5
3	5				9	1		
9	2		5	8	4	7	3	6
	9	3	8	2	6	5		7
2	1		4		3		6	9
8		6		5	1	3		4

Puzzle 73 - Very Easy

		1	6		2		4	9
7	6	2	9	4		3		8
5		9		1	8	6	7	2
9	1	6		8	3		2	7
		7	1	2	9		3	6
2	8		4	6	7		5	1
6	7	4			1	2	9	
	2	5	7		6	1	8	4
1	9		2	3		7	6	

Puzzle 74 - Very Easy

3	8		6	2	5		9	
5	1	9	7	4	3	8		6
6	2	4			8	3	7	
9	5	6					8	2
1	3			6		5	4	9
4	7	8	2	5	9			3
	6	3	4		1	9	5	
7	9	1			2		3	8
8	4	5	3	9	7	2	6	

Puzzle 75 - Very Easy

5	4	1	6	8	2	3	9	7
	8	6	4	7			1	2
7	9	2	1	5	3	4	6	8
4	2	8		3	5	1	7	
6		7		1		8		9
		5	8		7		3	4
8	5	3	7		6	9		1
2	6	4	5	9		7	8	
1	7			4	8		2	5

Puzzle 76 - Very Easy

4	6	1		8	5	3	9	2
9	5	2	3	6		4	7	8
3	7	8	2	4	9	1		6
	1		5	9	2	8	6	4
8	2	4	6	3	7			
6	9		8		4		3	
	3	9	4	7	8	6	2	
		7	1		6			3
	8	6		2	3	7		1

Puzzle 77 - Very Easy

4		6			8	1		2
9			6	1	5	8	3	
	1	8	2	4		6	7	9
2	5	3			7	4	9	1
1	6	4		5				3
7			3	1	2	6	5	
6	4	2	5	8	9	3	1	7
8	9	1	3			4	5	
		5		2	6	9	4	8

Puzzle 78 - Very Easy

9	3		6	1	5	4		2
1	4	6		8				9
8	2	5	7					
3	5	1	8	4		2	6	7
6	8		1	2	7	3	9	
	7		5	6	3	8		1
	1	2	4		6		5	
5	6	3	9		8	1	2	
4	9	8		5		7	3	6

Puzzle 79 - Very Easy

7	2		5	9	3		6	8
9	8	6	2	4	1		5	7
5	1	3	8	7	6	9		2
	9	1		3	8			5
8	4	2	9	5		6		3
		5	1	6			9	4
4		8	6	2	9		3	
1	6	7	3	8	4	5	2	9
	3	9	7	1		4	8	

Puzzle 80 - Very Easy

	9	7	2		8	5	3	
6	3	5	9	4	7		8	2
2	8		6		5	7	9	
5	4	2		8	3	6		
7	6		2			3	4	1
	1	3	7			8	2	5
			3	9	4		5	8
	5	4	8		2	9	1	7
	2	9	7	5	1		6	3

Puzzle 81 - Very Easy

		6	4	5	8		3	9
5	8		2	7	9	4	1	6
	4	2	1	3			7	5
3	9	7	5	1	4			2
4	2			8	7	3	9	1
		1		2	3	7	5	
7	1	4		6		9	2	
6	5	8	7	9		1	4	
2	3	9			1	5		7

Puzzle 82 - Very Easy

	6	1	2		3	7	9	
	5		7	4	8	1	3	
7	8		6	1	9		4	5
	7		1	2	5		8	3
	2	5		3	4	6		7
3	1		8	6	7	9	5	2
1		7	4	8	2	5	6	9
6				7		3	2	4
5			3	9		8		1

Puzzle 83 - Very Easy

8			5	6	3		9	
6	1	9		4	7			
3	7	5	8	1	9	2		6
2	9	8	6	5		3	7	4
5	4	3		7		6	1	2
1		7	3	2			8	9
4	5			8	2	9		3
9	8	6		3	5		2	7
7		2				1	5	8

Puzzle 84 - Very Easy

	8			2		4	3	1
	2	6	3	4	1	8	5	7
4		1	5	8	7		9	2
2	4	3	1	9				6
		9	7		5	2	1	4
1		7	4		2		8	3
	1		2		6	3	4	9
6	7		9	5	3	1	2	8
	9	2		1	4	7	6	5

Puzzle 85 - Very Easy

```
. 3 . | . . 8 | . 5 4
2 4 . | 3 . 6 | 8 . .
. 9 . | 5 2 4 | 7 6 3
------+-------+------
5 2 9 | 6 8 3 | 4 1 .
. 7 . | 9 4 . | . 3 2
3 1 4 | 7 5 2 | 9 8 6
------+-------+------
4 . . | . 6 5 | 1 . 9
9 . 1 | 4 3 7 | 6 2 .
7 6 2 | . . . | 3 4 5
```

Puzzle 86 - Very Easy

```
5 2 8 | 9 . 6 | . 4 7
4 1 . | 3 8 . | . 6 5
3 6 . | . 2 5 | 9 . 1
------+-------+------
1 8 4 | . 3 2 | 7 9 .
7 . 2 | 8 6 9 | . 5 4
6 9 5 | 7 4 1 | . 3 .
------+-------+------
. 1 6 | . 3 . | 5 2 8
. 7 . | 2 5 4 | 6 1 .
. . 6 | 1 9 8 | 4 7 .
```

Puzzle 87 - Very Easy

```
5 2 . | . 7 3 | 4 6 8
7 . 4 | . . 1 | . 9 3
. 3 9 | . 8 2 | 5 7 .
------+-------+------
4 6 8 | 2 . 7 | . 1 5
9 . 7 | 1 3 . | 6 2 4
2 1 . | . 5 . | 9 8 7
------+-------+------
8 9 . | 7 4 6 | . . 2
1 . 6 | 3 2 5 | 8 4 .
. . 2 | 8 . 9 | . 5 6
```

Puzzle 88 - Very Easy

```
8 6 . | 2 5 1 | 4 3 .
1 5 4 | 7 . 3 | 2 9 .
2 7 . | 8 4 9 | . 5 .
------+-------+------
. 3 5 | 1 9 . | . . 6
6 4 . | . 3 8 | 9 . 2
9 8 2 | 4 7 6 | 5 1 .
------+-------+------
3 1 6 | 9 2 4 | 7 8 5
5 9 8 | 6 1 . | . 2 .
4 2 . | . 8 5 | . 6 9
```

Puzzle 89 - Very Easy

```
3 9 2 | 8 5 1 | . 4 6
7 4 8 | 2 3 6 | . 9 .
1 5 6 | . 7 9 | 3 . .
------+-------+------
2 7 9 | 1 8 . | . . 3
. 8 1 | . . . | 4 7 2
5 3 4 | 6 2 7 | . 1 .
------+-------+------
4 6 . | 3 1 2 | 9 . .
. . 7 | . 6 5 | 2 . 4
9 2 3 | 7 . . | 8 1 5
```

Puzzle 90 - Very Easy

```
1 . . | 2 . . | . 5 .
5 7 4 | 6 8 3 | 1 2 9
2 . . | 7 1 5 | . 8 6
------+-------+------
. 2 7 | 5 . . | 8 9 1
4 6 1 | 3 . . | 2 5 7
8 . 9 | 4 . 1 | 6 3 2
------+-------+------
7 1 5 | . 2 6 | 8 4 3
9 . 8 | . 3 7 | 2 6 5
. 3 2 | 8 5 4 | 7 . .
```

Puzzle 91 - Very Easy

5	2	6	4	3	9	8	1	7
3		1		5	7	4	6	9
4	9	7		8	6			
		8	3	9	5	7		4
2		5		1			9	3
9			7	6		5	8	1
7	1	2		4	8	9	3	6
8	4	9		2	3		7	
6	5		9	7	1	2	4	

Puzzle 92 - Very Easy

		9		4		6	2	3
5	4	6		3			9	7
	3		6	8	9			
		7	2		8	1	3	9
9	1	3	5			2	4	8
4	2		9		3	5	7	6
2	6	4	3	9	1	7	8	5
8		5		6		3	1	2
3	7	1		2	5		6	

Puzzle 93 - Very Easy

8	5	4		6	3	7	2	9
6	2	3	7	5	9	8		1
1	9	7		4	8	5	3	6
	1			7		4	8	3
	4	8	5	3		6	9	2
	6				2		5	7
9	7	1		2	5	3	6	
4			3	9		2	1	5
		2	6	1		9	7	

Puzzle 94 - Very Easy

4	7	9	2	5	3	1	8	
2	5		6	1	8	4	9	7
6	1	8		4			2	
8	4	6	1	7	5	9		2
5	9	2	3	8	6	7		
	3		4		9	8	6	
1	2	5	8	6	4	3	7	9
3	6		5		1			8
	8	4	7	3	2		5	

Puzzle 95 - Very Easy

	2	9		3	4	1	5	6
	1		6	5	2	9	7	8
5	6	7	8	9	1			
3	8	2	9	4	7	6		5
		5	2	8	6	7	9	3
	9	6		1			4	2
	5	8	1	7	3	4	6	
	7	4	5	2	9	3	8	1
9		1	4		8	5		

Puzzle 96 - Very Easy

7	2	6	8	4	5		3	
		8	7	2	1	5	6	
5	4	1	9			6	8	2
8			6		3	1		2
4	7		5	1	2	6	9	8
6		2		9	8	7	5	3
1	8	5	2	6	4	3		
2	6	7		8		4		5
3			5	7		2	8	6

Puzzle 97 - Very Easy

	1			4	8	2		
9		5	6	1	2		3	7
	2	3	7	9	5	8	6	1
5	4		1	3		9	8	2
3	7		2	8	9	5		6
2		8		5	6	1	7	3
1	5	4	9	7			2	8
7	6	9		2		3		
		2	5	6	1		9	4

Puzzle 98 - Very Easy

3	7	5	4			9		1
	9	2	6	5	1	3		4
6		4			9	5	2	8
	6	4	2	7	8		5	
7	1	8	3		5		9	2
5	2	3	9		4	6	8	
1	3	9	8	2			4	5
	5		1		7	8	3	9
4			5	9	3		1	

Puzzle 99 - Very Easy

4	5	2		8	3	6	9	7
	7		5	4	9	2	3	1
	3	1	6	2	7		8	
5	8		7	6	2	4	1	
2	6	7	4	9	1			8
1	4	9		3		7	6	
	1	4		5	6	8	2	3
	2	8	3			9	7	5
3	9		2	7	8	1	4	

Puzzle 100 - Very Easy

6	5	9	4	2	8		7	
	8	4	6	1	7		5	
7		2	5	9		6	4	8
5	3	8	7	6		4	9	2
1	4	7	9	3	2	8	6	5
9	2				4			
			1	7		5	8	
8	7				5	9	2	
4	6	5	2	8	9	7	3	1

Puzzle 101 - Very Easy

	6	1				3	9	2
		5		3	6		7	8
4	3	8	7	2		6		1
8	2	6	9	4	3	7	1	
		7		5	1	2	3	
	1	3	2	6		9	8	4
1	5	9	4	7		8	6	3
3		2		9	5	1	4	7
6	7	4	3	1	8			

Puzzle 102 - Very Easy

9			8	3	2	6	7	
6	8	2		4	7	9	5	3
1	7	3	5	6		2	4	8
7		9	2	8	5	1	3	6
2	1	5	6			4	8	
			7	1	4	5	9	
4	2			7		3	6	5
5	9	7	3				1	4
	3			5	1	7	2	

Puzzle 103 - Very Easy

2	1		4	8	5	6		9
		9		3		4		
	4	3	6	2	9	7	8	1
	8	5	2	4		9		6
	9	2		7	6	1	4	8
	7				8	3	2	5
	2	8	7		4	5		3
6	3	4		5	1	2	9	7
7	5	1	3	9	2		6	4

Puzzle 104 - Very Easy

9	4	8	5	2		7	6	3
1	5	7	3			2	4	9
2			9	4	7	1	5	8
3		2	6	5	4	8	1	7
8		5		7	3	4	9	
6	7	4		1	9		3	2
		9	4	3		6	7	1
4		1				3	8	5
7	8	3	1	6			2	

Puzzle 105 - Very Easy

9	3	7		2			1	5
4	5	8			1	6	2	3
2			8		3	7	4	
5	4	3	6	1	7		8	2
7		2	5	9	8	3	6	4
	9	6	2	3	4	5		1
1	7	4	3	8	5	2		
	2		7	4	9	1	3	8
3	8			6	2	4	5	7

Puzzle 106 - Very Easy

3	7	8	2	1	6			9
2	5	1		3	4		6	8
6	9		5	8	7	1	3	
8	1	5	3		2	6		
4		6	7	5				
9		7	6	4	8	5		1
7		3	1		9	8		5
5	8		4	7	3		1	6
	4		8	6	5	3	9	7

Puzzle 107 - Very Easy

6	3		5	4	7		9	2
		7			9	6	5	3
5	9			3	2	7	8	4
9	6	4		5		8	7	1
	1	3	8		4	9	2	
8	5		9	7		4		6
1			4	2	5		6	
	8	5	7			2	4	9
2	4	6	3	9	8	5	1	7

Puzzle 108 - Very Easy

4	9	6	3	7			1	2
7	5		2	1	4	3	6	9
2		1	9	8	6	5	4	
3		4	7	9		6		1
1			5	6		4	2	
6		5	4	2	1	9		3
8	4		6	3	2	1	9	5
	1	2					3	6
	6			5	7	2	8	4

Puzzle 109 - Very Easy

8			5	9	4		3	7
3	1	4	8	7		6	9	5
7	9	5		6	3	2	8	4
4	6	1		8	5	7	2	3
2	5	7	4	3	1	9	6	8
9	8	3	6	2	7	4		
	4		3	5		8	7	9
6	3		7		8	5		2
		8		1	9		4	

Puzzle 110 - Very Easy

6	8	4	3	7	1	5	9	2
	3	9	4	5	8	6		1
1			6	9	2	3	8	
	1		8	2	7		5	
9		8	5			2	1	3
5	2	6		3	9	7	4	
7			2	6	4		3	
8	6		7	1		4	2	9
3	4	2	9	8	5	1		

Puzzle 111 - Very Easy

3	4	7	5	2			8	6
1	9	5	6	7	8			
8	6	2		4	3	5	7	9
2	3			8	5	6	9	
4		6	2	9	7	3		
7	5	9		1		8	2	4
		4	9	6	2		3	
		3	8		1		6	
6	2	8			4	9	1	5

Puzzle 112 - Very Easy

7	1		9		5	3	8	6
5	9		6	3	8	1	7	2
	6			1	2	9	4	5
6		5			1	2	9	3
1	3	7			9	4		
8	2	9	3	6		7	5	
4	8		1	5	7		2	9
	5	1		9	6	8	3	7
	7		2	8	3	5	1	4

Puzzle 113 - Very Easy

2		4	1	9		5	6	8
9	3	5	7	8	6	4		1
6		1		4	2	9	7	
3	9	2	6		8	7	4	5
1		8		5	7		9	
5	4		2		9			6
8	5	6		7				4
	2	3	8	6	4		5	9
4	1		3	2	5			7

Puzzle 114 - Very Easy

5	4	6	7	9		8	1	2
	3	7		8			6	9
		9	5	6	2	7		3
8	7	1				2	9	6
6		4	1			3	5	8
3	9	5			8	1	7	4
9	1	3	2		4	6		7
4		2	8	1	7	9	3	
7	5		9		6	4	2	1

Puzzle 115 - Very Easy

3	7	8	4		9	2	6	
6	2	5		7		9	4	8
1	9		6		2	7	5	3
	1	3	2	9	6	8	7	4
4				3	1	6	2	
7	6	2	5	4	8	1		
8		7	9	2		3	1	
2	3	1	8	6	5	4		
	4	6		1	7			2

Puzzle 116 - Very Easy

1		9	4	5	6	2	7	3
5	3	6	9		7	1	4	8
4	7	2	8	1	3	9	6	5
6		5	2	9	8		3	1
9	1	7	5	3		6		
8		3			1		9	4
	5			6			2	7
2		8	7	4	5	3	1	9
7			3	8	2	4	5	6

Puzzle 117 - Very Easy

4	7	6	2	5		9	1	8
1		8	9			3	2	
9	3		8	1	7	6		5
3	6		5	8		2	9	
		9	3	7		4	5	6
	1		6	2	9			3
2	8	1	4	6	5		3	9
6		5	7	3		1	8	4
7			1	9		5		2

Puzzle 118 - Very Easy

4	9	5			2	8	3	6
7	1	8	3	9	6	5	4	2
3	2	6	4				9	
6		7	2		1	4	8	9
2		1		7	9	3	6	5
	8	9	6	4	3	1	2	7
8	7	2	5	6	4	9		3
	5	3	1	2	8			4
1	6		9	3	7			

Puzzle 119 - Very Easy

3	8	1	2	7			4	5
7	6	4		3	5	9	8	2
5		2	8		4	3	1	
	7	5	4	1	6	2	3	9
	2		7	5	8		6	4
1	4	6	3	9	2	5	7	8
2	3		5			4		6
4	1		6			8		
6	5		9	4	3	7		1

Puzzle 120 - Very Easy

	6	7		1		5		
1			8	3	7	9	2	
	3	2				7	8	1
2	8	1	7	4	3	6	5	9
7		4		9				
3	9	6	1	5	8	4		2
6	7	3		2		8	4	5
4		9	3	8	5	1	6	7
5	1	8	4	7	6			

Puzzle 121 - Very Easy

9		2	5	6	7	1	3	8
3	8	1	4		9	6	7	5
	6		3		8	2		4
8	5	9		3	1		6	2
4	2	3	6	9	5	8		7
		6	8		2	9		
6	3	8		5			2	1
2	9	4		7	3	5	8	6
5	1	7	2	8			4	9

Puzzle 122 - Very Easy

	3	5	6	8		9	1	
		4	3	2	7	8	5	6
	8	6		9	5	3	4	2
6	7	1			9	2	8	3
	5		2	1	8	6	7	4
8		2	7		6			5
5	6		9	7	2	4	3	
4	2	3	8	5	1	7	6	9
9	1	7				5		8

Puzzle 123 - Very Easy

7	2	8	6			1		5
6	5			8		7		3
3	4	9	7	1	5	2		
		6		3	9	4		2
9	8	4			6	3	1	
	1	3		7		9	5	6
4	9	2	3	6	8	5	7	1
8	3	7	5		1			4
1			4	2	7	8		9

Puzzle 124 - Very Easy

4	5	2	3	1		9		8
8		3		2	7	4	1	6
	6	7		9		3	5	
6		5	8	4		1	2	
3	1	9	6		2	8		5
	8	4	9	5	1	7	6	
9			2	6	4		8	7
5	4	6	7		9	2	3	
7	2		1				9	4

Puzzle 125 - Very Easy

7	9	3	6	8	4	2	1	5
8	4	1	9	2	5	3	7	6
5	2			3		4	8	9
		2		5	8	7		
		8	1	4	7	6	9	
1	7	4	2	9		8	5	
4	3		5	1		9	6	8
		9	8			5	2	4
2	8	5			6	9	3	7

Puzzle 126 - Very Easy

5	3	6	2				8	4
	4		5	8	7		1	3
1	7	8		6	3	9	2	5
6		7	8	5	4	2	3	9
		4	1		2		5	6
		5		9	6		4	7
		9	7	3	8	4	6	1
	6	1	9	2	5	3	7	8
7		3	6			5	9	2

Puzzle 127 - Very Easy

8	9			6	4	2	7	5
7			3	9			1	8
1	4	2	8		7	6	9	3
	1		4	7	3		5	
	5	4	9	8	6	7	3	
3	8	7	5		1	9	4	6
			2	3	5	1	8	9
5	2		7	1	9	3	6	4
		1	6			5	2	

Puzzle 128 - Very Easy

	6	9		4	8	1		2
7	2	5	3	1		8	4	9
	4	8		5	2	6	7	3
9		6		7	3	2		5
2	5	4	1	8	9	3		7
8	7		2	6	5	9		4
4	3				7		2	6
	8	2		3	4	7		1
6	9	7	5	2	1	4	3	8

Puzzle 129 - Very Easy

2	9	5	6	1	8	3	4	7
1	6	3		7		5		8
7	4	8	5	2		1	6	9
9	2	4				8	7	
6		1	2	5	7	9	3	4
3		7	8	4	9	6	1	2
8	1	6		9	2		5	
	3	2	4	8	6		9	1
4		9						6

Puzzle 130 - Very Easy

	5	8	1	4	2	9	7	6
		6	7			4	8	
7		1	9	8	6	2	3	5
8	1	7	2		4	5	6	3
	3	2	8	1	5	7		9
5	9			6		8	1	2
	6		5	2	8	1	9	
1	7	5		3	9	6	2	8
2			6	7	1	3		4

Puzzle 131 - Very Easy

		1	8	5	3	9	6	2
8	2		1	9	6	7		4
5	9		7			1		3
4			9	8	7	3	1	6
1		9		6	2			5
	3	7	5	4	1	2		8
3		5	4		9	8	2	7
		4		7			3	
9		8	2	3	5	6	4	1

Puzzle 132 - Very Easy

		1	2			9	4	5
	2		7		5	1		
4	5	3	8	9	1	7	6	2
3	1	2			4	8		6
	6		3	5	2	4	9	1
5	4	9				2	3	7
2	7	5	4	6		3	1	8
	3		5		7	6	2	9
6	9	8	1			5	7	

Puzzle 133 - Very Easy

	9	5	7	6	3	4	8	1
1	7	8			4		3	6
	6			1	5	7		
8	2	7	1			3	4	5
5	3	6	2		7	1		8
		9	5	3	8		6	
	5	2			9	6	1	3
	4		6	5	1	8	7	2
	8	1	3		2		5	4

Puzzle 134 - Very Easy

	3	8	9		1	7		
	4		6	8	2	3		
2		9	3	5	7		4	6
1	2	6	5	3	9	4	8	7
8	7		2	6	4		9	
	9	4	1	7	8	2	6	3
	5		4	1	3	6	7	8
	8	1		2	6	9	3	5
3	6	7	8	9	5	1	2	

Puzzle 135 - Very Easy

2	1	9	3	4	7	6	5	8
	4	3				9	7	1
6		8	5	1	9	3		4
	6	5	4	2	3	8		7
7		4	8	5	6	2	1	
		2	7	9	1	4	6	
	3	6	1	7	4		8	
	2	7	9	3		1		6
	5	1	6	8	2	7		

Puzzle 136 - Very Easy

4		3	5	6		8	2	1
8	1	9	3	4	2	7	5	
6	5	2	8		1		9	
5	3		4	1			6	2
9	6	1	2	5		3	4	
	4	8	9				1	7
1			7	2	5			9
3		6			4	2	7	5
7	2	5		9	3	1		

Puzzle 137 - Very Easy

1	8	2	6	7	5			
9				4	2			
	5	4				2	7	8
3			4	5	7		2	1
7	2		3	1	8	6		9
4	1	8	9		6	7	3	5
8	3		2	6	9	5	1	
2	9	1	5	8		3	6	
5	4	6	7		1	9		2

Puzzle 138 - Very Easy

1	9			7	3		8	5
	4	3		6	8		9	1
5	8		1	9		4	7	3
		9	6	5	1	3	4	
	5			2	4		1	6
6	1	4		3		9	5	2
4	6		3		5	7		
	7		2	1		5		4
2	3	5	7	4	9	1	6	8

Puzzle 139 - Very Easy

3	4			9	1	7	5	8
7	1	9		5	8	6	4	
6	5	8	7	2	4			1
1		4	8	6	7	2		5
5	2		9	4	3	1	8	
	3	6	5	1	2	9	7	4
9	8		4	3		5		
		5	1			4	6	
			2	7	5		1	9

Puzzle 140 - Very Easy

6	2	7		5			4	3
9	4	3	7	2	1	6	8	5
1	5		6		3		2	
2		9	8				3	4
5		1		9	4	2	7	6
4		6		7	2	8	9	
	1	2	4	6	9		5	7
3	9	5	1			4	6	2
	6	4					1	8

Puzzle 141 - Very Easy

1			2	3	9		4	
	2	4	7	8				9
	3	7	4		6	5	8	2
5	9	3	6	2	8	4	1	7
7		6	9	5	1	8	2	3
8	1	2		4	7	6	5	9
2	7	1	5	6		9	3	8
				7	3	2	6	5
	6		8	9			7	4

Puzzle 142 - Very Easy

7	5	9		4	6	1	8	3
	8	4	5	1	9	7		6
6	1	2		7				9
9	7	6	1		5	3	4	2
	3	1		2	4	5	9	7
4		5			7	6	1	8
	4		7	6	2			1
2		7	3		1	8	6	4
1	6	3	4	9	8	2	7	

Puzzle 143 - Very Easy

1	6	2		5	7		9	8
9	4	8		6	2		3	
3	5	7			9		6	
			2	3		7		6
8	3	6	7	9	5	1	2	4
2	7			1		3		9
6		5	9	4		8	7	2
		3		2	1	9	4	5
4		9	5	7	8	6	1	3

Puzzle 144 - Very Easy

5	8	2	1	7	9		3	
6		7	5	3		9	8	2
9	4						1	5
	3	1	6	9		5	4	
		5	7	4	1		9	
4	9	6			5	1		7
3	6	8	4	1		2	5	
1	7		9	2				3
2	5	9	3	6	8	4	7	1

Puzzle 145 - Very Easy

4	5	6	3	8	2	7	1	9
2	3	8	9	7	1	4	5	6
9	1	7		4	6	8	2	3
3	8	4		2	9	1	6	
1	6	2			8	9		
5		9				2	8	
6	2	5	1		4	3		
7	4	1	8	3	5	6	9	
8		3	2				4	1

Puzzle 146 - Very Easy

2		6	4	1			8	7
1	5	7	3	8	2	9	6	4
9	8	4	5	6	7	2	3	
6	7	5	8	3		4	2	9
3	1	8	9	2		7	5	
	2	9	7	5	6			3
8	9		1	7	5			
5	4			9	3	1	7	8
	6	1		4	8	3		5

Puzzle 147 - Very Easy

4		6	9		2		7	3
8	7	9	1	5	3	6	2	4
3	2			4	6		9	5
2	9		5		8	3	4	6
	4	5	3				1	8
1	8			2		7	5	9
5	1				7	9	6	2
	3	2		6	1		8	7
7	6	8	2	9	5		3	1

Puzzle 148 - Very Easy

3	4	5	1	8	9		6	7
6	7		2	3	4	8	5	1
1			5		7	9	4	3
	9		3	7	8	1	2	
5	1	3		9	2		7	8
2		7	4	5				6
7	5		9	4		6		
	3		7	2	6	5	1	9
	6	2		1	5	7		

Puzzle 149 - Very Easy

5	4	9			6		2	1
8	6	7	1	4	2	9	3	
		2	5	7	9		4	8
	2		4		5	3	7	
		6	9	2	8	1	5	4
1		4	6	3	7	2	8	9
4		5	2	9			6	
6	8	1	7	5		4		
2		3			4	5	1	7

Puzzle 150 - Very Easy

2		4	6		3	5	7	1
6	8	7	9				3	4
5		3	4			9	6	8
3	2	9	5	1	7		8	
	4	5	3	9			1	2
	6	1	2	4	8	3		5
9	5	8	7	6		1	2	
4	3	6	1	2	9	8	5	7
1	7			3				9

Puzzle 151 - Very Easy

```
. 3 . | 7 9 2 | . . 5
8 9 2 | 6 . 1 | 4 3 .
. 5 . | 3 4 8 | 2 . 9
------+-------+------
9 6 3 | 4 7 5 | 1 . .
4 8 . | 2 6 . | . . 3
2 7 5 | 8 1 3 | 9 6 4
------+-------+------
7 4 9 | . 8 6 | 3 5 2
3 1 8 | 5 2 4 | 7 9 6
. 2 . | 9 . 7 | 8 4 .
```

Puzzle 152 - Very Easy

```
9 8 . | 7 . 3 | 2 6 .
6 7 3 | . 4 1 | 5 . 9
2 . . | 8 . . | 7 3 4
------+-------+------
. 6 1 | 8 7 . | 3 . 2
3 . . | 5 . 6 | . 7 8
8 9 7 | . . 2 | 4 5 6
------+-------+------
7 5 6 | 4 2 8 | 9 1 3
4 3 . | . 1 7 | 6 . .
. . 9 | 3 6 . | 8 4 7
```

Puzzle 153 - Very Easy

```
1 2 7 | 8 4 6 | 5 9 3
6 . 3 | . . 9 | 1 . 7
. . 5 | 1 3 7 | . 6 .
------+-------+------
8 5 2 | 7 6 3 | 4 1 9
3 7 . | 9 . 4 | 8 . 5
9 1 4 | 5 8 2 | . . .
------+-------+------
2 4 9 | 3 . 5 | . . 1
7 6 1 | 4 9 . | 3 5 2
5 3 8 | 6 2 . | 9 7 4
```

Puzzle 154 - Very Easy

```
. 4 2 | . 3 6 | 1 . 7
3 . . | . 1 . | 5 2 8
1 . 5 | 8 9 . | 3 . 6
------+-------+------
. 8 6 | . . 3 | . 5 1
4 9 . | 6 . . | 8 7 2
5 . 1 | 7 . 8 | 6 3 9
------+-------+------
9 1 8 | 2 5 4 | . 6 3
6 5 7 | 3 8 9 | 2 . .
2 . 4 | 1 . 7 | 9 8 5
```

Puzzle 155 - Very Easy

```
. . 2 | 9 8 1 | 3 . 7
. 9 6 | 4 . 7 | 2 8 .
. . . | . 6 2 | . . 1
------+-------+------
8 . . | 2 9 . | 7 . .
. 1 9 | 6 7 3 | 8 . 4
2 7 3 | 1 . 8 | 6 5 9
------+-------+------
. . 5 | 8 2 9 | 1 4 .
9 4 8 | 3 1 6 | 5 7 2
. 2 1 | 7 5 4 | 9 3 .
```

Puzzle 156 - Very Easy

```
. 3 6 | . 9 7 | 2 1 .
8 9 4 | 6 2 1 | 3 7 5
. 2 1 | . . 8 | 4 6 .
------+-------+------
1 8 9 | 7 . 6 | 5 2 3
2 6 . | . 3 9 | 8 4 7
. 7 3 | 2 . 5 | 6 . .
------+-------+------
3 4 8 | 9 . . | 7 5 6
. 1 2 | . 7 3 | . . 4
9 5 7 | 8 . 4 | 1 3 2
```

Puzzle 157 - Very Easy

8		6	3	1				2
5	1	4	6	8	2	7	9	
	7		9	4	5	8	6	1
	5	7	1	6		9	2	
1	3	9	7			5		
6	2		4		9	1	3	
3	4	1		9	6	2	7	5
9	6	5	2			3	8	4
7	8	2	5	3			1	

Puzzle 158 - Very Easy

8	5	6				1	7	3
1	3	2		7	6		4	8
4	9				3		6	
7	4	1		2	8	6		5
5			9			7	3	1
3	6		7	1	5	8	2	4
9	8	3	6	5	2	4	1	7
6	7	5	4					
	1	4	8	9	7	3	5	

Puzzle 159 - Very Easy

7	3	5	9	6		1	8	
	9	1		5	3	6		4
2	4	6		1	7	9	3	5
3	6		7	2	1	4	5	8
		8		4	9		1	3
1			3	8	5	2	9	6
4	8	7	1		2	5	6	9
6	1	2	5	9		3		
9		3	4			8	2	1

Puzzle 160 - Very Easy

6		2		5	7	4		3
5	4	1	6	3		2	7	9
3	7	8	4		2	1		5
	6	4		1	5	9		
1		5		4	9	8	2	6
9		7	3	8			1	4
7		9	8	6	1		4	2
2	1	3	5	7		6		8
4	8	6	9	2		7		

Puzzle 161 - Very Easy

		3		6		1		2
		8	3	2	1	6	4	9
	1	6	4	8	9	5	7	3
1	3	2					5	7
4	9	5	1	7	3	8	2	
6			2	5	4	9	3	1
		9	6	4	5	3		8
	6	4	7	1	8	2	9	5
	5		9	3	2	7	6	

Puzzle 162 - Very Easy

	7	1	2		5	3	4	8
2	8	3		1		6	5	7
	4	6		3		2	1	9
	9	8		5	6	7	2	3
6			3	8	2	4		5
3	5	2	7		9	1	8	6
7	2		4		3	8		1
8	3	9	6	2	1	5		4
1	6	4				9		

Puzzle 163 - Very Easy

	8		9		5	4	7	1
	9	7	3		1	6		2
5	6	1	7	4	2	8	9	3
	7		5	1	4	3	2	6
2		3	8	9	6	7	4	
6	5				7	1	8	9
	4	9	1	2			6	8
1	2	5	6	7		9	3	
8		6	4		9	2	1	7

Puzzle 164 - Very Easy

	6	7	5	1	4			2
	1	4	3		9	8	7	6
9	2		6	8	7	5	4	1
	5		7	6	8			9
	8		2	9	1			
2		9		3	5	1		8
1		2	8	7	3	6		
7	4	8	9	5		2		3
6	3		1	4	2		8	7

Puzzle 165 - Very Easy

3	6	2				1	5	8
8	7		1	3	5	9	6	2
5	9		8	6	2		7	4
	2	3	4	9		5	1	
6		8	2	1		4		7
4	1	9		5	7	2	8	3
	4		7	2	1	8	3	5
2		7		8		6	4	
1	8	5	3	4	6	7	2	9

Puzzle 166 - Very Easy

1		3	7		8	6		
7		8	6	3	4	9	1	2
9		6	2	1		8	3	
3			1			2	6	
2		9			3		7	1
5	1		9	6	2	3	4	8
4		2	5	8	6	1	9	3
6			4	2	9	7	8	5
8	9		3		1		2	6

Puzzle 167 - Very Easy

	2			5			8	9
	3	4	9	7			5	2
5	6		2		4		1	3
6	8		4	1	7	9		
4	1		8	3	9		6	7
9	7		5		2	1	4	8
3		6	7	9	8		2	1
1	5		3	2	6	8	9	4
2	9	8	1	4	5		7	6

Puzzle 168 - Very Easy

5			7	1	8		6	
	2	1	6	4	5	8		
8			2		3	5	4	1
1	3	6	4		2		9	
9	5	8		7	6		1	2
2	7	4		5	9	3	8	6
4	9	3	8		1	6	5	
	1	2	5	6	4		3	8
	8	5		3		1	2	4

Puzzle 169 - Very Easy

4	5	7	9	2		6	8	3
	8		7	5	6	4	2	9
	9		3	4				
	7	4			5	9	3	1
5	3	2	1	8	9	7	6	4
6	1	9		3	7	8		
9	2	5		7		1		
7	6	3	5	1	4	2		8
	4	8	6	9	2	3	7	5

Puzzle 170 - Very Easy

		9	6	8	3			
5	6	7	2	1			3	4
	8	2		4	5	6	1	9
8		6		2	7	4	9	1
2	4	1		9	8			5
7	9	5	1	6	4		8	3
1			8	5	6	9	4	2
6		4	9				7	
9			8	4	7	3	5	6

Puzzle 171 - Very Easy

1	9		6	2	8	7		4
	4		3		5		9	1
7	6	5		1		2	8	3
9		6		4	3	5	1	7
3	5		1		6		2	8
4	2		5		7	9		6
5		4	8	3	2		6	9
2	3	9	7	6	1	8	4	5
6		8	9	5			7	2

Puzzle 172 - Very Easy

2		1	7	6		9		5
3	7		5	8	9		4	1
5		9		4			7	6
7		5		1	4	8	9	
1	2	4	8	9	7	5	6	3
9	3	8			5	7	1	
			1	5	6	4	2	7
4		7		3	2	6	5	8
6		2			8	1	3	9

Puzzle 173 - Very Easy

5	1	2	9	3	6	7	4	8
	9			1		5	3	2
3	8	7		4	5			
2		1		5	9	6		7
9	7	5	8	6	1	4	2	3
6	4	8		7	2	9		5
1	2		5		7		6	
8	5	9	6	2	4		7	
7	6	4	1	8	3	2	5	9

Puzzle 174 - Very Easy

	9	1	8	3	2	5		7
7	8	3	5	6	4	9	2	1
2	4	5	7	1	9	6		
		8	1	2	3			9
		2	6	7	5			4
1	6	7			8	3	5	2
3	2		9	5	7	1	8	6
8	7	6				2	9	5
		9	2	8	6	4	7	3

Puzzle 175 - Very Easy

3	8	9	4	6	2	1		
		4	9	1			3	
	1	6			3	4	9	2
6	5	2	3	7		9		1
7		1	6	9	8	2	4	5
9	4	8	5		1	7	6	3
	2	7	8	5	6	3		9
		3	2	4	9	5	7	8
8	9	5		3	7	6	2	4

Puzzle 176 - Very Easy

		7	5		8			4
6	1		3	2	4	5	8	
4	8	5		1	7	2	9	3
9	6	4	1	7	2	8		5
2	5	3			9		7	
8	7		4	3	5		2	
5	9	6		8				2
	4	8	2	5	3	9		6
	3	2	9		6	7	5	

Puzzle 177 - Very Easy

9	7	3	6	2	4	8	5	1
2	5	6	8	1	7	4		9
4	1	8		9			2	6
		9		5		6		
3	8	4	2		1	9	7	5
5	6			8	9	3	4	2
6			1	3	2	5	8	
8		2	5		6	1		7
		5	9		8			3

Puzzle 178 - Very Easy

8			9	6	7		4	2
2	1	4	3	8	5	7	9	
9	6		2	4		8	5	3
3	8		4	9	2	5	1	7
5	2	1	7	3	8	4	6	
	4	9		1		3		8
1	7	2	8	5	9		3	4
4			6		3	9		1
6		3	1		4	2	8	5

Puzzle 179 - Very Easy

4	9	3	7	5	6	2	8	1
6	7	1	4	2	8	5	3	9
8	2	5	9				6	
		8	2		7	6		
2	3	4	5	6	1	7	9	8
		7	8	9	4		1	2
7	5	2		8	9		4	
3	8	6	1		2	9	7	
1	4	9	3	7		8	2	

Puzzle 180 - Very Easy

2	1	3	7	5				6
4	7		8	6	9		2	1
	9	8			2	5	7	4
1	8		3	9	7	2	6	
5	6	7	4		8		3	
3		9		1	6	4	8	7
		2	6	4	1	7		8
		6	2	8	5	9	1	
8	5	1	9	7	3	6	4	2

Puzzle 181 - Very Easy

	2	3	5		9	4	8	7
5	4	8	3	2	7	9	6	
	9	6	1	4	8	5		3
	7	4		5	3	2	1	8
3		9	7		2	6	4	
	1		6	8	4	3	7	9
	3	7	2	9			5	4
9	5		4			3		
4	6	1	8	3			9	2

Puzzle 182 - Very Easy

5		6	8	1	9	3	7	4
9	3			5	2	1	8	6
4	8		7	6		2	5	9
2				7		8	9	3
	9	8	6	3				2
1	5	3	9	2	8	4	6	7
	1			8	7	6	4	
	7	5	3	4	6		2	
	4	2	1	9	5	7	3	8

Puzzle 183 - Very Easy

1	7	5	8	9	2	4	3	6
	9	6	7	4			1	5
3							7	8
	4	1	3	8	5	6	2	7
6	3	2	4		7		5	9
5	8	7		6		1		3
2	6	8	5	3	1	7		4
7	5	9	6				8	1
	1	3	9			5	6	2

Puzzle 184 - Very Easy

4	3	1	9	2	6	5	7	8
7		5	1		4		9	
	9	2	5		7	1	3	4
2		3	7	9	8	4	6	1
	4	6	3	1		8		
	1	7	4		2		5	
	2	9	6	4	1	7		
1	7	8	2	5	9		4	
5	6	4	8		3	2		9

Puzzle 185 - Very Easy

1	4	7	6	9	2	8		5
	8		3	4	1	6	7	
3	2	6	5			4	9	
	9	3	1	7	6	5		4
4	6	2	8	3	5	9	1	7
5		1		2	4		8	
7	5	8	2		9	1	4	3
2	1	9	4	5			7	6
6	3	4	7	1				9

Puzzle 186 - Very Easy

1	8	4	5	2		6	9	
3	5	7	1	9	6	8	2	4
			9	8		5	3	1
4	3	6	7	5		9	1	2
5	7						4	8
9		8	4	3	2	7	6	5
8	4			6	1	2	7	9
7	2			8	4	1	5	
6	9	1	2	7	5		8	3

Puzzle 187 - Very Easy

4	8	1	3	5	9	6		7
6	9	7		2	8	5	3	1
5	3	2	6	1			9	4
		5		3	2	4	8	
2	4		7	8	6	9	1	5
8	6	9	1		5		7	3
9	5	4	8		1	3		
7		6	5	9	3			8
			2	6	4		5	

Puzzle 188 - Very Easy

9			4	8	6		2	7
4		5		7		1	8	9
	8		5	9	1	6		3
7		9	2		8	4	1	6
8	2	6			7	3	9	
3	1	4	9	6	5	2	7	8
5	7	2	8	1		9		4
6	4		7	2	9	8	5	1
1	9	8	6		4			

Puzzle 189 - Very Easy

5	3	4	2		7	8		6
7	6	2	4	9	8	3	1	5
1	8			3	6	4	2	7
6	2	8		5	4	9		1
3	1	7		6		2	5	
	9	5		2	1	6	7	
9	7	6			3		8	
8		3	6		2	1	4	9
2	4	1		8	5		6	

Puzzle 190 - Very Easy

	6	5	2	3		1	4	9
4	7		8		9	5	6	
9	2		4	6				8
7	9	4	3	2	1	6		5
3	5	2	9		6	7	1	4
1		6	7	5	4	2		3
6			5	7	3		2	
5	1	8	6	9	2	4	3	7
2		7		4	8	9	5	6

Puzzle 191 - Very Easy

	9	7	3	8	5	2	6	
6		1	7		2		8	9
8	2	5	1	9	6	4	3	7
	8	4	6	2	7	9		5
2	1		5	3	8		4	6
			1	9	3			8
7			8	5		6		2
	6	8	2		4	1	5	3
1	5	2	9		3		7	

Puzzle 192 - Very Easy

7	4		3			1	8	2
1		3	9		8		5	6
5	6	8	1	2		4	3	9
8		2	7	1	3			4
	3	7	5	9	6	2	1	8
9	1	6	4	8	2	3		5
6	9	4	8		1		2	7
2	7	1		5	4	8	9	3
	8			7			4	1

Puzzle 193 - Very Easy

2		1		8		5	9	3
5	7	4	1	3	9	2		8
8	9	3	5	2		7		1
4	3	7			2	8	5	9
6	1	5		9			7	
9		8	3				1	
		9	2		3	1	8	4
1			9	4	5	6	3	7
	4	6	8	7	1	9	2	5

Puzzle 194 - Very Easy

	2		9	1	4		7	
7	1	8	2	3		4	6	9
5		4	8	6	7		3	1
	4	2	7	5	3		1	8
8		5		9	1		2	7
1	7	3	6		8	9	5	
6	3	1	5	8	9	7	4	2
2	8	7	1		6		9	3
4		9		7	2	1	8	

Puzzle 195 - Very Easy

5	3	4	1		7	9	2	
9			5	2	6	8		
8	6	2	9	3	4			5
	8	7		1	9		3	2
3	2	5	7				6	9
		6	2	5	3	4		
2	1		4	6	5	3	9	7
7	4	3		9	2	6	5	1
6	5		3	7	1	2	8	

Puzzle 196 - Very Easy

6		8	7	3		5	4	9
2	4			5		6	7	
5	9	7		4	6		3	1
1	5	4			7		2	6
	6		5		1	3	8	4
8	3			9	4	1	5	7
4	8	6	2	1		7		3
		5	9		8	4	1	2
	2	1	4	7	3		6	

Puzzle 197 - Very Easy

		7	4		6			2
5	1				3	7	8	6
6	2		1	8	7	3	4	5
7	8	5	3	2		6	1	
		1	9		8		5	3
2	9	3	5	6	1	8	7	4
3	5	6		1	9	4	2	7
	4	8	7	3	2		6	
1		2	6	4	5		3	8

Puzzle 198 - Very Easy

	3	8	9		5	2	7	
6	2		7	1	3		5	8
5	7	9	2	8	6		4	1
2	5	6	1	3	7		9	
	1	3	5		8		2	7
	9	7		6	2	5		3
9	6		3	7				2
7	8		6	2	9	4	3	
	4			5	1	7		9

Puzzle 199 - Very Easy

1	8	9			3	2		7
4	5			1		6		3
7	6	3	2	4		5		8
3	1	8		9	5	4	2	6
	7	4	3	2	6	9	8	
9					4	3		
6	4		5	7		8	3	9
	3	7		6	8	1	5	4
8			4	3	1	7	6	2

Puzzle 200 - Very Easy

9	1	6			7	4	5	2
5		3	1	6	2	7	9	8
		2	4		9	3	6	1
1	2		6	7	3	5	8	
6		5		4	1	2		7
7	3		5	2	8		1	9
3		7	2	1	5	8	4	
2	6	1		3	4	9		
	5	8			6	1	2	3

Puzzle 201 - Very Easy

3		4	9	5	1	7		8
8	7	6	4	2	3	5	1	9
	5	1	7	6				3
7		9	6			1	8	5
	6	3	5	8	7	9	4	
5	8	2	1	9	4		7	6
4		5	2			8		
	1	8	3	7	9	6	5	4
6	9	7	8	4	5		3	1

Puzzle 202 - Very Easy

5	4		1	2		7	3	8
		1	7	4	3	2	5	
3		2	5		8		1	6
4	9	7		6	5	1	8	3
	5			1		9	6	7
1	3	6	9	8	7	5	2	4
9		3	8	1	4	6		5
8	1	5	6	7				2
	6	4	3	5	2	8	9	1

Puzzle 203 - Very Easy

5	4				2		8	7
7	1		5	6	8	9		2
2	6	8		7		5	3	
8		2		9		4		3
9	7	1	3	8	4			6
6	3	4					1	9
1	2			4	6	3	9	5
4	9		2		3	7	6	8
3	8	6	9	5	7	1		4

Puzzle 204 - Very Easy

1	7	2	3		5	6	4	8
5	3	9	8		4	1	2	7
4		6	7	1	2	9	5	
2	9		6		3		8	1
8	1	7		4	9			6
	6	5			8	7	9	
6				1		8	7	2
9	2		4	8		3		5
7		8		3		4		9

Puzzle 205 - Very Easy

6	9		1	4	2	8	3	
3	4		5			1	2	6
2	1	5	8	3		4		7
9	7	2	3	6	8	5	1	
	5	6	2			3		9
4	3	1	9	7	5	6	8	
7					9	2		
5	2	4	7		1	9	6	3
1			4		3		5	8

Puzzle 206 - Very Easy

3	4	7			2	8		6
9				7		1	2	4
	1	2	8		5	3	7	9
5	2	1	7	8	9	4	6	3
	9	8	3	5	6	7	1	2
7	3	6			4		9	8
2	6	3				9		5
1	5		9	6	8		3	
	7	9	5				4	1

Puzzle 207 - Very Easy

	2	7	8		4	3		5
	1	6			2	4	8	
	4	5	6		7	2	9	1
2		3	7	4		6		
4	6		5	8	3	9	7	2
5			2	6	9	1	4	3
6	8		3	2	5	7	1	9
1	3	9	4			5	2	6
7	5		1	9	6		3	

Puzzle 208 - Very Easy

7	6		5	3	1		9	4
9	1	2	6	4	7	5	3	8
4	5	3	2	9		7	1	6
	7		8			4	6	1
6	4			1		3	8	
		1	4		6			
3	2	6	9	8		1	4	7
5	9	7	1					3
1		4	3	7	2	6	5	9

Puzzle 209 - Very Easy

4	7		1	3	6	9	5	8
9	8	6			2		1	7
3	1	5	9	7	8	6	2	
1	3	4		8	5			6
		6	7		3	2	4	
5	2		6			3	1	
	4	3	5		9		8	2
	5	8	3	2	1	4	7	9
2		1	8	4	7	5	6	3

Puzzle 210 - Very Easy

6	4	9	8	1	7	2	3	
1	3	8	6	5	2	4	7	9
7		2	9	3	4	1		6
	1			9	6	8	5	3
2	8			7	3		9	1
9		3	1	8	5	7		4
8	2		5	6	9		1	
5			3				6	
3	9	6				5	4	8

Puzzle 211 - Very Easy

8	9	1		3	6	7	5	4
4		3	7	8	5	6		9
6	7	5	1	9	4	3		2
	8		5				4	1
1		9	8	6	3	2		5
2			4	1			3	6
7	6	2	3		1	4	9	8
5	3		9	2			6	
9		8	6	4	7	5	2	3

Puzzle 212 - Very Easy

1	3		6	9	4	2	8	
9	8		2		1	7	6	4
4		6	7					1
7	1	4		8	3	6	5	2
3	9	8	5			4		7
6	5				7	3	9	8
8	7	3	4	6		1	2	9
2	6	1	8		9	5	4	3
5	4		3			8	7	6

Puzzle 213 - Very Easy

8	6	9	2	4	1		5	7
	7	5	8	9	6	1		
1		2	7		5	6		8
5	3	6	4	1		7	2	
2	9	8	5		3	4		6
7	1	4		2	9		8	3
4	5	3	9	6	2		7	
9	8	1		5		2	6	4
	2	7		8		9	3	5

Puzzle 214 - Very Easy

6	7	8	2	1	4	9	5	3
	5	2	3			6	4	8
3	4	9	8	6	5	2	7	1
	8		1			4		7
9	3	1	7	4	2		6	5
		4			6		3	9
	9	5	6		8	7	1	
8	1	3	4		7	5	9	6
2	6	7	9	5	1		8	

Puzzle 215 - Very Easy

9	1	6		3		4	8	
2		7	6	5	8	9	3	1
3		8		1	4		2	
	3	1				7	5	
8	9				1	3	6	2
	7	2	3	8	6	1	9	4
4	2	5	1	6		8	7	9
1		9	8	2	7	5	4	
		3	4			2		6

Puzzle 216 - Very Easy

8	7		2	1	5		6	
	3	5		6	7	8		2
	2		9		3	4	7	5
		9	8	7	4	6	5	3
4		3	1	5	2	7	9	8
7	5		6	3	9	2	4	1
	9				8	1	3	
		1	7	9	6	5		
5	4	7	3		1	9	8	6

Puzzle 217 - Very Easy

3		8	6	1	4	9		5
	6	4	3	2	5	7	1	8
2	5	1	9	7	8	6	3	4
			1		3	8		2
5	1	2	7	8		4	6	
4	8	3			5	6	1	
		6	5	3	1	2		9
1		5	8	9	7	3		6
8			4	6	2	5		

Puzzle 218 - Very Easy

7	6				2		4	
4	2	9	7	6		3		5
8	5	1	4	3	9			6
1	8	7		4		5	2	9
		6	1	2				
2	3	4	9	8	5	7	6	1
	7	8	3	9	4	1		2
3	4		8		1		9	7
9		5		7	6	4	8	3

Puzzle 219 - Very Easy

3	8		2	4		7		1
6	1	2		8				9
	4	7	6	1	9		3	8
9	2	4		5	6		7	
7	3			4				6
1	5	6	8	7	3	9	2	4
4	7	3		9			1	5
8	6		7	3	1	4	9	
2	9	1	4			3	8	

Puzzle 220 - Very Easy

	6	4		3	1		8	
5	3		4	8	2	1		9
	1	8	9		7		4	
7			2	9	3	6	1	4
	2	1	8	7	6	9	3	5
			1	5	4	8	2	7
8	5	3	6	2		4		1
1	7		3	4	8	2	5	6
	4	2	7			3	9	

Puzzle 221 - Very Easy

	6		9	5		7	4	
3		7	1	4		2	8	5
8		4	2	7	3		1	
		8		6	1	7	4	
7	4	6	5	3		1	2	9
	3	9	4	2	7	5	6	8
6	1	2	3	8	9			7
		3	7	5	2	6	9	
9	7	5	6	1	4	8	3	2

Puzzle 222 - Very Easy

4	2		3	1		8		7
1	7		4	8		2	5	3
8		9	2	5			6	
		4		2	5	3		9
3	5	7			4	6		2
9	1	2	6	3		7	4	5
5			8		2			1
2	9	8		7	1	4		6
7	6	1	9	4		5	2	8

Puzzle 223 - Very Easy

	9	7		6	3	4		5
4	5	1	8	7		6	3	
	2		5	1	4	8	7	9
2	8	9		5	7	1	4	3
	3		9		8	5		7
6		5	4	3	1		2	
5	4	2	7	9	6	3	8	1
7		3	1	8		2	9	
	1	8		4	2	7	5	6

Puzzle 224 - Very Easy

		1			9	4		3
6		7		1		8	2	5
4	8	3	2				9	
9	2	8	3	5	4	1		6
7		5	1		6	2	4	9
	6	4	9	2	7		3	8
5		9		3	2	6	8	4
3		6		4		9	1	
8	4	2	6			1	3	7

Puzzle 225 - Very Easy

4	9		3	2	5	8		1
1	3	5		8	7	2		9
	7	8		1	9	3	6	5
5	2		8			7	3	6
3	8		1	7	6		5	2
	6	9		3			8	
		1	9	4	8	5	2	7
8	4			5	1	6		3
9	5	2	7	6	3	4	1	

Puzzle 226 - Very Easy

			3	5	1	7	8	
3	1		8	4	2	9	6	5
5	2	8	7	9	6	4	1	
1		4		8	5	3	2	7
6	7		9	2	4		5	8
8		2	1	7				
7	3	5	4	1	8			6
	4		5	3	9	8		1
9	8	1		6	7		3	

Puzzle 227 - Very Easy

	4	2	5	7	6	3		
5		9	1			6		4
3	6	1	2	9	4	5		8
4		6	7	2	8		5	3
9	8		4	3		2	1	6
2	3	5	6	1		4	8	
6		8	9	4	1		3	2
1	2		3	5		8		9
7	9	3	8	6	2		4	

Puzzle 228 - Very Easy

	6	5	2	8		4		7
	9	4	5	6		8	1	
		8	9		7	6	3	5
6	8		7	5	2	9		1
1	2			3	6	5		
4		7	8		9	2	6	3
9	3	1	6	2	8	7		4
8	4	6	1	7		3	2	9
5	7	2	3	9		1		6

Puzzle 229 - Very Easy

4	9	5			6		8	1
8		1	3	4	5	2		6
3	6	2				5		4
1	8	7	4	6		9	3	5
2	4	9	7	5			1	
5	3		1	8	9	7		2
7	1		5	2	4	8	6	9
6	2	4	8	9		1	5	3
9	5	8	6			4	2	

Puzzle 230 - Very Easy

	7		5	1	2			
9		1	8	7			6	
3	8	5	4			1		7
7	9	6	3	2		5	1	4
2	4	3	9	5		6	7	
	5	8		4	7	3	9	2
		2	7		4	9	8	1
8	1	9		6	5		4	3
	3	7	1			2	5	6

Puzzle 231 - Very Easy

6	9		7	2	3			
5	8	4	9	6	1	7		3
3	2	7	4			1	9	6
2		8	3		6	5		7
4	5	3	1	7		9		2
7		9	2	5	4	8	3	1
	4	2	8	3	7	6	5	9
8	3	6	5	1	9	2	7	
		5	6		2		1	8

Puzzle 232 - Very Easy

	9	2	8		6	1		
	1	6	4	2		8		3
4	7		9	1	3			
	5		7	4	8	9	2	1
2				9	1		8	5
1	8		2	3	5	7		4
9	6	5	1	7	2	4	3	8
8	2	1		6	4			9
7	3	4	5	8	9		1	6

Puzzle 233 - Very Easy

8	3	9	5	6	7	1	2	4
6	5		2	4	1		3	9
1	2	4	3	9	8	7	5	6
			4	5	6			3
3	6	1	7		9		4	5
4			1	3	2		8	7
5	8			7				1
			8		5	3	7	2
7	1	3	9	2			5	6

Puzzle 234 - Very Easy

7	2	4	6	5		9	3	
	9	1	2		3	6	8	
6		8	9				2	
1	4			6	9			2
2	6	3	4		7	1	9	5
9	8	7	1		5	4	6	3
3	5	6	7	9	1	2		
4	1	9			2	7	5	6
8				4	6		1	9

Puzzle 235 - Very Easy

9	6			7	2	5	3	8
2		7		3	6	1	9	4
1	3	5		9	4	6	2	7
	4		3		1	7	5	2
		6		4	9	3	8	
	1		2	5				9
3	5		4	2		9	1	6
6	7		9	8	3	2		5
4	9	2	6					3

Puzzle 236 - Very Easy

8		1	2	9	6	7	4	
7	4	5		8	3		6	2
2	9	6		4	5	8	1	3
	7	3	5		2	4	8	
	2		9	6	8		7	1
			4		7	2	5	6
3	5		6	2	4	1	9	8
4	8	9				6		
	6	2	8	7		5	3	4

Puzzle 237 - Very Easy

7	8	1	3	5	2	6	9	4
9	2	3	8		6	7	1	5
	5	4	7	9	1	8		3
3	1	2		7	8	4		
5	6			2	4	3		8
4	7	8	9	3	5			2
8	9	7	2			5	4	
1	4		5		9	2		7
2	3	5		1				6

Puzzle 238 - Very Easy

4	6	8	2	3		1		9
	2	5		9	4	7	6	
3	7		5		6	8		4
2		4	3	7		5	1	6
9			6	8	5	4	7	
6	5	7	1	4	2	3	9	8
7	4			5	8	6		1
5	1	6		2	3			7
8	9	3	7	6	1	2	4	5

Puzzle 239 - Very Easy

2	3	4	9	7	5	6		8
			3	2	8			4
5		8	1	4	6	3		7
4		7	6	9	3		8	2
9		2	5	1	4	7	6	
1		3	2	8	7		9	5
3				6	2		7	
		1	4	5	9	2		6
6	2	5	7	3	1	8	4	9

Puzzle 240 - Very Easy

3	5	7			2		6	
1	6		7	8		5		9
9	8		5		1	2		7
4				3		6		5
5	9	6	1	7	4	8	2	
	3		9	5	6	4	7	
8	1		6		7	3	5	2
6	2	5		1	9	7	8	4
7	4	3	8	2	5		1	6

Puzzle 241 - Very Easy

		1	5	8	7	3	6	9
3	6	7	4	1	9			8
	9	5		2			4	1
6	1		8	5	2	4	3	7
4	5		9	7	3	8	1	
7		8	1			5	9	2
9	8		7	4	1		2	5
5	2	4	6		8	1	7	
1	7	6	2	3		9		

Puzzle 242 - Very Easy

7	4			6	3	5	2	8
1		6	7	8	5	3		9
		8		9		1		6
9	8		5	7	6	2		
6	7	2	3		1	9	8	
	1	5	8	2	9	7	6	
2	6		9		8	4	5	7
	5	7		3		8	9	
8	9	1		5	7	6	3	2

Puzzle 243 - Very Easy

6		9	8	7	1	5		4
	8		6		5	3		
1	5			3		8	9	
3	1	5	4	6	9	7	8	2
8	4		7		2	9	3	5
	7	2	3	5	8	4		
5	9	8	1	4			7	3
	2	1	5	8		6		9
4	6	3	9		7	1	5	

Puzzle 244 - Very Easy

7	9	4	6	8	1		5	
5	1	8	9		3		4	
3	2	6	4	7	5	1		
4		1	3			5	6	
	5		1	6	4		9	
6	3	9	8	5	7	4	1	
1			3	6	9	2		
9	4	3	2		8			5
2	6	5			9	8	3	1

Puzzle 245 - Very Easy

6	7	3	9	4	2			
8	1	5	6		7	4		2
9	2	4	1	5	8	3		7
		7	4		1	2	5	8
1	4	9	8	2		6	7	3
2	5	8		6	3	1	4	
4	3		2	7	9		1	5
5	9	1	3	8	4	7	2	
	8	2			6	9		

Puzzle 246 - Very Easy

9	5	1		8		4	6	7
2		8	6	5			1	3
	6	4		9	1	5	2	
	1		9	2	8	3		4
	2	9	3	4	5	6	7	1
4	3	5	1	7				2
	9	3	8	6				5
6	4		5	3			8	9
5	8	7	4	1			3	

Puzzle 247 - Very Easy

5	3	9		4		1	2	7
8	6	1	7	5	2	3		4
4	7				1	6	5	8
3			2	8		9	6	1
2		8	1		5	7	4	3
7		6	9	3		5	8	2
6	8		4	1	9	2		5
9	2	5	8	7		4		6
1		7	5		6	8	3	9

Puzzle 248 - Very Easy

5	7	2	6	3	8	9	4	1
1	8		4		7	3	6	2
3		4			1	5	8	7
6		7	1	9	2		5	
4	1	8	3	6	5	7		
2		5		8	4	1		
9	5	3	2	7	6	4	1	8
7	4		8	1	3			
8	2	1	5		9	6	7	3

Puzzle 249 - Very Easy

1	8	7		5	4	9	6	3
2		3	7	6		8	1	4
4			1	8		5	2	7
6		1	9	7	8		4	
8			6	3	1		9	
	3	9	5	4	2	6	8	1
3			8	9	5	2		
		8	4	2	7	1		9
9		2	3	1	6			

Puzzle 250 - Very Easy

8	4		2	5				9
9	7		4	3	1	2		5
3	5	2		6		4	7	
4	8	3	1	7	2	5		6
	1	7	5	9		3	4	
5	6	9		8	4	1	2	
	3	5		4	8	9	1	2
1		8	2	5	3	7	6	4
		4	9				5	3

Puzzle 251 - Very Easy

		2	9	7	5	3		
8	4	7	1	2	3		9	
5	9	3	6	4	8	7	2	1
4		9		3		1		2
2		8	4	6		9		3
7	3	1	5	9			8	6
9	2	5	3		4	8	6	
1	8				9	2	3	4
	7	4			6	5	1	

Puzzle 252 - Very Easy

	4	9	2	8	6	5	3	7
8	3		7	9		2	1	4
5		7	1			6	9	8
		4	5	2	9	3	8	6
	6	2				7	4	
3	8		4	6	7		2	1
		8	9	4	2		5	
4	9	1	6	5	3	8	7	
2	5			7	1		6	

Puzzle 253 - Easy

7	5	4	8		2			9
1						8	7	3
8		9	7	1		5	4	2
			5	6			9	1
	9				3	7		
5	6		1				3	
		2				9	5	
		8	3		5		2	
			4				8	

Puzzle 254 - Easy

		4	7		1		5	
7		1	3	6		9	2	4
		8	5	4	2			
3	8	6		2	5			
4		5	6	8			3	
	1			3		5		
			9	5	6			1
				1		3		
		9	8					

Puzzle 255 - Easy

5	9		1	8	3			2
	8		9	2		4		3
					5	1		8
6	2		8	7		5		4
8	5	9				6		1
	4			5		8	2	
	6	3			8		4	7
9	7	8	3		6			
	1	5						

Puzzle 256 - Easy

			1	9		2		7
	8		2				9	
	7	2			3			
8		1	6		5	4	2	
	6	4			9			
2	9	7	3					8
4	2	8			1	7	6	
6	5		7		2			1
	1				8			2

Puzzle 257 - Easy

	2	3	8	4		1	5	
	4	6	7			2	8	9
		8						
	9		5	1		3	2	
6	8	1	2					
	5					6	9	
2		5			8	7		
1						9	6	
		9	7		6			4

Puzzle 258 - Easy

7	5	2		1		9	8	4
		1		9	8	5		2
4								1
			4				9	
	9	4		6		7	2	5
5	7			2	9	4		6
		8	6				4	7
6		7	9					8
2	3	5	7					9

Puzzle 259 - Easy

				6		9	4	
		6				5		8
2	1			5		6	7	3
9			1		5		3	
1		7		4	3	2		5
	2		6			9	1	
		2				3	5	
	7	9		3		1	2	6
	5	1		9	6	7		

Puzzle 260 - Easy

5	3	9	7	6		1	2	8
2	4		3	5				7
8	7	6	2		9			
		2	5	7				1
			4	9				2
3				8	2			
				2			9	6
	2	7			6		5	3
	9		8	3		2		

Puzzle 261 - Easy

	7		1	4		3		
3				8		9		
8	4			6		5	2	
9		2	4					7
7	6	5	3		9		2	4
	1	3	6	7	5			
	3	4	1			7		
6			7	9		2		
1		7			3	8	4	5

Puzzle 262 - Easy

			5			8	1	
3	1			9				
2	6		8	3	1		9	
		3			5	6	2	9
			6			3		5
		9	4	2	3		8	
1			2	5	7			
7		2		4	6	1	5	
3	4			1		2	7	6

Puzzle 263 - Easy

1	9			5				
2	8	4					9	1
7	5					2		
	6			8	3		5	4
		7		6	5		8	
	1		2			3	6	7
		5		7	1	4	3	6
			5					2
	7					8	1	

Puzzle 264 - Easy

			2	8				6
	2			6	5		3	
			4					
4		3			1		7	2
			3	7	8			1
7		6						9
3	7	9	8	5	6			
8	4	5	1	2		9		3
2	6				4	8		

Puzzle 265 - Easy

				1			5	
4					3	1		
		1	5		6		4	
5	8	9	3		1	6	2	
6		3	9		7			8
7		2	8		5			
			1		4		6	5
9			2	5	8			
	5	4		3		8		

Puzzle 266 - Easy

2	3				6	4	8	9
6	5		3	8	9	2		7
8				2				
5	8		4			9	2	
9	4	1			2			5
	2			5				
1		5		3				4
	7		2	9	5			
		6	1		8	5		2

Puzzle 267 - Easy

3				7		1	8	5
			3			2		
8	2					3	9	7
		6			3			1
			6	9		8	5	
		8			5		3	
			5	3	7	9		
	3	9		8	4		7	
	8		1		6		2	3

Puzzle 268 - Easy

6		8		9		5		3
					4	1	2	
1	4	2		5			6	
2	8		6			7		
					3			1
4		7		8	6			2
8								6
7			9		1	3		
3	5		4	7	6		1	

Puzzle 269 - Easy

		8	6	7				
		2		1		8		
1		9	8	3	2		5	7
4	2	3	1		5	7	9	
6			3		7		4	
	9	7	2					8
9	3	5				1		2
8		4		2				
2		6				9		4

Puzzle 270 - Easy

		3						
		8					2	4
2				1		8	7	
	3		6	1		2		5
	2			5		7	1	6
	6	1		7	4	9		
	1		3	9	7			8
	8			4		6		7
7	9	4			6	1		

Puzzle 271 - Easy

2					6			
			9		1		7	2
7	9		3	2	5	1		
9	7	5		1				
6	3	2		7		9		
	1				3	2		7
	5	7			4	6	9	
							3	
		6	1	5	9	7		4

Puzzle 272 - Easy

4			8		3	2	7	6
	2	3	1		6	9		5
	6			4		8		
6		1	9			5		8
5			6	1		2		
	3			5		6		
	8		3					
	7	9			4	8		
3	5	6				4	9	2

Puzzle 273 - Easy

	9	7	5	4	8			1
		8	2			5		7
					7		8	6
	4	6	3					
		3	8	2				4
	5	2	4		9	6	1	3
					4			
2	6		9		1		3	
7		9		8				

Puzzle 274 - Easy

8		4	1					9
7			6					
5	2		4			7	8	
	7			5		9	2	
	5				4	8		
3		8		2	9	6	4	5
1		5	2	4	8			7
	8	7				1	9	4
6	4	3		1			5	8

Puzzle 275 - Easy

4					2		1	6
6	2		4	8			3	5
3	5		6				9	2
2		9		4			7	
	1	5		9			6	4
	6	4			8		5	9
		3	1	2		6	4	
5	7				4	9	2	
1				6			8	

Puzzle 276 - Easy

	9						7	
6					9		4	
		8	5	6	7		3	
7		5		1	3			
	3	1		8	2	5		4
	8		4		5		1	7
3	6	9	2			4		1
8	2	7				6		
1	5	4		3		7	8	2

Puzzle 277 - Easy

	7			4			2	
	6		9	2	1		4	7
2					6	1		
3		2				7	1	9
			7	5				8
		5		1	7		3	4
6	2		5	3	4		9	
4	3	1			9		7	2

Puzzle 278 - Easy

5			7		8	6		
		8				1		4
					1			8
8	2			1	9			
9	6	7	3	8	4			1
4	5	1			2	3		9
			1	2	5			
		5		3	6			7
		2	8		7	4		5

Puzzle 279 - Easy

		8			6			
9		7	4					8
	2	6	7	8		4		
		9			2	5	3	
5	7	2	3					6
	6					9		4
6	1	4			5			
7		5		6				2
2		3	1	4			7	

Puzzle 280 - Easy

1			5	3		8	6	9
	4							7
	3		1		6	4	5	2
	8	4	3	2	7	9		
2	9			5	1		8	
7			9				2	
				9			7	3
4		6	7	1		2		
3		9	8			5		

Puzzle 281 - Easy

	9		1					
	4	8	3	5			2	
		3	9	6		4	1	
3		7	8		4			
9	6	5		1			4	7
	1				5		9	
5	7	2	4		1			9
6		1			9			
		9	5	2	6	1		3

Puzzle 282 - Easy

2		5			4	7	1	
	4	3	6			5		
					8	4	3	
			4	8	1			
3		2					4	
4	8	6	3		5		7	
		8	1	7		6	5	
	2	1			3			7
5		4	8	9	6	3		

Puzzle 283 - Easy

```
5 . . | . 1 . | 3 . 6
. . 3 | . . . | 1 8 .
. . 2 | . . . | 4 9 7
------+-------+------
8 . 6 | 5 . 4 | 9 . .
. 5 . | 9 2 1 | 7 . 8
9 . . | 6 . . | . . .
------+-------+------
. . . | . 9 . | 8 . .
1 . 8 | 7 4 . | 6 . .
7 4 9 | 8 . 5 | 2 . 1
```

Puzzle 284 - Easy

```
7 . . | 2 1 9 | . . .
1 . . | 8 . 5 | . 2 7
. . 3 | 4 . . | 1 9 8
------+-------+------
. . 9 | 6 5 . | . . 1
. 1 . | . 2 3 | . . .
4 3 . | 7 8 . | . 6 .
------+-------+------
5 8 . | . 9 2 | 6 . 3
. 6 . | . 4 7 | . 8 9
. 7 . | . . . | . . 4
```

Puzzle 285 - Easy

```
. 9 6 | 7 . 1 | 5 8 4
. . . | . 6 . | 9 3 .
. . 5 | . 9 . | 6 . 7
------+-------+------
. . . | 6 5 . | . 7 1
. 5 . | . . . | 3 9 6
6 . . | . 3 2 | . 5 .
------+-------+------
9 6 7 | 2 . 3 | . . 5
2 3 . | 5 8 . | . 6 .
5 . . | . 7 . | 1 . .
```

Puzzle 286 - Easy

```
. 8 9 | 2 . . | 6 4 . 5
```

```
. 8 9 | 2 . 6 | 4 . 5
. 5 . | 4 1 9 | 2 8 6
2 4 6 | 3 . . | . . .
------+-------+------
. . . | . . . | . . .
. 7 . | 6 2 1 | . 5 .
8 . . | . . . | . 1 7
------+-------+------
. 2 . | 9 . 4 | . 7 8
. . 4 | 1 8 5 | . 2 3
. 3 . | 7 . 2 | . 4 .
```

Puzzle 287 - Easy

```
. . 1 | 2 5 3 | 7 4 6
4 7 2 | . . . | . . .
6 . . | . . 4 | . 8 1
------+-------+------
. 6 . | 7 8 2 | . 5 .
. 8 3 | 6 . . | 1 7 .
. 2 . | 1 3 9 | . . .
------+-------+------
. . . | . 6 . | 5 2 .
. 5 9 | 4 2 . | . . .
2 1 6 | . 9 7 | 8 . .
```

Puzzle 288 - Easy

```
. 2 . | . 1 8 | 9 . 6
8 . 7 | 2 . 6 | . . 1
9 6 . | 5 . 7 | 2 . .
------+-------+------
. 8 6 | . . 9 | 3 . .
. . . | 6 5 . | . . .
7 . 2 | . 8 3 | . 5 .
------+-------+------
. 9 . | 8 . . | . 6 .
6 . 4 | 1 3 2 | . . 8
. 7 8 | . . . | 1 3 5
```

Puzzle 289 - Easy

						4	8	5
5				6		9	1	
		9				6	7	3
7						8	2	
6	1	5		7		3		
		4		9	3			
4	8	7		3		1		6
9	5	6			7			4
	3	2		4		7		

Puzzle 290 - Easy

6		1				5	9	
	2							7
	7		5			4		2
		2						4
			1	7	4	3		9
		7	6			3	1	5
		8	4	3	7	9		
	5	3	9	6		2	4	
9	4			8				

Puzzle 291 - Easy

		4	2					1
6		7	5	4		3		
9	5		3	6			2	4
	9	3	1		4	5	6	
		8	9				1	7
			7					
	4				2	1		6
3			6		7			5
			4	9	5		8	

Puzzle 292 - Easy

3	9			4				1
	7	1		2	8			
4	5	2		6	1			9
8			2			6	1	7
			7	6			3	
	6		4		3		2	
7	4	6				1		
5						7	4	
2	1	9		3		5		6

Puzzle 293 - Easy

6			3	1		7	2	8
							5	
5	8	7				3		1
3			8	7		1		
9			3	2			4	
4	6		5				3	
		2		5		4		6
	5		7	2	4	9		
7	3	4	1	6	9	5	8	

Puzzle 294 - Easy

7			1		5	2	6	3
	6	9			8		1	5
			3	6				9
	3					7		
4		7	2		9			6
6						2		
	7		6	5			4	2
5	2	6	4		7	9	3	8
1	4		9	8	2			7

Puzzle 295 - Easy

		8	3					2
3		2	6	7	8		9	
		1		9	2	3		
	1	3			9		2	
		9	1		5	7		3
	4	5	7	3	6	9	1	8
			9	6	3		5	
8				5		4	6	
5			4		7			1

Puzzle 296 - Easy

4	6					5		
5	2			7	8			
		8	9			1		
		6	7				3	4
8				6	1		9	5
				8	4	2	1	6
	1		8	9				
	7			1				9
	8	3	4	5	7	6	2	

Puzzle 297 - Easy

			5	1				9
	9	3	2				7	
	5	4	3	9		1		2
	8		6			4		1
5				2	1	3		8
9			4	8	3	7	6	5
4	6	1		7	2			3
				4		9		
8	7		1	3				6

Puzzle 298 - Easy

	1				8	7		
5		8				9		6
	9						4	
8	7						3	4
	6		8	3	4		7	2
		4	6	5		1		
		7	2	8			9	1
4	2	6		1		3	5	8
1			5	4		2		7

Puzzle 299 - Easy

8					4		6	1
6			1		8	7	4	
3			6	7				9
4	1	9				8	3	
7	6				9	4		
5	3	2	8	4			9	6
1			9		2		7	
2	7	3	4		1		5	
			7	5				

Puzzle 300 - Easy

			6				1	9
		4	9	7		6	5	
					1		2	
7	3			2	5	9		
1						2	8	5
	5	2	1					
6	2	7		5		1	9	
3		5			9		6	7
4			7		6	5	3	

Puzzle 301 - Easy

	2		3	4				
4	1		2	8	6			3
3		8		9		2		
2				3	9		6	
	5	6		1		8	3	
8		3			2	7		1
				6	8		4	7
	9		1			3		8
5	8	4			3	1	2	6

Puzzle 302 - Easy

		1	9		4		5	3
		8	1					
5		3		2			9	1
		9		5	6	4	8	7
4		6			7	1		5
				8		6		
8	3			6		5	1	
	1					9		8
	7	5		1	2	3		6

Puzzle 303 - Easy

					5		1	
		4		1	8			
						6	3	8
							6	3
9	5	8	3	6	2		7	1
	2	3					8	5
5	8			4				
4				2		8		
7	6		8	5	1	3	4	2

Puzzle 304 - Easy

				5	8			
	2		4			7		
8		3	2		1	9	5	
2		4		1		5	6	
3	1	6	5	9	4		8	
9	8	5	7	2		1	4	
	3			5	7	6		
								8
		9		8	2	4	1	

Puzzle 305 - Easy

	1				9	7	8	3
7						9	6	
		6	5				1	
			8	1	5		9	
5		1			2	8	7	4
8	9	2	4		6			1
3	5		9	6			2	
		9		3		6		
		7			1	3		

Puzzle 306 - Easy

			9			6		5
	5	7	8	4		2	9	3
			5	6		1	4	7
	2		1		5	9	7	
7						5	2	
5				7	4	3		1
4	7			2	6			9
	6	1		5		4	3	
		9	4		8			

Puzzle 307 - Easy

1		6				2		9
	2	4		8		5	3	1
9					2	8	7	
8	6		5		1			
	9					1	5	3
5	4			9		7		
3	5			4	6			7
					7	6	8	
	7				8			

Puzzle 308 - Easy

	1				2			3
		5		4		7	6	8
8						9		
	3	2	6		1	8	9	
1		9	7	3	8	5		2
	8	7				3		
3			1			6		
	4	1	8	7			3	
6			4	5	3			

Puzzle 309 - Easy

8		9	4	2	5	6	1	
7				6				4
			3	7		5		
	5					2		1
	4				6			5
	8		5			3		6
5		2	8	9				3
	9		3	7	1	5	2	
3	7							9

Puzzle 310 - Easy

	4		8	7		5	9	
5	7		2			8	1	4
	8	9				3	7	
			6					
6			7	2	3			
	3		5	8		6	2	9
4			8		2		6	3
	1		6	7	5	2		8
	6	8		9		1	5	

Puzzle 311 - Easy

	4		2	5			3	1
3	2	7						
			9	3		4	6	
		2	9		4			
4	7	3			1	6		8
	8	5		6		4	1	2
2	3	6	8	1		7	4	
8		4						3
7	5	1		3			6	

Puzzle 312 - Easy

		6			9	3		4
	4	3		7				
8		9		3	4	1		
	5		3	9	1		6	
6		7	5	8	2	4		
1	9	2	4		7			
							4	2
9	6				5	8		1
		1				7	9	

Puzzle 313 - Easy

7	3		4		1			
4						3		6
		6	3		8		5	7
3	1	8	5		7		6	4
	6							
				1	2		8	
6	2	3			9			5
8		7		4	3		2	9
	4	1	2				3	

Puzzle 314 - Easy

	2	7		4	6	5		
	1		5	3		7	2	8
	5							
2	8							1
			8	9				6
1	4		6	2		8		
4	7	8	9	1	2		5	3
	9	1	3					
3	6					9		7

Puzzle 315 - Easy

2			5			1		
3	5	7						8
1					7	5		
	8				5	7	4	
7			8			9		3
5			7		4	8	2	
8			1	4	9	3		
		3			6	4		
4	7	6		5		2		9

Puzzle 316 - Easy

9		3	8				2	7
	1		3	2	7	5		8
		7	5		9		4	
4	2	1	9			8		
	6			3			7	
3		5			2		1	
			1	9	6		5	
	6		2	5	3			
	9				8		3	6

Puzzle 317 - Easy

7		4		8		5	9	
	6	5		9	2		1	
9		1	5				6	4
3						9		5
			3	5	9		7	
6			2	7	4			1
2		6		4				7
	1		8	2	5	6		
				1		4	2	8

Puzzle 318 - Easy

	7	1	4		5		9	
	3			6	1		2	5
	6						1	
1	5	4	2	8	6	9	7	3
			3	4				1
		2	5	1	7	8		6
5	2					1		
7			1		8		6	9
		6			4		3	

Puzzle 319 - Easy

7		5	2	1		8		4
8			5	7				2
			8	6	9		7	3
		1				2		
		9	3			6	4	8
3	2	8		4				1
			1	3				
	5	3			8			6
4	8			5	7	3	1	9

Puzzle 320 - Easy

6	9							3
3	2		7		5	1		
	8		3			2	9	
4		9	5		8	6	3	
		8		7			1	
			4		3		7	8
	3	6						1
	5	1	8				6	
	6	2			9			4

Puzzle 321 - Easy

	8		4	3	9			6
	1		5	7	2		8	
	2					4	7	5
		1				7	9	2
		5			4	6		
6	7	3				8		4
	6		8				1	
	5		9		3	2	6	
				1		9		

Puzzle 322 - Easy

	2	5	4					6
		8	2	6	7	9	4	
	6	4	5		3		1	7
2			1	7				
5			9		8	6		2
	8	6		5		1		
	3		8			5	2	
1			6	3	5	7		
	5	9	7				6	3

Puzzle 323 - Easy

3	4							
						4	1	2
		1	5		4	3	9	
1	7	2		9		6		
6	3			2				
		5	4	7		2	3	
5	9		7				2	
4			2	1			7	
7	2			5		1	4	9

Puzzle 324 - Easy

5			9		2	4	1	8
			3					9
			1			5	3	6
4		9	5	3	8			7
		8	1	2				
7	3		4		9		5	2
3	4	6					9	
8			6	9	1	3	2	
				4			6	8

Puzzle 325 - Easy

3	1		6	5				
2		9		8	1		6	3
5	6			2		7		
8				9		4		
		3	2			8		6
4		1	8	6	7	9	3	5
			5		8			
	8			4		3	5	
	3	5				6	8	4

Puzzle 326 - Easy

9	4		2		3			8
	5	2	9		7			
		1					4	
5	2	4		7		8		1
3	7			9		6	5	2
	6	9			5	4		7
			6	3	2			5
6	8		1			7	2	
2						9	6	3

Puzzle 327 - Easy

		2	4	5	6	9		
9		5	7			4		
1		3	2					7
	9	4	5		7			2
	5		6		4	7		9
			9			6	5	
	8	9	3	7	2			6
7	2	6		4		8	9	
	3	1	8	6				

Puzzle 328 - Easy

1		2	8	4	6	3	7	
	4	6	5	3	7	2		
7		3	9			5	6	4
2		1	6			8	4	
	9		3					6
	7		2	5		9		
			4		8	1		3
8				2				
			7	9	5		8	

Puzzle 329 - Easy

				9	6	7	3	4
9	3		8			6	5	
		7	3	5				
8	2		6	4				5
		3	1	2		4	6	
7			5					
			4	6		5	8	7
3		4			5		2	
6	5	8	7		2	9	4	

Puzzle 330 - Easy

3			2		4		7	
5			8	6			1	
		9		5	7			
8		5		7			4	
1	2	7			9	8	6	
	9	3		1			5	2
	1		4	3	6	2		
	3		7				1	4
2		4					3	

Puzzle 331 - Easy

					5	1	6	3
6			8	4		9		5
	5	7	6		1	8		
4	2	9			8		5	1
3	7	5		1		6	8	2
	6		2	5		3		
		8					9	6
1	9		5	8				
5		6	4			2		

Puzzle 332 - Easy

	6		5			7		1
1	3		8	7		9		
	2		1				8	3
	1	5		2				
2			7		5			4
			9	6				
3	4			8		2		7
				4		1		8
	8	9	2	5	1	3		6

Puzzle 333 - Easy

3	5	2	7	8		9		1
6			2	5	3	4		
	8		6				5	
4	2	5				1		
8	3		9		6	2	4	
9	1		5		2			7
			8		7		1	
				3				
		3	1			6	8	4

Puzzle 334 - Easy

8						7		
9	1			4	3		6	
	3			8	7		2	1
5				1	4		7	
1			6	7				5
							8	
3	4		8		1	6		
	7	1				8	9	
2	5	8		6	9			4

Puzzle 335 - Easy

6		5				2	1	
4	9						6	
2	7				5	4		
				6				
7			3			9		
3	2	6	4			5		
5	3		1				4	
	1	4	5	3	6	7		2
8	6		7	9	4			

Puzzle 336 - Easy

4							3	9
3				8	2	6		4
8		6	3			1	5	2
9	4			2	5	3	1	6
6		5				2		
2			6	9		5	4	8
		8			9	7	6	
7	6	4	1	3				
1				5			8	

Puzzle 337 - Easy

			5			3	1	
						8		
2	3	1		4	9	7		
9		3			5	2		7
	8	7	6	2	3			9
		4		8				
4	9				8		2	
3	5		1		4	9	7	8
7	1		3			5		

Puzzle 338 - Easy

			5			1	9	6
6		1		7			4	
5		4	3	6	1		7	2
	8				9			1
9		5		4	3		6	
	3	7	6			5		
	6	2	4				1	
1		9	2	8		6		
				3				

Puzzle 339 - Easy

9			7			6		
2	8		9	4	1		7	3
				8	5	4		9
	5	8	4			1		7
	1	2		3				
4	6			5			8	2
			2		8	7		
	2			7	9	8		6
8		3		6		2		

Puzzle 340 - Easy

4				7	5			
1				6	2	8		
		6	4			9	5	
9	1			4	3			8
	3	2		8			6	
8	6				7		3	5
3							9	2
		1		5				
5		7	8		4	6		3

Puzzle 341 - Easy

	4	2						9
7	3	9	2	6		5	1	
6	8		7		9	2		
2	5			7		9		1
	9				1	7		5
4	1					2		
9		3	1				5	2
5				9	3		6	
	7	1	6			2	4	

Puzzle 342 - Easy

2		9			6	7		
	7			4		9		
4						5		2
7			6	2		3		9
9			4		3		8	
	6			5			7	
6	4		5	2	9	8	1	7
1	9		3	8		6	2	
5				7				3

Puzzle 343 - Easy

			7					9
2	4	9	6	3	1			8
		5						
	8		1		4			
7	3			9		5	1	4
1	9						2	7
				1		4	6	
		3		4	8	1		2
	1		2		6		8	3

Puzzle 344 - Easy

5	7	4		1	6	3	8	9
8		6		3				7
2	3		5	8			6	4
				5				2
						5		8
4				9			3	6
	9				2			
1			9	7	5		4	
		3	8			9	2	

Puzzle 345 - Easy

2	5	7	4			1		
8	4	3				7	2	9
6		9		2		5	4	
		5	8	3	4	9		
1		4		7	2		8	5
9	8	2	1		5		7	
					3			
3		6						7
	2			8		3		4

Puzzle 346 - Easy

1						8	2	3
	8	3	2	7		9		
2	9		3	1	8		7	5
9	5			7		3		
7	1	4	5	6		2		
			8	9	4	7		1
				2	1	9		
3				8				
	2			1				7

Puzzle 347 - Easy

1		4	3	6			8	9
			4		7	1	5	
9			1		5	6	4	3
2	9					5		4
6	4	8			9		7	
								8
		6	9				2	
7	8	1		5				6
4					6	3		

Puzzle 348 - Easy

8	1	6	9				5	7
	7		1					
4	5	3		7	8			9
5	9			6	2		4	8
3				8	9			6
6	8		7			3		
7					3			
		5	8	9	1	7		4
	3	8	6	4				5

Puzzle 349 - Easy

```
2 8 4 | 1 7 . | . . 6
1 5 3 | 6 8 . | . 7 .
7 9 6 | . . 4 | 8 . .
------+-------+------
4 . . | 3 1 . | . . 9
8 7 9 | 2 . . | . . 3
. 3 . | . 9 8 | . . 5
------+-------+------
5 . . | . 4 . | 3 . .
. . . | 7 5 . | . . 4
9 . . | . 3 . | 5 1 7
```

Puzzle 350 - Easy

```
2 4 9 | . . . | . . .
. . . | . . 8 | 9 1 2
. 8 . | . . 9 | 5 4 6
------+-------+------
. 2 1 | 9 6 . | 4 5 8
. 9 7 | . . 4 | . . .
6 . 4 | 8 2 1 | 7 . 9
------+-------+------
. 7 2 | 1 . . | . . .
9 . . | . 4 6 | . 2 .
. 3 . | . . 2 | 1 9 5
```

Puzzle 351 - Easy

```
. 9 . | 1 . . | . 7 .
. 1 7 | . . 4 | . . 3
3 . . | . 6 . | 2 . 9
------+-------+------
5 2 1 | 4 . . | 3 9 6
. 7 . | . . . | . . 2
4 . . | 6 2 . | . . 5
------+-------+------
. 6 3 | 9 4 . | 8 . .
9 8 . | . 1 2 | . 6 .
1 5 2 | . 8 6 | 9 3 4
```

Puzzle 352 - Easy

```
7 9 . | 8 1 . | 4 . .
2 . 1 | 6 . . | 5 . .
4 . . | . 2 . | . . .
------+-------+------
. 4 . | 9 3 2 | . . 1
. 7 . | . 5 . | 2 8 6
. 1 . | 7 6 . | . . .
------+-------+------
. 2 . | . 7 . | 3 . 5
6 3 4 | . 8 . | . . .
. . 7 | 2 4 . | . 9 .
```

Puzzle 353 - Easy

```
3 7 4 | 5 2 . | 9 . 1
. . . | 9 6 . | 7 3 2
. . . | 1 . . | 8 . .
------+-------+------
1 . 3 | . 5 . | 6 . 7
7 . 9 | . 8 . | . . .
. 4 8 | 3 6 . | 2 . 9
------+-------+------
. 3 2 | 7 4 . | 1 . .
4 . 7 | . . 9 | 3 2 .
9 . 1 | . . . | . . 4
```

Puzzle 354 - Easy

```
. . 9 | . . 8 | . 5 4
. 8 . | . 5 . | 7 . .
. . . | . . . | 2 . 8
------+-------+------
8 . . | 4 . . | 1 . 7
4 . 3 | 1 . 5 | 9 . .
. . . | 1 7 . | 3 . .
------+-------+------
6 . 7 | . 4 . | 5 . .
9 . 5 | . . 2 | 6 7 .
3 . 8 | 5 7 6 | . 2 9
```

Puzzle 355 - Easy

	1		7			9	8	
		8					7	2
		4	8	9	1		3	6
3				7			4	
6		5	4			2	9	
	4			1	9			
			3		6	8		9
4	6					8		1
2	8				7		6	

Puzzle 356 - Easy

7			4			9	8	
	6	1				5	3	4
		8		6			1	
	2	3	8	9		4		
		4					5	
9			1	4		8		3
			5	8		1		
5	4	2		1	7	3		8
1							9	5

Puzzle 357 - Easy

	8	6		3	5	7		1
4	7	3	6				5	2
5	9	1			2	6	3	
		6		9	3			
1			5		8	3		
8	3		4	6	7			9
9	1		3			8		6
3			2			1	9	
						5		3

Puzzle 358 - Easy

5	2		8	9	4	3		7
1			6		7			4
	7							5
	1	8		7		5	3	6
		7	1	3	6	8		
9			5		8		7	
7		5		2		4	6	
						9	5	
				5			2	8

Puzzle 359 - Easy

		2			9	4		5
	4	6	7	3			2	
	9	5		8		6	3	
		6		9	1	5		
				5		2	9	
	5	9			4			3
		3						2
		8	5					4
5	2	4	9	7		3	6	

Puzzle 360 - Easy

5	3		4		1			6
6		9			8		3	
8	7	4		9			5	
		3		1		6		
9	8	7		3				5
	6			5	9			
3	2	8				6		
	4	5	2				9	
			8			4	2	

Puzzle 361 - Easy

```
7 4 9 | . 1 . | . 3 .
8 2 . | . . . | 4 1 .
3 6 . | . . . | . . .
------+-------+------
. . . | . . 9 | 2 . 5
. 7 . | . . 5 | . . 3
. 5 . | 6 3 1 | 9 . .
------+-------+------
4 8 7 | . 9 2 | 6 . .
. . . | 1 5 7 | . 8 4
. 1 . | . 4 6 | 7 2 .
```

Puzzle 362 - Easy

```
6 2 7 | 4 3 9 | . . .
. 5 . | 7 . . | . 2 9
. 1 9 | . . . | 6 3 4
------+-------+------
2 . 4 | . . 5 | 7 1 .
3 . . | 1 2 4 | 9 . .
. . 1 | 6 . . | . . .
------+-------+------
9 5 . | 8 . 2 | . . 7
1 . . | 9 6 7 | 4 . .
. . . | . 3 . | . 9 2
```

Puzzle 363 - Easy

```
4 . . | 2 9 . | 1 5 .
. 7 . | 5 . . | 2 . .
5 . . | . . 1 | 9 . .
------+-------+------
. 8 2 | 6 . . | . 9 1
. . 9 | 7 8 . | 4 6 .
. . . | 1 . 9 | . . .
------+-------+------
. 9 . | 3 6 . | 8 1 .
. 4 5 | 9 . . | . . 2
3 . . | 4 . . | 5 7 9
```

Puzzle 364 - Easy

```
. 4 . | . 6 8 | 2 . 9
3 2 8 | . . . | 6 . 1
. 6 . | 1 . 7 | . 3 .
------+-------+------
. . . | 8 3 . | . . .
6 5 . | . 7 1 | 4 8 3
. 3 4 | . . . | . 9 7
------+-------+------
. . . | 8 3 . | . . .
5 . 3 | 6 1 . | 7 4 .
2 8 7 | . . 9 | . . 6
```

Puzzle 365 - Easy

```
. . . | 9 8 . | 5 3 .
2 8 . | 3 . 5 | . . 7
. . . | . 7 . | . 8 1
------+-------+------
4 . . | . . 7 | 3 . .
. . . | 9 . . | . . .
7 2 6 | 5 . 3 | . . 4
------+-------+------
9 . . | . 6 4 | . . .
3 4 . | 8 5 . | 6 . 9
1 . 8 | 7 . 9 | 2 . .
```

Puzzle 366 - Easy

```
. . 8 | . . . | 4 . .
9 . . | 2 . . | 1 8 .
. . 6 | 4 7 . | . 9 .
------+-------+------
. . . | . . . | 3 5 .
8 . . | 7 . . | . . 6
3 5 . | 1 . . | . 4 .
------+-------+------
5 . . | . . 2 | 7 9 .
7 6 . | . 8 1 | 2 3 4
4 2 3 | 6 . . | 8 . 5
```

Puzzle 367 - Easy

		9		3		1		
8	3	2	1	9		4	6	
	5			8	2		7	9
	8	4		1			9	
9			8			7	5	
	1	7	3		9		4	
		8		4		6	2	
		1		7	8	5	3	4
4				6	3		1	

Puzzle 368 - Easy

1	6			8	5	2		
		2	1		9		3	
				6	4			5
		7	3				6	
	3			2		7	5	
5	2	4	6	9		3	8	
6					2	8		3
	7	8		4	3	1		
3	4	1	9	6				

Puzzle 369 - Easy

		5	9			3	2	6
7	1	3	8			5		
			5			1	7	8
8				9	3		1	5
							3	2
6	3	1		2	5	9		7
			3	7	8	2		
3	2			5	1		9	4
			4	9	7	6		

Puzzle 370 - Easy

1			4		6	5	8	
2		8	1	7	5			3
	5	6	3		2	1		4
	9		5			6		8
7				6			4	
	6						1	5
		1	7	2			9	6
	2	4	9		1		5	
	7	9						1

Puzzle 371 - Easy

		8		2		7	9	
4			6	7	5	3	1	
1			9		3		2	6
9	4	7	2		8	5	3	
				5		9		
8					4	2	6	
			8	1	9	6	7	2
			1		6		4	
			9	7		2	1	3

Puzzle 372 - Easy

1	7	5	6	3	9			4
	9				8	1		5
4		8	5		7		9	
		6	7			5	4	
5					6			2
7	8	2		5				
				8			3	1
3	4	9		6		2		7
8	6	1				4		

Puzzle 373 - Easy

3		1	7				4	
	7	8			5	2		
	2	4	1				7	
					9	5	6	
	8		4	6		9		
	6	7	2		3	1	8	
7		6	5		2	4		8
8	5	9		1			2	3
		2		8		6	5	

Puzzle 374 - Easy

			2		4			1
2				7		9	4	
4					8		2	
			5	3	9	2		
			7	4			9	6
9			8		6		3	7
	1		4	8	5	3	7	
3	5		1		7	8		2
8	9			3			5	

Puzzle 375 - Easy

	9		8	6	1	7	3	
4	7			3		9	6	
6		1	9	4			2	
3		9		5			8	
					4	3	5	9
	8		3	2	9		1	7
9		3		7	5			
		7		1	3			6
			4		8	2		3

Puzzle 376 - Easy

9				3				
4	5		8					
3	8	2			4	7	6	5
	9	4		8	3	1		6
	2					3	7	
	3						4	9
2	7	3		6		5		
5		8	3	2		6	9	7
1			7	5	8			

Puzzle 377 - Easy

	7		3			5		
9	5			2	6	1		
			5				9	4
					5	8	3	
1			8	6		4	5	
5	8			4		7	2	
3	1	8		7	2	9	4	
		7	5				1	8
			9	8	1	6		

Puzzle 378 - Easy

							6	5
3		6	8	9	4			
				3			8	9
1	8		9	6	3	5		2
6	7	2				1		
9			1					6
2				1	9	7		4
	9	7	2	4	8			1
		3						8

Puzzle 379 - Easy

			7		9	2	3	8
	1	9	3	8	2			
8		3		6	4			
		7		5	3	1	2	
4						5	6	
	5		9			3		
3		2	6	7		9	4	5
1		6		3		8		2
5	7		2		8			3

Puzzle 380 - Easy

9		7	8					6
		4			2			7
1	2			3				9
			5	7	8		3	4
	5	1	4	6				8
4	8			9	1		6	5
2	7	6	3		9	5	4	1
8					5			3
		5	6			8	9	

Puzzle 381 - Easy

			7		5	1		9
9	8	1			6		2	
7	6	5	2	9		4	3	8
5	3		1		4	9		
	1	6		5	2			
		9			7	6		
	9		4		8		5	
		8					9	1
			5					6

Puzzle 382 - Easy

			7		9	3		
7	9			3	5		1	8
6	3	5	2				4	9
		4	3			2		
3							5	
	2			7		1	3	
1		3				9	7	2
		8		2	3		6	1
2				1		5	8	3

Puzzle 383 - Easy

	6	8	1		5		3	
					4	8		7
3								5
			3					
9			6	7	8			
5		3		1	2	9	7	6
			9				8	1
8	5		7	4		3	6	
6		1	8				5	4

Puzzle 384 - Easy

	6		2			5	9	
3		9			6		2	
5	8			1		4		
6			8	3		9	4	1
					4	7	6	
4						8		2
	5		4					9
9		4	5	2		6	8	
7	1			6	9	2		4

Puzzle 385 - Easy

9		6	5		2		4	
5	2			1				
	7			9			5	2
1		3		8				
			3	7	9			4
	5	7	6			3		9
				5		9	6	7
	3			6	8			1
2			1	4	7			

Puzzle 386 - Easy

8		2	9	6		1	3	
			2	7			4	
1	4			5		2		9
6	8		3	4	5	7	9	2
	3	4	7	2			1	8
	2			1				3
4		3		8				
2	1	9	4	3				
	5			9	2			

Puzzle 387 - Easy

		3	5	4	8	9	6	1
	9		7	6			8	2
1		8	2				4	
9		5		2		8	3	4
	4		8	3			7	
			9			2	5	6
		2	4					
4	8			5			1	
		7	3	8	9			5

Puzzle 388 - Easy

		7	1		2			9
		3	6	4				
5	9				3		1	
7		1	5	8	6	2	9	
9		8			4	5	7	6
	5	6		2				8
8		5	2	9	1	4		
3		2					8	1
	6		4	3				

Puzzle 389 - Easy

	4	3	5	1			7	
8	5			2		6		
7		6	3		9		4	1
	9	8	6		5		1	2
	1				3		8	
				4				
1	8	4				7	5	
	3				8		2	
2		7			4		9	

Puzzle 390 - Easy

7		2	4		5		6	9
			7	9	8	5	2	1
9				6			7	4
8		4	9		3			
1		6		8		4		
5	3	7			4			
	5						8	7
	4	1	8	7	2		9	
	7			5	9	1		

Puzzle 391 - Easy

```
2 . . | 9 . 4 | . 5 .
3 7 5 | . . . | . 4 6
. . . | 6 3 5 | . 2 .
------+-------+------
7 . . | . . . | 6 8 .
. . 3 | 4 . . | 7 9 2
. . . | . . 7 | . 3 .
------+-------+------
. . . | 5 4 . | 2 . .
4 5 1 | 7 2 . | 8 6 .
8 2 9 | . 1 6 | 5 7 4
```

Puzzle 392 - Easy

```
8 . 9 | . . . | . . 6
. . 3 | . . . | . 2 1
2 6 1 | . 9 . | 3 5 .
------+-------+------
. . . | . 6 . | . . 2
6 1 4 | 9 . . | . 3 .
. 2 8 | 5 4 . | . 1 9
------+-------+------
4 9 . | 1 . 5 | . 6 .
5 3 7 | 6 2 . | 1 9 4
. . . | 4 . . | 2 7 .
```

Puzzle 393 - Easy

```
9 6 7 | . . 2 | 4 . 1
1 . 3 | 4 9 . | 8 . 2
. . . | . 6 3 | . . .
------+-------+------
3 . 8 | 9 2 6 | 5 . .
7 . . | 3 4 8 | 1 . .
4 . . | . . 5 | 3 . .
------+-------+------
6 . . | 5 8 . | 2 . 7
8 4 9 | 2 . . | 6 5 .
5 . . | . . . | 9 . 8
```

Puzzle 394 - Easy

```
. . . | . . . | 6 . .
4 6 2 | 8 9 5 | . . 3
. 9 . | 6 . 3 | 8 2 4
------+-------+------
2 . 5 | . 4 . | . . .
9 7 6 | . 2 . | . 4 .
. . . | 7 . . | 2 5 9
------+-------+------
7 . 8 | . . . | . . 2
3 2 9 | 1 . . | 4 . 8
. 1 4 | . . . | . . .
```

Puzzle 395 - Easy

```
8 6 . | 7 . 4 | . 9 1
. 9 . | 8 2 5 | . . 7
4 5 . | . . . | . . .
------+-------+------
3 8 . | 9 7 2 | 6 . 5
. 7 . | 4 . . | 9 1 .
. 4 . | . . 1 | 8 . 2
------+-------+------
. 2 . | 6 1 . | 7 8 4
. . . | . . . | . . .
7 3 . | . . 8 | . 2 6
```

Puzzle 396 - Easy

```
. 5 6 | . . 8 | . 4 7
3 . 4 | . 2 . | 6 9 .
. . . | 1 4 6 | 5 8 .
------+-------+------
. . . | . 9 2 | 7 . 8
. 6 . | . 7 1 | 4 . .
7 . . | 6 8 3 | 1 . .
------+-------+------
6 3 . | . 5 4 | . 7 .
8 7 5 | . . . | . 1 4
. . . | . . . | . 5 .
```

Puzzle 397 - Easy

				4				5
2		8	6			1	4	
	4				8			
6		1	8		3	2	9	
3		4			9			6
9	2	5	1			3		
4		7	2		5	8	6	1
8	3	6		7	1			2
5			9	8	6			

Puzzle 398 - Easy

		5					8	1
9			7			4	5	
					8		6	9
			2		7			
		4			1	2		8
2	3	1		8		5	7	6
4	5				9	8	1	7
3		7		4	6	9		
8			1	7	5		3	4

Puzzle 399 - Easy

2	8	1					7	3
			3	7	9	8		
7	9			1		6		5
		9	1	8	3		2	
8		7		9				
						3	9	
	1	2		6	8	7		4
5	4				1	2	8	9
9		8	5	2	4	1		

Puzzle 400 - Easy

7		4		1		8	9	
			7				3	
	3			9	4	2		
	4		6			5	2	8
	5			4			7	
6	7				8		4	9
4	8	5	2	3			1	
3	9						8	2
	1		9	8		3		

Puzzle 401 - Easy

	4	9	7	8				3
7				5		4	9	8
3			9		6			7
	7			9				
	9		6	2				5
6	8		5		4	3	7	9
	6	4	2		5	7		1
2	3	1						6
8	5			6		9		2

Puzzle 402 - Easy

2		7					5	
		4		5	6			
			7	9		1		
	3			4	1	7	8	2
4				8	7		6	9
7		8		6		4	1	
9				2	8	5		
3		5				8		1
	8		4	3		9		

Puzzle 403 - Easy

4					1			8
6		5				4	2	7
8		2	6	4	7	1	5	
		6	1		4	3	8	
2	7	3				5		
			3			9		6
			8	9		7	1	5
7		9		5				
			3			2		4

Puzzle 404 - Easy

9			8		5		1	
5	3							
		7		2		5	9	3
1		3	6	9	2		4	
	9			4		3	6	1
6			3	8		9	5	
	6	8		5			7	
	4		2	1			3	5
3	1			6	9		8	4

Puzzle 405 - Easy

9						3	1	8
	7			3			2	
3	1				5	6	7	
					6		3	7
	6	7		4	2			1
5	3	1			8	4	6	2
7	2			8	4	1		
		6			1			
1		5	2		3	7	4	

Puzzle 406 - Easy

			5	6	3			
	5		7		1		8	
1		9	4		8	6		7
5	7	1						
8		3		7			6	2
2				8		7		4
7			6				9	8
	6			1		2	7	5
9	2	5	8	3	7	1	4	

Puzzle 407 - Easy

3	6	9		2			7	
		4	6		1		3	
		1	7		3			8
				3	2		1	9
		7	4		6			
2	3			1			8	
9	7	3	2				5	
8		6				3	4	
			3	5	8		9	

Puzzle 408 - Easy

2					6		5	
5		9	2	4	7			3
	7			5	8		4	
7		8			3			6
		6		2				
		2	6	8				4
	9	7		6	2	5	3	
	2	1					6	
6	3	5	4	1			2	

Puzzle 409 - Easy

8	1						4	3
	4		6		1			8
	7		8	3	4	5		2
7								
6				8		9		4
4	9	8	2		5		6	
1	6		3	4	8	2		7
3		4	5	6		1		
	8	7	9		2			6

Puzzle 410 - Easy

	1					4		9
4	8	6		9	5			
2		7	4	8				
9		5	1	7	2		6	
3	6	2			8	7		5
1	7					2		4
8	3	9	7	2		6		1
			8			3		2
	2	1		5		9		

Puzzle 411 - Easy

4					3		1	7
5	8	1				3		6
			1	5				
9		6	5		8	1	7	
			9	3	1			8
3			7		6	2		
		5			9	7	8	4
	6	4					3	
8	2	9				6		

Puzzle 412 - Easy

8	7		4	9				
		4	6				9	
6		9	2			8	7	
9	4	3		5	2	7		8
7	6			4	8	9		2
	2	8		6		4	3	
1	9				3	6	4	
			9					
	3	6				1		9

Puzzle 413 - Easy

			4					8
7			6			4		
		5	1	7	9			2
	4			9	7	1	2	3
3	1	8			6	7	9	
2						8		
	3	7	5		8	6	1	4
	6	2	7	3	4		8	9
			9			2	3	7

Puzzle 414 - Easy

			6			2	1	3
			8		5	6		4
6	7			2			5	
		8					3	
	3				6	7		9
7	6		9	4	3	1		
		8	2		9			
5		7	3		4			1
3			1		8	5		

Puzzle 415 - Easy

		6	1				4	7
						2		5
2		4	5		8		1	9
			4		1	7	9	3
			2				8	1
5	1	3			7			
4		1		2	6		5	8
3	2			4	9	1	7	
6		8	3					4

Puzzle 416 - Easy

						3		5
			6	2		1	9	
			9	7	3		2	6
	3		7			8	5	9
7	5	8		2	9	6	1	3
	1	9						7
5			3	8	7		6	2
2			6	5			8	
8			2	9	4			1

Puzzle 417 - Easy

3		1	2	4				9
	6	7	1	9	5		3	
9				6		5	1	
	7	5					9	
2			4	7				5
1		6			2			
	3			2		9	5	1
						3	4	
	1	4	9		3			7

Puzzle 418 - Easy

	2	4	1	9	5			7
			4	7				
5				3			9	4
	4		6		7	5	1	
6			9			4		2
	1				3	8		9
8	9	1		2			5	6
	5							
4			7	5	1	2		

Puzzle 419 - Easy

8	1		9					
	3		7	9		1	5	
					6			3
6		5	3	8		2	7	
9			4	2	7	5		1
	7	2				3		
		1	2		8	4	9	5
	2	8	9		4			7
7	9	4	1			3		2

Puzzle 420 - Easy

3	4	6		8	2			9
9	5		6	3		4	8	
		8			5		3	6
7	8	3					4	5
	9							
	1		2	7				3
	6	7	3	2			5	
			7		9			8
8					1		2	7

Puzzle 421 - Easy

		4		8		5	1	
	7		6				8	9
8	3	2	9			4		6
3			4			9		
	4		3	9			5	
7	8	9		2	6		3	4
6			7		3	9		
				5				1
	5							7

Puzzle 422 - Easy

9			1	8		4	5	
5	4	1		9	3	2		
	3	8	5	7		1		
			2		9	7		8
8	2			1	6			4
		9	8				2	
7	8		9		1	3	4	5
3							1	6
1		4		6				

Puzzle 423 - Easy

7	3		8	2		1	5	4
	8	4		9			2	
			7		4			
4								
2	9	6	1	5	3	4	7	
8		3		4				6
1						3		7
3	2				9	6		
9	6	8			5			

Puzzle 424 - Easy

			7	6				
1						9	8	
2		7	8		1	5		
		2				1	9	4
	1			7		3		
	9	5	1	3				6
5		4	3	1	6	8	7	
3	7	1	9			6		
	8		4			2	3	1

Puzzle 425 - Easy

	1	3	6			8		
4	8			3		5	1	
2		7	5	8	1		3	
7	4		8				9	1
9		1						
8			1	7	9	6		
		5	3	2		1	8	
	2		9	6			5	3
3						9		

Puzzle 426 - Easy

						6		
	5	9	8	6			7	4
8		4			7	1	9	3
	7					5		9
5	8			9	2		1	7
			6		5	8	3	2
4	2		7	3		9	5	1
6	9	3		5				8
			9	4	8			

Puzzle 427 - Easy

```
. 3 . | 2 . . | 8 . .
. 6 9 | . 7 . | . . .
. . . | 6 . . | 1 4 .
------+-------+------
2 5 . | 8 . . | 4 . .
3 4 6 | 7 2 5 | . 8 1
9 . 1 | . . . | 7 2 .
------+-------+------
. 9 . | 5 8 . | . . .
. 1 2 | . . . | . 9 .
5 7 4 | 9 3 2 | 6 . .
```

Puzzle 428 - Easy

```
7 . 6 | . 8 . | . 5 3
3 . . | 4 6 5 | . . .
5 . . | . 1 . | 2 4 .
------+-------+------
9 . 3 | . . 8 | . . 2
. . . | 2 4 . | . 9 .
6 . . | . 9 . | . . 5
------+-------+------
2 5 9 | . 3 4 | . . 1
8 . 7 | . . 9 | 5 . .
. . 1 | . 5 . | . . 9
```

Puzzle 429 - Easy

```
1 . . | . . 4 | . . .
9 . . | . 5 . | 8 1 3
2 6 5 | 1 . . | . 9 .
------+-------+------
8 2 6 | 5 . . | . . .
. . . | 3 2 . | 6 . 8
. 1 3 | . . . | 5 . .
------+-------+------
3 4 2 | 9 . 6 | . 8 5
. . 1 | 2 8 3 | . . 4
6 8 . | . . . | 7 . .
```

Puzzle 430 - Easy

```
. 1 . | 5 . . | . 2 3
. 8 6 | . 9 2 | . . 5
5 . 9 | 3 . 1 | 6 . .
------+-------+------
. . . | 4 . . | 7 5 .
. . . | . 5 . | . . 6
. . 1 | 6 . 3 | . . 2
------+-------+------
. 3 5 | 9 1 . | 2 8 .
. . 8 | . . 4 | 5 . .
. 4 2 | 8 6 . | . 1 9
```

Puzzle 431 - Easy

```
. 2 . | . . . | . . .
. . 6 | . 3 . | 4 . 1
9 . 1 | 8 5 . | . 7 6
------+-------+------
1 8 5 | . . 3 | . . 4
7 . 3 | 4 9 2 | 5 1 .
. . 2 | . . . | 3 . .
------+-------+------
. 4 8 | . 2 5 | . . 9
2 . . | 9 . . | 3 . 5
6 . 9 | 3 1 8 | . . 2
```

Puzzle 432 - Easy

```
. 4 . | 8 2 3 | . . 5
8 . . | . 5 . | 7 . .
1 . 5 | . . . | 6 . .
------+-------+------
6 . 1 | 2 . . | 4 . .
9 2 . | . 7 4 | . 1 3
3 7 4 | . . 8 | 5 . .
------+-------+------
. . 9 | . 3 . | 5 2 .
2 6 3 | . 5 . | . 8 9
5 . . | . 6 4 | . . 1
```

Puzzle 433 - Easy

				4		5	6	7
3	6	5	8	2				
					9	8		
8		7	6					
		3		5		6	7	8
6	2	4		7		3		9
4	1						2	5
			9	1		4		
9		2	4		5			6

Puzzle 434 - Easy

5	1	9	7		6			
2	3	7	5	9		1		
8			2			9	7	
			8	2			3	
		6			3		9	7
		2	6	1	7			4
9	7			6	5			2
4	5		9		2		1	6
6		3	1			7		

Puzzle 435 - Easy

		9		3		4		
7		8	1		6		9	3
	6			9	2	1		
8	5			3	4	7	1	
2		1		5	7	3		
9	3		6			2		
1			4	7	5	8		6
6				2	1	9		
3				6		4		1

Puzzle 436 - Easy

1				8		4		
				9	2		1	
7	2		1		5			9
	7		8				3	4
	1	2		9		7	6	
4		9				5	1	
		3				2	6	
	5	7	6			9	3	
9	6	1	2	8		5	7	

Puzzle 437 - Easy

			6		4			
4		5		1		9	7	
		6		9	5		4	8
6	8						9	
5	1	7	9	8				4
9				6	7	5		
			3		8	4		
2	3					8	5	7
8	5	4	7		9	6	1	3

Puzzle 438 - Easy

3	2			6			4	9
	9			8				
		4			9	8	2	5
	4		6		5		8	3
		6	1	9	8			4
	8	5		4			9	
8	6					4	7	
9	1	2		7	6	5		
4	5		8				1	

Puzzle 439 - Easy

5				9	1	2		6
	2	6	7			5	4	
	4		2	6	5			7
2			5	3		4		8
3				2	9		6	
8	6	9		4		3	2	5
			3	1	4	6		
		8				2	9	
6						1	7	4

Puzzle 440 - Easy

4			6	7		9	1	
		1	2		3	6		7
			9	1	8			
			8	6	7		3	4
6		8	1	3	4		9	5
1				9			8	6
	5			8	1			9
3	2		4	5				
	1					6		

Puzzle 441 - Easy

			1					3
						9	7	5
4		2	7	9		1		8
		5			3		9	
8		7				3		1
		1	9	4	7	5		
9	5		6	8		3		
2					9	5		4
	8	3	4			6	2	9

Puzzle 442 - Easy

		2		4		9		
2	6			1				4
1			9	5		7		3
3	4		8	2		9		6
	2	8	5		6	1	3	
	7					2		8
5	1							9
8		6	4					5
	7			9		6		

Puzzle 443 - Easy

	4	9	1			7		
	6	3	5		7			
		7	4					5
	3	8		5		6		
7	2	6	8		9			4
		1			2	8	3	
			2		6	7	1	
		4			5		9	
3		2	9				4	

Puzzle 444 - Easy

		2	7			6		3
	6	7	9		3		2	
		8	1			9		
	7			6			4	
						5	3	6
8		6	3		4	2	9	
7			6				5	8
6		5	2		8			9
		9	7			3	6	2

Puzzle 445 - Easy

	8	3	2					
				4			8	5
7		5	6					
5	7		3	2	4	6		8
4				1				7
6			9				2	1
8		6	7		3	2		
9		4						3
3		7	4	8		1		

Puzzle 446 - Easy

	9	7		4	1		5	6
	1		9		6			
	6	5			7	9	1	2
	3						4	
5			6			7	2	1
	2		5					3
	5		1	6				9
1	8	3	4		9	2		
		9	3			1	8	4

Puzzle 447 - Easy

	8	4	7	6		5	9	
1	5				4			2
			2			1	4	
	1		4	3				
3	6	5		1				
9	4				5	8	3	
5	3		8	9	6		1	
	9		2			6		
4			1	5			8	9

Puzzle 448 - Easy

7	8	9			3	6		1
		5	8			7	9	
	6		7		2	4		
	7			4	9			2
	9		7	5	8	6		
			2	8	9	3		7
5			9					
			4				7	
	4	7			6	5		9

Puzzle 449 - Easy

			6			8	1	
				8		5		2
	8	5	7	2	1	3	4	6
					5	6	8	3
	9			7				
			1	3	6			
8				1	7			5
5	6	1		9	2			
2	7	4				1		

Puzzle 450 - Easy

		7				3	8	
	9		6	8				5
	5			2	4			
7	6		1	2	8		4	3
4	8			5				7
	2		4	6	7		9	
		2	8	1		3	5	4
6	1	4	3	7				2
			2	4		7		

Puzzle 451 - Easy

	5	8						6
	2	9	4	1		7	5	3
4	3			2	5		9	
2			8		4		3	9
3	8		2	5	9	4	1	7
				3	7		8	2
7	9		3	4	1			5
	1		5	7		9		

Puzzle 452 - Easy

7	6	8	1	5			4	
		4	2		8	9		6
2	3		7	6		8	1	
		5	6		7		2	
	2	7				5	8	
4	9	1	5	8		6		
				6		2		1
8		6		2		4		
			3					8

Puzzle 453 - Easy

4			6	5	9		2	8
		2	3				1	4
8		3	2	4	1		9	
	2		1			7		
		1				8	4	9
	8	4			3	2	5	
2		6		1			8	5
	4	8		3		1		
	7	5	8					2

Puzzle 454 - Easy

		5	9			8		
	7					2	4	9
1				3		6	5	
7			6	9		4		2
	4	9			7		3	1
							7	6
5		7		9	8			
4	8		7	2		3	9	
		3		5			2	8

Puzzle 455 - Easy

				9		4	1	
1	5	9	7	2	4	8	3	6
		6	8	3		2		
4			2		7	9		
	2			5	3		8	
3			4		8	6	7	2
						3	4	8
8		2					6	
5					6	7	2	

Puzzle 456 - Easy

4	5	3			9		2	7
6	8		5	1	2			
		2			3	8	6	5
3					7	5		2
8		4				7		6
				8	6			4
2			7		8			9
5		6	9				3	8
9			6	5	4	2		1

Puzzle 457 - Easy

```
7 . . | 4 . 2 | 6 9 .
5 1 6 | 3 . 9 | . . .
9 . . | . 1 6 | 3 . 7
------+-------+------
2 5 7 | . 3 8 | 4 . 6
. . . | . 2 . | . 7 .
8 . 4 | 7 . 1 | . 3 2
------+-------+------
. . 1 | 6 . 5 | . . 3
3 . 9 | . . 7 | . . .
. 7 . | 1 . . | . . 4
```

Puzzle 458 - Easy

```
3 5 . | . 6 . | 9 7 .
. 7 . | 1 . . | . . .
. . . | . . 5 | 4 2 .
------+-------+------
. 1 . | . . 8 | 2 . .
. 2 9 | 4 1 6 | . 3 7
. 6 8 | 9 . 3 | . . .
------+-------+------
2 . . | . . 1 | . 8 6
1 8 6 | 2 . . | 7 9 .
9 . 5 | . 8 7 | 1 4 2
```

Puzzle 459 - Easy

```
8 . 6 | 2 3 . | . . .
. . 9 | 6 4 . | . 3 .
. 3 4 | . 1 . | 9 . 2
------+-------+------
5 4 . | . 9 . | . 8 7
3 . . | . 7 6 | . 9 4
6 9 . | . 8 . | 2 1 3
------+-------+------
. . 5 | 8 . . | 7 . .
9 2 . | . . 4 | 3 5 .
. . 3 | 9 5 1 | . . 6
```

Puzzle 460 - Easy

```
6 7 . | 8 . 9 | . . 4
. . 9 | . 1 . | 2 . .
3 1 5 | 2 4 7 | 8 . .
------+-------+------
. . . | 7 8 1 | . . .
1 . . | 6 . 2 | 4 8 .
. 9 8 | . 5 4 | . . .
------+-------+------
7 6 1 | . 2 . | . 4 .
. 8 . | . 6 3 | . 2 1
9 . . | 4 . . | 6 . .
```

Puzzle 461 - Easy

```
6 4 5 | 7 3 . | . . 1
. 2 . | 6 9 1 | . 5 3
. . . | . 4 8 | . . 7
------+-------+------
5 . . | 4 . . | . . .
. 9 7 | . 1 5 | 3 4 6
4 . 1 | . . 7 | . . .
------+-------+------
1 . . | . 7 9 | 6 3 .
9 6 . | 1 . 4 | . 7 .
8 . 2 | . 5 . | 1 9 .
```

Puzzle 462 - Easy

```
. . . | 8 . . | 2 . .
. . 3 | . . . | 4 7 .
. . . | 3 1 . | 8 . .
------+-------+------
. . 7 | 2 . . | . 5 .
2 9 . | . 4 . | . . .
5 . . | . . . | 7 2 9
------+-------+------
6 . . | 1 2 4 | . 3 .
7 1 9 | 3 6 5 | . 4 2
3 2 4 | . 9 . | . 6 .
```

Puzzle 463 - Easy

2	4	3	7		6		8	
8					5	2	3	7
	5	7				6	4	1
	2		8					9
		5			4		7	6
7	6			5		1		
6		2		3	9		1	8
5					1			
	3	4		7		5		2

Puzzle 464 - Easy

3		4	5			8		
			4			7		9
8				9			5	4
1	6							
		7	3	2	5			
2	4	5		7	6	3	9	8
7			5			6	8	
5					1	9	4	7
4			7	6		5	1	2

Puzzle 465 - Easy

5		4			7	1		
			9			2	5	4
	8		5		4	3	7	
2				3			4	
3	9	5	4	7	8			1
	4	7	1	5		8	9	
4			8	9		5		
					5		6	
8			1	7		6		3

Puzzle 466 - Easy

	1		9					3
8	2			4		1		7
6	3	5	8	1				9
		7		9		6		2
9	6		5	2	1			
4		2	7	8		9	3	
		8		3		7		
3							2	5
	7	1				3		8

Puzzle 467 - Easy

	4	3			1		9	
6		9	8	5	4	2		1
2			4	6		3	5	7
3				5	7		8	4
	9	4						
8		7		4		9		3
		2			3			5
				7			3	8
7	3		5	1	6			9

Puzzle 468 - Easy

	8	1				5		
	7	6						1
4	2		3					6
5		2	1		7	4		8
	1	4			3		5	
		7		8	5			2
	6		8	4		1	9	
		8	5	3		6		
2				7		8	3	

Puzzle 469 - Easy

1	4	2	3		5	9		7
						8	3	4
7		8	4	9				2
		9	5		1	4	8	6
		1		2			9	5
			8					
6						1		9
	2	4	9			6	5	8
9		5		4	8		2	3

Puzzle 470 - Easy

	8			1	3	2		
4				5		6	8	1
		2	9	4		3	5	
			5		2			
2	5		3	6	4	1	7	8
3						4		
	2		7	3	1	5	9	4
	4			2		7	6	
9			4		6			2

Puzzle 471 - Easy

6		2		1	7	3		4
1			9			2	6	
				2		9		
	3	7			4		2	
5		6	2					
4	2	9	7			1		
	7			4				2
					1		5	
2	6	5		7	9	1	4	

Puzzle 472 - Easy

		6		8	5		3	2
	7		9				4	
4	5	3		2	1		6	
				9	2			
	6	9	3	1	8		2	
			6	7	4	9	1	8
6			8		7			4
		7	1	4				6
9		4				6	3	

Puzzle 473 - Easy

		6		4		8		7
4				8	2	5		
	5	2	6	9			4	
3	4	8	2		9		6	5
5				7				
	6					2	8	
		4	7		1		3	
							7	4
		8	1		3	9	5	2

Puzzle 474 - Easy

6	1	2	3			9		4
					8		1	6
			4	2	6			5
4	7			1				8
2	6	1	7				4	
		9			4			
		9		7	5	4		1
7						3	5	
			6	4			8	2

Puzzle 475 - Easy

5				3		2		
9		1			8	7		
	4		2			5	6	
		3	7		5		8	4
	9			4	3	1		
	7		9	8				5
7			3				1	9
2		8	1		4		7	
1			8		6	4	5	2

Puzzle 476 - Easy

		2		3				
	3		1			6	7	
5			4		7	3	8	2
3			9	6	2			1
		8	3		4	7	2	
	6		7	5		9		
	8	3	2		9	1	6	
1			6			4		7
6	9	4	5	7	1			

Puzzle 477 - Easy

5			8			7		1
	6	3	1	5	7	8	2	
	7	1	9			5	3	
		6	4				9	8
		5		2			4	
7			6	8	1	2		
	2		7				1	
3							8	
	5		2			4		6

Puzzle 478 - Easy

6	5	1	7		9			2
		4			8			
	8	2	1	4		9		3
	3	8	4					1
4	7	5				3		
	2		8			4		
2		7				5	1	9
		3		6		2	8	
8	4		5	2	1	7	3	6

Puzzle 479 - Easy

		7			3		5	
	6		1					4
		3			5	7	1	
		6			9	8		3
3	5		7			4		
	2	8	3	4	6	5	9	7
	3	1		8		4	6	
9	7		6	3	4		8	
		4		5			7	9

Puzzle 480 - Easy

	1			8	6			
9	6	7	1			8	4	
		5		2		1		9
			3	1		5		
5			7		8		9	1
	4			9	2			
		8			3		1	4
		1	2				7	
	7		8			9	2	3

Puzzle 481 - Easy

			5	7		1		
		6		3	2	8	5	
5	1	9					2	
6		1	9	8	4			
			3				7	4
	3			5				
	5	2	8	9		4		1
	9	7	4			3	5	
		8				7	9	3

Puzzle 482 - Easy

	9	4	7	2	5			1
1	3	7	8			2		4
	2		3					
7		2			8			
			2		7	6		3
		5	1			9	7	
		8	4	9	1			
9			6				2	8
		3	5	8		4	1	

Puzzle 483 - Easy

4				2	8			
6	2		1	4	9		3	
		3		6			4	
5	9					4	2	6
3	4		9	5		7		
					6	3		
1			2			6	3	
		9	5				1	
2	3	4		9	1			5

Puzzle 484 - Easy

7	1	2	4		9	6	8	3
6		8				5	1	
5					6		4	
3		1		2				
	6	5		7	3		2	
8	2	9			4	3	7	5
2			3			1		
				4	8		5	
							2	3

Puzzle 485 - Easy

	9	5	2		8	7	3	
2				6	9	5	4	
		4		3		1	9	
	6	9		1	2			4
1					7	2		6
7	5	2	4	8	6		9	
9		1						
4			1	9	3	6	2	5
	2							9

Puzzle 486 - Easy

		9		2	6		3	4
8		6				9		2
	5	3		1				
				6	7		9	3
6	7			9		4		1
3		8		4		6	5	7
9	8			3			7	5
	3		9					6
4	6	2	1	7	5		8	

Puzzle 487 - Easy

4	2			6		5	1	9
1				3	2			8
8	6			5				
3	4				7			
			3	4		7	9	2
	7	2			6	8		3
2	8	4		7		9		1
7	3				9			5
		1	6	8		2		

Puzzle 488 - Easy

		5					3	2
3	7		6	9		1	8	
	9	8		3				7
	8		5	4		3	9	1
2	1					7	5	8
5			1	8	7	2		
	2							3
4			3	2			1	
	5		8	7	1	4	2	

Puzzle 489 - Easy

							1	2
1		4	5	9	3			7
6	9			2				3
7	4	6		3		1	9	
				7	8		4	5
5		3		4				6
8	5			6	4			1
	6	1	3				7	4
	3		8	1	7		6	9

Puzzle 490 - Easy

	2	8	6		7	1	3	5
			9				7	
		4		3	8		9	
4		6		8	5	9	1	2
	1	2			9			
	9	5	3	2		4	6	
		9	8	5		7		
6		1					5	
						6		8

Puzzle 491 - Easy

3	1	5				7	8	
	8						6	
7	6	9	8		1	5	2	
1	4	2	6	5			7	
		7	2	8	9			
	9			4		6		
8	7					1		6
9				3			4	
		2	6	1	9	3		

Puzzle 492 - Easy

5			1	6		2		8
	8		5					3
6	7	2			9	5	4	
		7	2		3		5	4
2						8	6	
	1	5		4	6			2
	5	6			8			9
		4			2	7	8	5
7			3		5			6

Puzzle 493 - Easy

	8		4	6			1	7
9	3				1		5	8
	4		3		8	2		6
5		1	8	2	4	6		
		4	6				8	
		3	7		5		2	4
	1	2		3		8		5
4		9	1	8				2
		8						1

Puzzle 494 - Easy

	5			4	1	9	3	2
	4		9	6				5
3	1			2				
		2	3			5	6	
5	8				6	4		1
		6			9		2	
9			3	7				
		5		1		3	7	
7		1			8		4	6

Puzzle 495 - Easy

2	8			3	4			6
		9	2		5			3
1	5	3		8				
		8	5		6	9		7
5		2					3	4
9	6		7		3		8	
8		6			9	4		5
	3		1			7		
					8	3		2

Puzzle 496 - Easy

	6		7	3	5			4
3				4	8		1	
			9	6				7
2	1	3		5		4		
9	8	6		1		7		
5		7	8	2			6	1
	3	5		8				9
	2	4	1	9	3		8	5
	9			7	2			

Puzzle 497 - Easy

		1		9		6		
8	2	6	3					4
		4	6			7	1	
2			1		4			5
	1		8		6			
	4					3	8	1
	5	3		4	8			6
4	6			9	3		5	7
9				6		4		8

Puzzle 498 - Easy

	9	4	1	2		5		
						4		
				4		7	9	2
8		9			2	6		7
		7	9	1			8	
4	5	3	6		7	9	2	
2	3					1	7	
5	4	1		9	8			3
9	7	6	2				5	

Puzzle 499 - Easy

8	1			5			6	
9		5	1	7				3
		2	6		3	5	9	1
			8		6	9		5
	8	6		9				
7	9		5		4			
					1		4	2
			2	6		3		9
2			8	9	4	5		1

Puzzle 500 - Easy

6					1	5	3	
			5		9	1	8	6
				2		7		4
	7	3			2		4	
		1		9			7	
5		6	1	8				
	1	9			8	4	5	7
		4		3				
	5		7	1			6	3

Puzzle 501 - Easy

	6	9			2	7		
	1		3		7	2		8
	2		9	5				6
2	7	3	5	4	6			
8			2	7			3	4
4						5	7	
	3			2	4		1	7
		7	8					
1	8	2		9	3		6	5

Puzzle 502 - Easy

3	6	8	1		7	9	5	4
	4	8						
		5	3	6	4		2	8
1	5					8	6	
	9		7	8	6	3	1	5
		3	9	5			4	
				4		5		
	9			3	8			1
	1				9	4	8	2

Puzzle 503 - Easy

8							3	
6	5		9	2		8	1	
	3	9	8	4	1		6	5
5	7				2	4		8
9	8		4	7			2	
			1		9	6		
		8		3		9		
7		3	6		4	5		
		5				3	7	1

Puzzle 504 - Easy

		2						
1				5				
7	8			6		9		5
5	3		7		9			8
8	9					5	4	3
4	2		3		5		9	7
		8	5			4		9
3	4		8		6		5	1
2	1					8	3	

Puzzle 505 - Medium

3		2	8	5				
9		6					2	3
5	7	4		9		6		
7				8	2	3	6	
			1					9
4					9		8	
2			7	6				
		9		4	1		5	2
			9			8	4	

Puzzle 506 - Medium

	9	2	5	6		3	1	
7			3		2	5		
	5	3		1				2
		5		7		4	2	
	2		6				3	
			3	8				
2		9		8		6		
	6				9		8	
	8		7	2		9	4	

Puzzle 507 - Medium

1		7	2	3		5	6	
	4		1	7	9	3		
		8	6	4			1	
								8
7	6	2	9	8		4		
		9		2			7	
		3	8	9			4	
	2	1		5			3	
					7	2	9	5

Puzzle 508 - Medium

6	7		1					2
	4	2	5	9			8	
	1		8		2	6	4	3
2	6	1		8		4	7	
	9					8		
	8	7	4	1				
		4		1	9	9		
						5	1	
	2		9			3	4	

Puzzle 509 - Medium

	1				9	8		2
2		8	1	4		9		
			3	2		4		
			5	8	9			4
9		2						
		5	2					
1				7	4			5
5	9		2		6	3	1	
7		3	1	4			6	9

Puzzle 510 - Medium

1		7	6	2		3	5	8
9	3	8		4	5		1	
2	5						9	
5				9			7	
6		2	4	3				
				5		6	4	
7		1	5	8				4
	2							
	9	5					2	

Puzzle 511 - Medium

8				5	7		6		
2		4		3	8		1	9	7
6	7	9			4	2	8		
	1						7	3	
3		7		5	1	9		4	
	2								
	9				7				
7	4				3				
5		3				1	4		

Puzzle 512 - Medium

	7	6			1		4	3	
1	9	3		5		4	8		
	8	2						9	
2	4							6	9
				8			7		1
		5				3			4
8					9	7	6		3
		4		6		8			
9	6				4				

Puzzle 513 - Medium

	2				6		5		
3	8	6		9			7	1	2
5		1				7			6
8					3				1
		4							
2		3		5	7		6		
6		2		7				5	
9					1	6	2		
4		8		2	5	3			

Puzzle 514 - Medium

4	8			1	5		2		7
				8			4	5	
2		7					3		
9	1								
	2	8		6	4				
6		3		7		5	8		
5				3			7	4	
		6		4			5	9	
	3	4			9				

Puzzle 515 - Medium

	6			9	7		2		
		4			8	3			
	8				4	1			9
	2			8	7	5			4
					4		2		1
	4						9	5	
				7		6	5	9	
6	5							1	
	9	2		3					6

Puzzle 516 - Medium

				2	7		4		
9				8	1	6	3	5	
		6			4		7		
6	4				9			2	3
7		1			3			8	
	3	2				1			
	1							3	6
		9			8				4
					5	7			

Puzzle 517 - Medium

	7		5		3		6	1
		3		4				
	1	8			6			4
			6	1				
2	4	6					9	5
	8			5	9	4	7	
1	2		3		5	8		9
		4	8					
		9	4			6	2	3

Puzzle 518 - Medium

8	1	6			4	2		
	5		8	2	1	6		
		4		6		7	1	
4			6			3	5	
				5		8	7	
6	7	5				4		
	3		4	1			8	
2	6				7	1		
	4		3		2			

Puzzle 519 - Medium

4				7		1	8	
			9			3		
6	3		8	4		7		
		8	4	6		9		3
	5		3	2				4
		4	7		1		6	5
8						6	9	
				3	8	4		7
	7				4	5	3	

Puzzle 520 - Medium

	7	5		4		6	3	1
					7	4		
		4	1		9			
5		2			6	1		9
	4		5	9	1			
7		1			2			
	7	9	6				2	
3		9	8			7		
	2		7	1	3		6	

Puzzle 521 - Medium

		9						
		1		5	3			
2			9	3		7	8	
1	9	7				2		3
	2		5	7		4		
	4	8				6	7	
			7			1	2	
7	3	5	2		1	9		
8		2	4	9				

Puzzle 522 - Medium

4		8				5		
3		5			2	7		1
2				5				8
	8			3	5			
5		9		1	6		7	4
		1			4			
	5		1	6			3	
1	6		5				8	9
	4		2	9	3			6

Puzzle 523 - Medium

				5	9			6
		4	7	8	6		2	
6	3	9		2				
4		8		3		9		
	6	2	8					
	7	3		1		8		5
		6	9					
8		1						
	9	5	1		3	2		8

Puzzle 524 - Medium

2	5				8		1	
9			1					2
1	3	6	9	2	7	5		4
		5	6	1			3	
3				8	5		9	
	6	9	3	4		1		8
			8					5
	9	1			4			
		8				7	4	

Puzzle 525 - Medium

8	3			2	1	5		7
	1	7	5				9	
	4	5						
	7		3			2	5	
3	6	1	7	5		4		
5		2			4	1		
		6		8		7	4	
				6			1	8
			1				3	

Puzzle 526 - Medium

1	8				5		3	7
	9				3			
		3						6
	4			2			6	
2	7		6	1				3
	6	8	7			2	1	
			5	9		4	8	1
			4				9	5
9	5	4	8					

Puzzle 527 - Medium

9		8	2		3			7
	6		1	5		9	4	
	4							8
2				6			7	
	5		3			2	8	
		1		2	9	4		6
				1	7			
		6		3	5			
7	9	3		8		6		

Puzzle 528 - Medium

		8		7		5	3	6
	5							
		6		1		2	4	
8	7	2		6	1			
		1			3	8		
3	6		2	8	5	1		
1		4					5	9
7	9				2		1	
	8			4				3

Puzzle 529 - Medium

	3					2	6	
	8	5						
	7		3	8	9		5	
	4	7	9		6	8	1	
				5	1			
	1		2		8		4	3
	5	2		1	3	9	8	4
						1		5
		8				3		6

Puzzle 530 - Medium

6	3			9	7		2	
	7	5			3			
	9		4	8				7
	6		8	2	4		3	5
	5	3		6				
		7		3		8	9	
		6					4	
		9		4	8		7	
	1	4		7	2	6		3

Puzzle 531 - Medium

1	9			3		7		
							8	
	2	6			7			
	1	2	8		4			
9	5					2		8
6		3			1	5		
5	6			4		8		
2	7	9		8	6			1
4				1		6		2

Puzzle 532 - Medium

			9				7	
1	5	6	3			9	2	
				4	2		1	3
3				9				
9	4		7	2				
2		7	8		1			
				8		6	5	2
5		4			7			
6	9	2	1	5		7		

Puzzle 533 - Medium

	9							
1	5		6	4	9			
8		7	2		1		3	
						2	5	8
			6	2	3			7
7			8					6
2			4					
9		8	2	6				5
	1		5		7	8		

Puzzle 534 - Medium

	3	1				9		
				8		5	6	
			3					
			1	3	9			
2			4		5		8	9
9	6	3		2				
	5	6	9	4	1		7	8
		4	5					
		6		2		4		5

Puzzle 535 - Medium

```
2 . . | . . 3 | . . .
. . . | . 6 8 | . . .
. . . | 1 5 . | 3 2 .
------+-------+------
. 7 6 | 9 2 . | . 3 8
. 9 3 | . 4 . | . 5 2
. . 8 | . . 6 | . 7 .
------+-------+------
7 1 . | 4 3 . | 5 . .
. 8 4 | . . . | . 9 3
9 . . | 6 8 . | 7 1 4
```

Puzzle 536 - Medium

```
. . . | . . . | . 7 .
. 3 1 | 9 . 5 | 8 6 2
. . . | . 6 . | 9 . .
------+-------+------
1 . 3 | 7 . 6 | 2 8 .
9 . . | 2 . . | . 3 5
. . . | 4 3 . | . . .
------+-------+------
. 6 2 | . . . | 3 . 9
3 . . | 8 . . | . 4 6
. . 5 | . . 4 | 3 2 .
```

Puzzle 537 - Medium

```
. 1 3 | . . 6 | . 2 .
6 . . | . 4 1 | . 5 9
. . 8 | 7 . . | 1 3 .
------+-------+------
. . 4 | . 6 . | 2 . .
2 6 9 | 1 . . | 4 7 .
. . . | . . . | . . 5
------+-------+------
. 4 5 | 6 . . | . . 2
7 9 2 | 5 1 . | . . .
. 3 . | . . . | 5 4 7
```

Puzzle 538 - Medium

```
9 . . | . 5 . | 2 . .
7 . 4 | . 2 3 | . . .
. 5 3 | 9 . 6 | 4 . 7
------+-------+------
3 7 . | . . 1 | . 5 .
. . 2 | 8 . . | . 6 1
. . 6 | 3 9 . | . 8 4
------+-------+------
. . . | 6 . . | . . .
1 . . | . . . | 6 4 .
6 . . | . 8 . | 5 . 3
```

Puzzle 539 - Medium

```
. 7 9 | 2 6 . | 4 . .
2 . . | 7 . . | . 8 .
. . . | 9 8 6 | 7 . .
------+-------+------
. . 5 | . . 7 | . 3 6
7 3 6 | . . 9 | . . .
9 . . | . . 5 | . 1 .
------+-------+------
4 5 8 | . . 3 | 2 6 .
. . 1 | . 7 . | 3 . 5
3 . . | 5 . 6 | . . 1
```

Puzzle 540 - Medium

```
. . . | 4 . 7 | . . .
. 5 . | . 2 . | . 7 3
1 . . | . . . | . 4 9
------+-------+------
. . 5 | 3 9 6 | 4 8 .
. . 3 | . . . | . 5 .
. . . | 5 . . | . 3 1
------+-------+------
6 . . | 7 . 4 | . 9 .
. . . | 8 . 9 | . 2 4
. 4 9 | . 5 . | . 6 7
```

Puzzle 541 - Medium

```
. . 6 | 4 . 1 | 7 . 5
. . 7 | . . . | . . 2
2 . . | . 3 . | . 4 .
------+-------+------
7 8 1 | . 9 . | . . .
. 2 . | . 4 . | . . 7
5 3 4 | . . 8 | . 6 .
------+-------+------
9 . . | 6 . . | . 5 4
. . . | . . 9 | . 1 .
1 5 . | 3 8 4 | 6 . .
```

Puzzle 542 - Medium

```
2 . 7 | . . 8 | . . .
. . 1 | . 2 . | . 4 .
. . 3 | 5 1 . | . . .
------+-------+------
. 3 4 | 8 6 7 | . . 2
5 7 . | 9 . . | . 3 .
9 8 . | 3 . 1 | 4 . 7
------+-------+------
. . 5 | 4 . . | . . 1
. . 9 | 2 . . | . . .
4 . . | . 3 . | 7 2 5
```

Puzzle 543 - Medium

```
. . . | 3 9 8 | . . 1
. 9 . | . . . | . . .
. 4 8 | 7 1 . | 3 5 9
------+-------+------
3 1 6 | . 2 4 | 7 . .
. 5 . | 6 7 . | 9 4 .
. 7 . | 5 . . | 1 . .
------+-------+------
. . 5 | . . 6 | 2 3 .
9 . 4 | 2 . 7 | . . .
. . . | 3 . . | . . .
```

Puzzle 544 - Medium

```
. . 9 | . . 7 | 8 1 .
. 4 . | 9 1 . | . 2 .
. 1 8 | . . 2 | . . 6
------+-------+------
8 . . | . . . | 2 6 7
. 2 . | . . 6 | 5 . .
. 5 7 | . . . | . . 1
------+-------+------
. . . | . . . | 6 . 5
3 6 5 | 2 7 . | 1 . .
. . 1 | 6 3 . | 4 . .
```

Puzzle 545 - Medium

```
. . . | . . . | . . .
8 4 5 | 6 3 . | 9 . .
. 3 7 | . . 9 | 8 . .
------+-------+------
. 7 . | 9 6 . | . 1 3
. 6 . | 2 1 3 | 5 8 .
. . 1 | . . . | 4 . .
------+-------+------
6 . 2 | . 7 8 | 1 4 9
. 8 4 | . . . | . . 2
9 . 3 | . 5 . | . . .
```

Puzzle 546 - Medium

```
9 4 . | 2 . 7 | . 5 .
5 . . | . . 3 | . 8 .
. . 8 | 5 . . | . 4 1
------+-------+------
7 . 9 | . . . | 8 2 .
. . . | 9 . 2 | . 7 .
1 5 . | 8 7 . | . . .
------+-------+------
. . 4 | 7 6 . | 5 1 .
. . . | . . . | . 6 7
6 7 . | 4 . 1 | . . .
```

Puzzle 547 - Medium

			6	5				
7		8		3		5	4	9
3		4		7	8			
1		7		6		2		
		9						
4	2		1		3		5	7
		3					1	
	4	2		1		9		
		5	4				7	6

Puzzle 548 - Medium

	2				4			
8		5				6		
		9		5	8	1		3
		8		6		4	5	
6	1	4	8					
2				7	9		1	
		2		8	7	5	9	
9		7					6	
5	4			2	9		7	3

Puzzle 549 - Medium

6	2	4			5	7		
		7	6		1			
9	1						5	
					4		9	
			2	6	9			8
				7		1	2	
	6	9		2		5		4
2				9	6			
3	7	1						9

Puzzle 550 - Medium

		6		5				8
				4		5		
5		7		2		1	6	
3		4			5	6		7
			3		6			
8	6					3		9
	5		2	6	4	7	3	1
	1	3			8			
6			1					

Puzzle 551 - Medium

		5	6	9				
				3	8			5
7	6			8			9	1
8				4				3
	7	6	9					8
	3		1	8	6	7		
	1	3			9	5	8	7
	4				7	9		
9					4			6

Puzzle 552 - Medium

7			5		3	6		4
4	5	3			6			
		8			9	1		
9		1	3			6		
3				5		9		
		2	9		8	7	4	3
8	4			9	7			
			1			5	7	
	7	6	5			8		

Puzzle 553 - Medium

	4			7			2	8
	2		3			7		
	6		8	9			3	
2			5		7			3
			1				5	
5	9	6	2				1	
8	1		7			3		
9	5			1		6		
	3		9					5

Puzzle 554 - Medium

		4		9	3			
		8	6		7			
		3		1		7		
3			4		5			
9	4	7	8		1	6		
5				9			3	1
	1	9				2		
4		2	1					9
	3						6	4

Puzzle 555 - Medium

4				5		7	8	
5	8			7		4		1
						5	9	2
2		5	3		6	8	4	
					7		5	6
8		1					2	
	5	2			3		7	4
6								
	9					6		

Puzzle 556 - Medium

3			7	8	1		6	
9	5	1				2	4	
				5			3	
					5		1	
	6			1		3	9	
4	1		2		3			5
1	8		5	2		7		
7			1			8	9	6
2			9	3	7			

Puzzle 557 - Medium

			2		5	4		
			6			3		5
5		3	4	9	1			8
8		6			7	4	1	
	4	5		6				7
			8	1				
			1			7	6	3
2		9	7	4	6			
	6		5	8				

Puzzle 558 - Medium

	9				7			
	6		9	3		7	2	
					5			4
	4		1				9	3
5	8	9		6	3			
2				9		6	5	
9		2	4		6		1	
1	3	8				4	7	
	5		3	7		2		9

Puzzle 559 - Medium

					2	3	7	1
2				1	5			
		9			6	4	2	5
3	1		6			7		
			2	5				
6	7		3				9	8
9			1		4	5		
		4		7	3			
		1		2	9	6		

Puzzle 560 - Medium

	1	3		8	4	6	2	
			1	2			3	5
			9				4	7
		1	7					8
		5		4	8	7	6	
8	7	6		2				
						1	9	6
	9							
6	8			9				3

Puzzle 561 - Medium

1			5			2		
	3		1	2	7	5		
	5	6						1
7			2	9				
3	2	4			1			5
	1			3		6	4	2
						2	5	9
5	3				2			
	7	2		4	6	3	1	8

Puzzle 562 - Medium

	9			3	7	2	1	6
1								5
		7			1		8	
8					5		6	
						3	2	8
	3	2			8	5	9	1
4	8		2	5				
			7					
9		3	8				5	2

Puzzle 563 - Medium

4	2	7					5	
	3			5			1	
1	9							
			1	4	7			5
3		2			9			
5	7		3	6		1	9	8
2			4		8			1
	4	1				2	8	9
8				7		4	6	

Puzzle 564 - Medium

		8			9			
	3		4	7			5	
9	6		3	1	8	4		
3						7	6	
			6	7	2			
6	2		9	3	1			5
		3		8				4
2	7			4	5		8	3
4				3		2		

Puzzle 565 - Medium

```
. 8 . | . 2 . | . 5 .
. 4 . | . 1 7 | . . .
7 1 . | 4 . 5 | . . 6
------+-------+------
. . 4 | . 8 . | 6 7 .
. 7 . | . 4 . | . . 2
1 . . | 2 5 . | 8 . 3
------+-------+------
3 5 . | 9 . . | . 6 .
. 6 . | 3 . . | . . .
. . 1 | . 6 8 | 9 3 .
```

Puzzle 566 - Medium

```
. . . | 3 . 7 | 2 . 8
. 7 . | . . . | . . 5
. 3 . | . 1 5 | . . .
------+-------+------
2 . 3 | 9 . . | . 8 7
5 1 . | . . . | . . .
4 . . | 5 2 . | 3 . .
------+-------+------
7 . . | 4 . . | . 1 .
. 4 2 | 7 . 8 | 6 . 9
. 8 5 | . . . | . 2 4
```

Puzzle 567 - Medium

```
9 5 . | 8 7 1 | . . .
3 . . | . . . | . 8 .
6 . . | . 3 2 | . 9 4
------+-------+------
1 . 7 | 3 . . | . . .
. . . | 1 9 7 | 8 6 .
. . 9 | 2 8 . | . . 3
------+-------+------
. . . | . . . | . . .
. . 1 | 7 2 3 | 6 5 .
. 9 . | . 1 5 | . 7 .
```

Puzzle 568 - Medium

```
. 4 . | 7 . . | . . .
7 . . | 3 9 . | 1 6 5
9 . . | . 5 8 | 7 4 3
------+-------+------
. 7 . | 4 8 . | . 3 9
. . . | 6 1 7 | 4 . .
6 . . | . . . | 8 7 .
------+-------+------
. 7 6 | 8 3 1 | . 9 .
1 . . | . . . | . . 6
. 9 . | . 6 . | 3 . .
```

Puzzle 569 - Medium

```
. . 1 | 7 . 5 | . . 3
7 2 . | . . . | 8 . .
3 9 . | . 6 8 | . 4 7
------+-------+------
. . 2 | 8 . . | 6 . .
. 7 6 | 4 . . | 3 . 8
5 4 . | 6 . 7 | . 1 .
------+-------+------
. . . | 7 . . | . . 1
. . . | 8 6 . | . . 9
. 8 7 | . . 9 | . . 6
```

Puzzle 570 - Medium

```
8 1 . | 9 . . | . . .
. . . | 1 . . | . . 4
. 4 . | 2 3 . | 9 5 1
------+-------+------
. 6 . | . . 2 | . 7 .
. . . | 1 7 . | 3 4 .
3 . 4 | 5 . . | 9 1 6
------+-------+------
. 5 8 | 4 6 . | . . 7
6 2 7 | . . . | 4 . 5
. . 9 | . 2 . | . . .
```

Puzzle 571 - Medium

1	7		5					3
4				9			7	
		5		3		1		
7	2	3	9	8	5		4	
8		4	6				2	
	9				2	5	3	8
		9				2		4
	1	8	2			9		
			5	9				

Puzzle 572 - Medium

	3				4		1	6
	9	7		5				
1				2	9		8	
6	8	1	9					7
2	5			6	7		3	
			8				6	
		8				6		2
4	6	5		7				3
3		9	4				7	

Puzzle 573 - Medium

1				4				3
9	7	3						4
4	2	6	3		9	5	1	8
	5			6			3	
3			4			2	8	9
			2			7		6
			8	2				
		2	1			6	7	
6	4		7		5		9	

Puzzle 574 - Medium

8	6						9	4
		7	9		5			
			6	4	1			
7						1	5	
		5	4	1				7
1	8		5				4	
	9		1			3		8
3		8		5	9			6
	1	2						

Puzzle 575 - Medium

3	8		6					
					9			
		6	3	2	1	4	5	
		5				6	4	1
		6	1	5	7	2		3
2	1						7	5
		4		7				9
		8	4	9	3	5		
7	5		2	1			6	

Puzzle 576 - Medium

		1		5				
			3		7			4
5	3	7		1	2	9		8
4				3				
	2		6		1	4		3
	8		2	4		7		
		4				2		9
9		3	5	2		6		
		2	7					

Puzzle 577 - Medium

```
. 8 2 | 1 . 4 | 7 . 6
. . . | 8 . . | . 3 9
7 1 . | . . . | 2 . .
------+-------+------
. . 8 | . 7 . | . . .
1 7 . | 6 5 2 | 9 . 3
. 3 . | . . 8 | 6 7 .
------+-------+------
8 5 . | 2 6 . | 4 9 7
4 6 . | . . 9 | . . 5
9 . . | . . . | . . .
```

Puzzle 578 - Medium

```
9 3 . | . 7 4 | . 1 6
. 7 . | . . 8 | . . 9
8 . . | . . . | 2 . .
------+-------+------
. 2 . | . 5 . | . . .
. . 3 | . 9 . | . . 2
. . 7 | 6 . . | . . 3
------+-------+------
. . 9 | . 1 . | . 2 .
7 5 1 | 3 . . | . 6 4
2 8 . | . . . | 3 . 1
```

Puzzle 579 - Medium

```
4 7 . | 9 5 . | 6 . 3
. 6 . | 2 8 . | . 5 1
. . . | 3 . . | . 2 .
------+-------+------
8 . . | . . . | 5 1 .
2 5 . | . . 3 | 9 . .
9 4 6 | . . . | 8 3 2
------+-------+------
5 1 . | 4 9 . | . . 7
. . . | 7 . 2 | 1 . .
7 . . | 1 . . | . . .
```

Puzzle 580 - Medium

```
. 8 . | 1 . . | 5 . .
6 . . | 9 . . | 4 7 1
. 1 . | 7 . . | 6 . 2
------+-------+------
3 . . | 6 . . | 2 . .
. . . | 3 . . | . 9 5
. . 2 | . . . | . 6 7
------+-------+------
8 7 . | 4 . . | 1 2 .
9 . 3 | . 8 . | . . .
. 1 . | 9 . 7 | . . .
```

Puzzle 581 - Medium

```
8 . 2 | . . 1 | . . .
. 5 . | 6 2 . | 3 8 4
. 7 6 | 8 9 3 | . . .
------+-------+------
. . . | . 3 . | 8 . .
5 8 . | . . . | 9 . 2
. 2 . | 8 . . | . . .
------+-------+------
2 . . | 3 1 6 | 7 . 8
. . 1 | . . . | 5 . .
. . . | . 7 . | 9 . .
```

Puzzle 582 - Medium

```
8 . . | . 5 . | . 2 3
. . . | 3 . . | . 6 .
. 9 . | . . . | . . .
------+-------+------
. 8 . | 2 . . | 7 5 .
6 . 4 | 7 9 . | 3 . 1
. . . | 4 3 8 | . . 2
------+-------+------
7 3 . | . . 9 | . 1 .
1 . 8 | . 4 . | . 3 .
. . 2 | 1 . . | 5 . .
```

Puzzle 583 - Medium

```
1 6 . | . . . | . . .
4 5 . | . 2 9 | . 8 7
8 . . | . 6 4 | . . .
------+-------+------
. . 1 | . . 5 | . . 8
. . 6 | . 8 7 | 5 . 1
. . 8 | 2 1 6 | . 4 9
------+-------+------
. 7 9 | . 5 2 | . . .
6 . . | 8 . . | 2 7 .
. 8 . | 4 . 1 | . . 6
```

Puzzle 584 - Medium

```
. 1 6 | 9 4 3 | . . .
. 9 2 | . 7 5 | . . 3
. . . | . 8 . | . 9 6
------+-------+------
. . 4 | . 1 . | . 2 .
. . . | 4 . . | . . .
9 . . | 2 8 7 | 3 1 4
------+-------+------
. . 1 | . 5 2 | . 3 8
. 3 7 | 8 9 . | . 4 .
. . 9 | . . . | 4 2 .
```

Puzzle 585 - Medium

```
. . 8 | 6 . . | . . .
. 5 1 | 8 . 4 | . 6 2
6 4 . | 2 5 . | . . .
------+-------+------
. 8 5 | 4 1 . | 7 . .
. . . | 5 9 2 | 8 4 .
. . . | 6 . . | 1 5 9
------+-------+------
. . . | . 4 . | 3 9 .
3 9 . | . . 6 | . . 1
5 . . | . 8 . | . . .
```

Puzzle 586 - Medium

```
. 3 . | . . 6 | 2 4 7
. . . | . . . | . 8 1
8 2 4 | . 1 5 | . . .
------+-------+------
2 5 . | 3 . 8 | 7 . .
7 . . | 1 2 . | . 5 .
. . . | 7 . 9 | . . 3
------+-------+------
1 . . | . . . | . . 4
. . . | . 6 . | 8 7 .
3 . 2 | 4 9 . | 1 6 5
```

Puzzle 587 - Medium

```
. . 8 | 3 . . | . . 5
9 . 3 | 6 . . | . . .
. . . | 1 8 2 | 6 . 9
------+-------+------
1 7 . | . . . | . 2 6
5 . 9 | . . . | 8 4 .
. . 2 | 4 . 5 | . . .
------+-------+------
. . 1 | . . 9 | . 5 3
. . . | . . . | 9 6 .
8 . . | . 7 . | 4 1 .
```

Puzzle 588 - Medium

```
3 7 . | 4 9 . | 1 . .
. . 4 | 5 7 . | . 8 2
5 . 2 | . . . | . 9 .
------+-------+------
6 9 . | . . . | . 3 5
2 . . | 1 . . | 8 6 .
8 3 7 | . 6 . | . . 1
------+-------+------
1 5 . | . . 4 | . 2 .
. . . | . . . | . 1 .
. . . | 3 . . | 9 7 8
```

Puzzle 589 - Medium

		7	3					1
			9	4				
	6		1	5		8	9	
							3	
6	3		7			4	2	9
9	2		3			8	7	
			8			9	4	
			6			3		8
7		9	4		3	6		

Puzzle 590 - Medium

3		6	9			8	5	
1	2			5				
4	5				6	7	1	3
			7					5
			3	6	2			
8	4			1	9	6	3	7
		7			4			
			6		5			9
	8	1	2	9		3		6

Puzzle 591 - Medium

1	5		3					
3	8			7	5		2	9
	7			4	6		1	
	4		2		8	9	7	
					8			
		1	4	9		3		
7	1				4			
	6	2						
9	3				2	6	8	4

Puzzle 592 - Medium

		1				3		7
				9			1	
2	6		3			4		8
9	2			3	1		5	
	1	8				7		2
5	3			2			8	9
		2			8	9	4	
	4							
	8	9			3	2	6	5

Puzzle 593 - Medium

						7	3	5
			6		2			4
7		9						
6		1			8			2
4			2	5			9	
5		8		4				7
	5						7	6
		4			7		2	3
2		6	8			4	5	9

Puzzle 594 - Medium

		3	4	7		2		
8			5		9	6	1	
	2			6	3		8	4
3	5			4				9
			8		1		4	7
		7		3				8
	4		2		5			
6						8	5	1
		9		1	6		7	2

Puzzle 595 - Medium

5						2	6	4
7		1	6	4				
	4		9		5			1
			7	6	3	4		8
		9	4					6
			5		9		1	7
	3	7			4			5
4	1				7		3	
	5			1				2

Puzzle 596 - Medium

4		8						
6			8			2	9	
9	1			5				
2	8	9				4		1
	6							
	3	1		4	8		2	6
1		5			3	9	6	
		7		2		5	1	3
3		6	1				4	7

Puzzle 597 - Medium

2	9	7					1	
	1	8						9
	4		9		2	5		
9	6		1	2	8		7	
		4	7	3			9	
					8			
6		2		9		5	8	
					5	9		6
	5			8			2	1

Puzzle 598 - Medium

	6	7	5		9	8		
				4				
4	9	8		7	2	3	6	
		6			1	9		
	7	5		8	6	2	3	
9	2					6		8
		9				4		
5			7	9				6
7			8	6	4		9	

Puzzle 599 - Medium

	9						8	
1	8	5				6		
		6				3	1	9
	5		1	7			3	
7	6	4			8		5	1
8						2		
9	4	8	6		7			3
6		1	9		5	7		
		7	8		3	1		

Puzzle 600 - Medium

		6	1		3			
7					4			3
1	3			8	6	7	2	9
	9	7		2				1
			4	1	7	8	9	
8	1		6			3		4
			7				3	
	7	3			2	9		
		1				2	8	7

Puzzle 601 - Medium

3	5		6	2		9	1	8
8			7		9	5		
		1			8	4	7	
		9	8	7				6
7		2						
		6						9
	6	8	5			1		
	7	3		6	2			5
	4	5			7		6	2

Puzzle 602 - Medium

5					9			
	4			1		7	6	
			6	3		1		2
	7	2	3	4	8	6		
4			2		7			1
	3			6		4	2	
		4				5	8	
3		7					1	
6	8		5		3	2		4

Puzzle 603 - Medium

4		7		9		8	2	1
2		5	3		1	7		9
		9						4
			8	1				
						1	9	2
5	9					6		
				6			1	3
3	7				5	9		
			9		3	2		7

Puzzle 604 - Medium

5	7		9				8	
9		4	8		1			
6	8		3			4	2	
	5	2		9	3			
3		9	7					8
4	6	7	5	8	2			
1			2	7	8			
2							1	7
						8	9	

Puzzle 605 - Medium

	6		8	1	3		2	
				4			7	
3	5				2		8	6
6			3		8	5	4	
8		4	9		7			2
5		2	4					
		6						
2								9
7		5		3	6	8		

Puzzle 606 - Medium

	2	1	3	8			6	4
5	3				1		7	
	6							2
			4	5			8	1
4			7					
8	1	5	2	9	6			
			1			6		
	4				9			5
2			5	6		1		

Puzzle 607 - Medium

	9		8	6				
7		3				6		
4		8	1	7		2		9
		7	2	1			9	
2	8	5			6			4
3		9						
			6	8			4	
			3			1		6
	7		5		4			8

Puzzle 608 - Medium

			3		6	9		
	9			8	7	6	1	4
				1	9			
	3	5	2			1		
	4	6					3	
2			6	9			7	
	5		9			8	2	
		9				7		3
8	2	3	7		4			

Puzzle 609 - Medium

1	3	9	5					
		8	9		1	5	7	
5	4	7	8		3	9		2
			3	5				7
		5			4	2		
3			1		6		8	
	8	6				3		
	9	3		1	5	8	6	
4	5							

Puzzle 610 - Medium

	6		9				3	1
2	4	9	1	6		8		5
7		1		4				6
	2				6			
			2					3
1		6						7
	9	2		1				
4	5		2	9		1		8
	1	7	6		4		9	2

Puzzle 611 - Medium

			1				8	4
1				3	4		6	9
			7	9		3	2	1
	6	5	3			8	7	
	1		6	8	5	9		
8				4		1		
		1	4	6		2		
3			5	2				7
6			9	7				

Puzzle 612 - Medium

	1					6		
4		2		8			3	9
		6	5	3	9	4		2
1	8		4		3			
3				9		8	6	
		4		5	2	1		3
6				1		9		
		1	3				7	
					6		2	

Puzzle 613 - Medium

				9	5	7		
		3	2		4	1	5	
	8						4	2
				3				
3		8		4				5
		1	6	5	7	8		
	6	7			1			
	3	4	7	6	9			
8				3		4	6	

Puzzle 614 - Medium

	1		9		4		2	7
6			8	2	7			
	7							
5				3		9		
4	2		6		9		7	
7				4		3		8
	5	4	2	9				6
8		9	1	7			5	4
			4	6				

Puzzle 615 - Medium

						9	7	4
6			1	9	5	8		
9				8			6	5
3			9		6	7		
	9		8		1	4	2	
	1	6						
	7						1	
5	6			1	9	2	4	3
	3		2	6			8	

Puzzle 616 - Medium

2			9	6		7	8	
4			1		3		9	
9				5	8	4		3
			3				5	
7	8	3		4		2		
							7	4
	4						3	7
	1		5		4	9	6	
	2	9	8	6			4	5

Puzzle 617 - Medium

		6		1	7			
4	3	8		2				7
	1	2	3		8		5	9
					9	2	7	8
8				7			4	
			8		1			
9			6	8		1		
3			1				9	6
	2				4		8	

Puzzle 618 - Medium

5								7
7			6	5	2	8	1	
2			7	9			6	4
		7		1				9
			8	6		7		
6	2			4			3	
3		8				9	2	
1		2	5		9			
	6	4				3		

Puzzle 619 - Medium

	6	3			9		7	
	5		1				2	8
			2			3		
4		6					8	7
	8	7	6					3
					9			6
		5		8	3	4	6	2
	4	8		6				9
	2				1	8		5

Puzzle 620 - Medium

8					3	4	5	6
1	3		4					
		4	6				1	
	6	1	7		9			
	9				6	7		
	4	8					9	
3		5					6	9
4		9	5	6		1	3	7
	1			9		5		8

Puzzle 621 - Medium

		2	7		4			
1	5		3					7
		7				4		2
	1		6	8	2	7	9	4
9		6	1			8	5	
7					5		2	
				7				8
	8	5	2					
				1	8	3		

Puzzle 622 - Medium

1	7				2	5		
	6		3	7				1
2	8		6					9
6			4		5	9		
	4	8			6	1		2
	1	9			3			
8			1			6	9	3
7	3						2	
4		6				7	1	

Puzzle 623 - Medium

4	1	8			9	6		7
			7					1
	6	7		1				
			9	5	3			
	2	1			8			9
	8	3				1		6
8			3		2		1	4
7		2	5	8				
1	3	5		6	4	8		

Puzzle 624 - Medium

2				7				
5		7						
		6	8		9			7
8		2	4					
	6		7	5	2	4		
	4		9	6	8	2		5
					7	3		8
3			6		1			
	9		3	8		6		

Puzzle 625 - Medium

4	5				3	9	6	1
	7	1				2	5	4
6		2			1			
9	3						1	
	8	6				5	7	9
				6	8			
	6	7		1				8
8			7	4	6		3	
1			3	8				

Puzzle 626 - Medium

7	3				6			
2				3	8		1	
1	5		7		2	6	3	
	8				9			4
			2			5	9	1
	5							
	2			4		9	6	5
	3		9			7	1	2
6								

Puzzle 627 - Medium

		6					5	
3	1		4		5	9		6
	5			6	3		7	4
			7					1
		1		2		6	9	5
			5				3	7
	3				7	5	1	
9			6	3		4	8	
			2					

Puzzle 628 - Medium

1	2			6				
5	7	6		9				
			8		2	6	5	
2	8		6			7	9	3
	6	5	9				4	
		4		2		5	1	
6		2	3		5			
		7		6				2
4	9							5

Puzzle 629 - Medium

5			2					
			8	9		5		2
			5	6				
6		1	3			9		
	3	5	9	8			7	
4	9		7	1				5
		2						
	5	9		3		7	4	
8	4			5				3

Puzzle 630 - Medium

			9		6	3	1	
				4	1		5	
	1		3				4	9
8	3	6	4			5		
1		5					9	4
	4	7		1		2		3
3	6			9	4			
	7	1	8		2			
		9	1				2	

Puzzle 631 - Medium

		1		5	8	4		7
	2		1		4	8	9	
	6		2			5		3
2	4			1				
		7	3				5	4
		9	8	4	6			
		6			3		4	
9				6	5	7		
		8	7			9		

Puzzle 632 - Medium

1		7		9	6			
8	5					9	7	3
2			8	5	7		1	
	2				8			9
		1						5
4		9		1				2
7		2		6	9		4	
			1			2	3	
9				3		5	2	6

Puzzle 633 - Medium

					6			
	9			4	3	5	1	8
3		8	7			4	9	
6					8			4
8		2		7	9	3		
	3		4	5		6		
	7			9	4			5
9		1						3
5			3	6		9	7	

Puzzle 634 - Medium

9							5	8
2	7		9	5			4	
				1	4		9	3
	2			9			3	7
	5	7	8		3		1	
					5	8	2	
								5
	9	8		3		4		
	6		2	4	9			

Puzzle 635 - Medium

			8	7		9		
7			2					8
			6		9	4		
4	2	9	6					1
	1					6		
6			1	2	9		5	3
		7	9	1	8		6	
		3				1	8	
		1		5		7		

Puzzle 636 - Medium

		2				7		
7		9	6	8	4	3	2	
3						5		6
	6			3	8			9
8			9				5	6
9		7	5					
4			2		1	3		
		6					1	2
2			3			4	8	

Puzzle 637 - Medium

							5	
	9	4						
	6					9		7
		8	5		2			9
	7	9	3			6		
2	3				7	1		5
8				9	6			
9	5			8			7	
		7	1		5	8	9	3

Puzzle 638 - Medium

9								
	7	1	3			8		
	5		1	4	2		9	7
		9		8			5	3
1		7						4
		5			9	1	8	
4	1	8		3		6	7	2
							3	
5		3	2				1	9

Puzzle 639 - Medium

		1	6	5		4		8
2			8	3	4	9		1
3			1				6	
	3	5			7	6		9
7		4	2		5			
8		9			6	5	4	
	8	7						
				2	8	7		
	1		7					4

Puzzle 640 - Medium

	8		5		1		3	4
5			8			1		
	4			3	9		5	8
8		3		2	5	6		1
2			1					5
								3
		7	4					
9		8		5			6	2
4	6			1		3		

Puzzle 641 - Medium

		4	9					5
		7	3					8
2		9			7	3		6
8		1	4		5			9
						1		4
9		3	8			7		
	7	2	5	1	9	8		
6								1
			8				2	

Puzzle 642 - Medium

5							3	
8	4	3	7	9		5		2
	6	2						
			9	4	7	2		
2			1					
9	7			2		6	4	3
	5	6	2				7	9
7		9						
3	2	8	6	7		4	5	

Puzzle 643 - Medium

		7	5	8				6
		1				2		
	3	8			4	1		
8			7	2	3			9
3	9		8					
		4	3			5		
		1	7				3	
4		3		2	5			
		9			1		8	4

Puzzle 644 - Medium

			8	2				7
9		8	6		3	1	4	5
	3		1		9	2		8
		5		6	7	4		
		3		9		8	1	
	6	2			1			4
3				8		5	2	
8	1	9					7	

Puzzle 645 - Medium

	8						7	5
9		5	2					4
	6	4			9		2	
8			9	4		3		
1				3				8
			2	8	5	1	6	
			7		2	3		
	2			5				
6		9	4			2		7

Puzzle 646 - Medium

	4					9		
2		7		6				8
	5	8			4	3	7	
6	1			4		7		3
8	2	3			9	5		4
5	7			3	8			
	8	1	7				2	
	2		1					9
	6	5	4					

Puzzle 647 - Medium

			8	1		4	6	
			7	3		9		2
5	7	6					8	
						6	9	
8		3	7				2	
2					1	8	7	5
	5	8		3	7			
3		7		4			1	6
			5		2			8

Puzzle 648 - Medium

8				6		4		9
9	7				8	6		
	4		5		2			
5						9		
6	8	9	3		5	7		
			6	2	9			
	9		7	4	6	1		
	3						7	6
			8			5		4

```
1 . . | . . . | 3 . .
. . . | . 8 3 | . . 1
. . 7 | . 4 1 | . . .
------+-------+------
. . 5 | 4 9 . | . 1 6
4 . . | 1 . . | 5 . .
. 6 . | . 5 7 | . . 3
------+-------+------
6 7 . | . . 9 | 1 5 4
9 . 4 | . 1 5 | 3 . .
. . . | 7 . . | . . 9
```

```
3 . . | . 6 8 | 5 . .
. 1 8 | . . . | . 6 .
6 . . | 9 . 7 | 3 . 8
------+-------+------
. 4 . | 3 8 9 | 7 . 5
. 7 3 | . 5 . | . 8 .
. 8 . | . 2 . | 4 . 1
------+-------+------
. . . | 1 . . | 9 . .
. 3 . | . 6 . | . 7 2
. 9 1 | . 4 3 | 8 . .
```

```
1 . . | 6 . 8 | . 7 .
4 2 . | . . . | . . 3
. 5 6 | 2 . . | 9 . .
------+-------+------
. 8 . | . . 6 | . . 1
3 7 . | . . 2 | . 8 5
. 1 . | . . . | 3 . 4
------+-------+------
. 4 . | . . . | 2 5 .
2 . 9 | 5 . 4 | . . .
. . 1 | . . 9 | 4 3 7
```

```
. . 4 | . 8 . | 2 . .
2 6 7 | 3 . . | 4 8 5
. . 9 | . 2 . | . 7 .
------+-------+------
. . . | 2 3 . | 5 . .
1 7 3 | 6 5 4 | . 9 2
6 . . | 9 7 . | . . .
------+-------+------
. . 6 | . 9 . | . . .
. . . | . . 7 | . . 3
. 2 5 | 8 . 3 | . 1 6
```

```
1 3 5 | . . 2 | . . 6
. 6 . | . . 1 | . 8 .
7 . 8 | 3 . . | 5 . 2
------+-------+------
. . 6 | . 3 . | 1 . 8
. . . | . . 5 | . . 4
. . . | 6 9 . | 3 . 7
------+-------+------
. 7 4 | 5 . 8 | . . .
. . 9 | . 6 3 | . 7 .
. 8 . | . . 9 | 4 6 1
```

```
8 . 5 | 6 4 . | 7 . .
3 . . | . . 7 | 1 5 6
9 . . | . 1 . | 3 . .
------+-------+------
2 3 . | 4 . 9 | . . 1
5 . . | 7 6 . | 2 . .
1 . . | . 2 8 | . 9 .
------+-------+------
. 8 . | 2 . 4 | 5 1 3
. . 3 | . . . | . . .
7 5 . | . . . | 6 4 .
```

Puzzle 655 - Medium

3		9		8			1	
6	7			3			4	
			9		5	7	3	
			8				5	
								9
1			6	2		8	7	3
4		2	3		7	6		
		8	4	6				
7			5	9			2	1

Puzzle 656 - Medium

8	1			9				
			8	6		4		7
		4			7		9	8
4							8	
9	5		8					
	8		3	7	9		5	
1	2	8	7	4		9		
7	9		2		5		4	
	4				8		1	

Puzzle 657 - Medium

			7	1			6	5
7		5		9	6	1	4	
	6	1		3				
1	9				8			
5		8		2	9			
6	4	3				9		8
4					3		5	
			1		4		7	3
3	1			5				

Puzzle 658 - Medium

		1				3	7	
	4		8			5	2	1
6	5							
		3	6		8			
8		4	2		1	6		3
		6	3				8	4
4	3		1			2	6	
	2	8			6		3	5
						8		

Puzzle 659 - Medium

		6	4	5				7
4			3				1	9
5						6		
		1	2	8	7	3	4	5
			1		3		9	8
7		3			4		2	
	7	2	4					
					9	1		2
	9	5		1		4	7	

Puzzle 660 - Medium

1	7			6			2	5
							6	
				8				3
	1				8	3		
7	9		1	3	2	8		6
		8	3	5		2	4	
4	6	7			9			2
9	3	5			6		7	
			7	5				9

Puzzle 661 - Medium

2	1			5		7	9	3
						5	8	4
8						6	2	1
	3					2		
5					6			7
		4			8		5	
	2			6	5			
4			3	8			6	
	8			4	1		3	5

Puzzle 662 - Medium

5			2			6		
6	7		9		3		2	
	3		8		5		7	9
	1	4					8	
	8		1		4			
7	2			8			4	
		7						4
	6			5				7
1			6	7	2			

Puzzle 663 - Medium

3								
		2						
1	6	9		8				2
7	3	8	1			4	5	
2	1					7	9	
	9		4			1		
5				6		1	9	7
	7				2		8	4
	8		7		1		2	

Puzzle 664 - Medium

	3			6				
	4	9	5		8			3
8		5		7	3			
4	9						5	1
1		2	3					9
					9			
5						9		
9	1	4			7	2		
3		7	9		1			8

Puzzle 665 - Medium

	1		3			7	6	
		7			1	5		9
	6			7		3		1
6	7			1				
			2	4	5		1	7
2			7				8	4
				5		4	6	
	3	5		6			9	
		6	4	2			3	5

Puzzle 666 - Medium

4				8	7	9		
			9	5		6		
9	5	6	3		4			7
	4					5	9	6
		9				1	3	
	3		5		1	7		
3	2	5						
8		7						5
1	6		2		5			

Puzzle 667 - Medium

			7		1			3
2				3				4
9	3	7		2			5	
4				7	2			8
			3		5	4		9
1	5	9	6	4			7	
	9		4	5		6	3	1
	1	6	2					
						8		5

Puzzle 668 - Medium

	9		8					3
	6	1				8	7	
5								
	5			3		4		
3	4	6			2			
8			5			1		2
9		8			7			
6			9	1	3		5	
	1		2	6	8			9

Puzzle 669 - Medium

4	3						5	7
5			2	9			8	
1		2		3				
8					1	4	7	
7	4	1			8			
	2			4	7	1		
	5			8	2			6
			4					
	1	4	6	7	3	8	2	5

Puzzle 670 - Medium

					1	9	7	8
1			3			2	5	4
8		5		7	2		4	9
2	1				8			
7				4		5		
	9	8	7			4		6
5		6			4			
			2	8	6		9	5

Puzzle 671 - Medium

7		9		6	3			
	5		7	1			3	
					9	1		
8		7				5	9	
1				7	8	2	4	
2			1	5			8	3
	7			9				
9		2	4	3	1			5
			2					4

Puzzle 672 - Medium

8						9	4	
6	3						8	2
							1	7
2		8		5	7			9
7	4	9		8				5
1	5		2	6	9		7	4
9	1				8			
4			7		3	6	2	
			5				9	8

Puzzle 673 - Medium

7		2		6		3	5	
4		9			5			1
	3	8		7			4	6
2	4		5	1	6			
	9	7						
	5	6	7	9				4
			2	4	1			9
			9			7		
		4	6					3

Puzzle 674 - Medium

	1			2	3		4	8
						1		3
			1	5	7	2	9	
			4				1	
6	7		3			9	2	4
	2	1	7				3	
2	4							1
9			8	1	4	3		
		7	5					9

Puzzle 675 - Medium

7		8			3			
	3	5	7		6			1
4	1					3		
3			9				5	
	8		6					
		1	3		8			
		6		3	2	9		4
1					9	5		
5	9	4	8	1		2		3

Puzzle 676 - Medium

7					2	3		1
2				4		6	9	
		4				5	2	7
	2		9		1		3	6
	3	1		6				2
		7	2		4			
			3			1		9
		8	6	1	7			
			4	8	9		6	3

Puzzle 677 - Medium

	2	8						
		9	2		1	8		5
1	5	6		7		9		
6	7		3		8		1	9
	3		6	1			5	
		2		9		3		4
	6				7		8	
		1		8		6	2	7
2		7			6			

Puzzle 678 - Medium

	5	6	2		7	9		
9				3		2		
				8				4
8		5					9	
7			5			3	6	
3			6	9		7	1	5
		7	1			8		
	8			3			4	1
	3	2	6					9

Puzzle 679 - Medium

3	4						2	7
1		5			2		4	
			4	3	8	1		
	2		5				3	
		3		8				6
		4						2
4				1	7			3
8			9	6	4	7		1
7				2				

Puzzle 680 - Medium

	9			6	7	5	2	4
7		6	1	5	4			
				2	3		1	7
	3	1		9				
5	7						3	2
	6		2		5	8		
	8		3	1				5
	5				9	1	8	
						7		9

Puzzle 681 - Medium

	1	7	5	9				
	4	3	1			7		
9	5	2						
2	9				7	3		1
			4	5	3			
3			2			8		4
4				8	5			2
		9	6			5	3	
		1		7				

Puzzle 682 - Medium

6		2			4		9	3
			6		3			
	3					4		6
4	2			1	8			
3			5		9	2	8	1
	8					9		
8	6		4	7			2	9
1				2	5	7		
		7	8			1	4	

Puzzle 683 - Medium

7	3		1	4			5	8
			2		8		9	
8				5				4
6					2	9		
	4			9	3	1		5
	8					2	6	
		3	5	8	1	4		6
	5			2		8	1	9
	7						3	

Puzzle 684 - Medium

6	3	1	5	2		7		4
	9			7	6			1
	7			4				
		7	2	1	3	8	6	5
			6	9				7
5	1	6				3	2	
3			1	5			4	
1						5		2
		4		6				

Puzzle 685 - Medium

4		8			7	9		
	1	6	8	3				4
	2	7			1	8		
8	6	9	1	5	3			
		2						9
	7	5	2		4	6		
7	5	4	3				9	
		3	7					8
			4		3			

Puzzle 686 - Medium

			5		6	8	3	
5		6						4
		4		1		7	5	6
			4					
		1	7					
4	9	8		2		6		5
	4	5	2			1	8	
	6				8			
9	8	2					6	7

Puzzle 687 - Medium

	8			6			7	
4		7	8			6	9	5
9	6					3		
5				8		1		7
				2	5			9
8	3		9			1	5	2
		6		9		4		
1			3	5	4			6
			2				5	8

Puzzle 688 - Medium

					3	4	2	
		5	6			8	3	
8	9		4	5	2		6	
5	7	1		8				6
9	3			4		7		2
	4			7		8		
	5						9	
	9		6				1	
	6		9		4			

Puzzle 689 - Medium

	7		4	8			3	1
		2			1	7	4	8
1				7				6
	8	9	6			3		2
2		5		1	4	6		
				3	9			
					8		7	9
						4	6	
9						2		5

Puzzle 690 - Medium

9				4			8	
					6	3		
4	1	6			5		2	
		9		5		1		
		1	4		9		6	
	4	2		6	7		3	
				8	2			
7		5		1	4			3
2			5	9				1

Puzzle 691 - Medium

	8		5	4	2	6		9
6	7	9	1	3			2	
2		5				8		1
4	2	6		9		1		7
8	9		7			2	4	
5					6			3
	5	4						2
					4			
9	3			5				

Puzzle 692 - Medium

			5	1	3		6	
		6		8				5
4				9				
	5			4				9
	2		8		1			6
3		1						
1			6	5				
5	9			3	4		1	
7			1	2	9	5		8

Puzzle 693 - Medium

2			8		6			9
			4				1	8
		1			4	7		
4	2			7			5	
	8				2			
7	6		5		9			
9		7	6	5	8	1	2	
			4	3			7	
		6				8		

Puzzle 694 - Medium

						2	4	3
6		5						
2						5	9	
	7	4				9		1
5		9				7	6	
						8	9	7
7		3				1		
	9	4	2			3		
1		6	7	8		2	4	

Puzzle 695 - Medium

		2				8	7	
8			9			5	6	3
		9	8	7	6	4	1	
3		7		4				
2	1				8	9	3	
				6		1		
	2					1	7	4
	5	1	4	2	7			8
		8	3		5			

Puzzle 696 - Medium

			4		8			
8		1	9	7	3	2	4	5
7		4	5		6			
			1	5	2			
2	9	8	6			3		
6			8			7	2	
		7	2	9	1		6	
	2		7		4			9
1	8							

Puzzle 697 - Medium

	8		4					3
	2			7	8	6	4	
3				6			7	
2	9	7	8			5		
		4	6			8	3	
	3					9	2	
4				8	3			
			7	5	1		8	
					6	2		

Puzzle 698 - Medium

			9	3	5		2	8
5		3						
7				6		3		9
6	9	5	3					2
3	4			1			5	6
		8		5		2		7
		7			4	8	6	
	3		7			5		4

Puzzle 699 - Medium

	2	5				1		
	3		7					
			2	5				7
			9	1	7			
3	5	1		7				9
	7		5				1	2
		3		9		2		
	6	9	3	4	5	7		
8	4		6	2	9	3		

Puzzle 700 - Medium

4			9			7	8	1
2	8	1		3	6		9	4
	9	7						6
3	1		5	4				7
7				9				
		9		8	4			
1			9				4	5
	2			5				9
9		5			1	6		2

Puzzle 701 - Medium

	9	4			1	5	7	
	3			7		9		1
6				5	9			
	6		2		8	4		
		7		6		1		
	1			9			5	6
						1		
		8	1	4	3			5
1	5	6						

Puzzle 702 - Medium

			7	3	8			
	4		6				1	
	6	8			4	2		7
8			3				7	
2		3	9			1		8
	7	4						3
	8					7		6
		9		7			8	
	3	2	4			5		1

Puzzle 703 - Medium

	7	1		4	5		8	2
6		3		7			4	
		8			9	5	7	3
	9	6				4		
8	2		9	5	1		3	
	3		4			2	9	1
					3			
3	6			8		9		
				2		3		7

Puzzle 704 - Medium

8	7		9	1			2	
9	5	6			7	4	3	1
	2	4		6			8	
	9	8			5	2		7
	1	7	4	9	6			3
				2		1	9	4
5								
	8	9					7	
		1	6					

Puzzle 705 - Medium

	2	9	5			1		
	8	7		3		2	4	
3		4			2		6	
			8	6	4	3	9	2
			7	1	9		8	6
4			1					7
		2	3				5	1
7	5	1		9		4		8

Puzzle 706 - Medium

			3		5		4	1
	4	7			1			6
		1	6					9
	5				6		9	
		9	1		2			5
2	1					4		
	6			1	8		2	7
3	2			6	9	5	1	
	7		4					

Puzzle 707 - Medium

5				1		8		
1				5		2		
	8		9				6	1
3		1		9	6	7	5	8
6		5				9		3
		8	3	4	5	1	2	
8			1	7			9	
		6			8		1	
						3		

Puzzle 708 - Medium

9			5	1	3		4	
	6		2	9				3
5		3	6				7	9
7			1		9		8	
1	8			2		7		6
6	3	2			5		9	
		1	7				2	
		9		2	1			7
	7			3				

Puzzle 709 - Medium

		9	7	3				
3		8	1			9	6	
	4		9	8	6			5
				4		6		1
		5			2		8	
4	8			9				
	2		4	6	3			
8					7	3		
1	6				9			7

Puzzle 710 - Medium

	4			8	9			
	2				4			7
	6		3		2		5	
9	1		8		7			
	3		5		1		9	
2			9	4	3		6	
		8	1		6	9	7	
		2		9			1	5
3			2					

Puzzle 711 - Medium

7		5		3		6	8	
	4					1		7
		1		4		9	5	3
5			3	8			4	
	6		5			3		8
					4			
		2	1	6		8		9
9								
1			4	2		6		

Puzzle 712 - Medium

	5		8			9	4	2
			6					
9			4					3
	6				9		5	1
2	1		3	6	8	4	7	
7	4	9			1	8		
			2			3		
		6			3		2	4
	2	4			6			8

Puzzle 713 - Medium

4		7	5	9				
	2				6			
	8					9	3	
2	4		7			3	9	8
			6					1
	5	8		2				7
8		4		5				3
5	6	9	4	3	1		7	
3					7	4		

Puzzle 714 - Medium

7		9	2	8	3	4		
6	3						1	8
			5					
			7	5		6	3	2
	5	6	1	3	2	9	8	7
					8			
2	6				1			9
	9		3	2	7			1
			9		5			

Puzzle 715 - Medium

```
9 7 2 | . . 4 | . . 1
8 . . | . 9 . | . . 7
3 6 . | 7 . . | . 9 4
------+-------+------
. . . | 5 . 6 | 3 . .
. . 8 | 3 2 9 | 7 . .
6 . 9 | . . . | 5 4 .
------+-------+------
. . . | 9 . 2 | . 7 .
. 2 . | 4 5 . | . 6 8
. . 6 | 1 . . | . . .
```

Puzzle 716 - Medium

```
2 . 6 | 8 . . | 1 . 5
9 . . | . . . | . . 4
. . . | . 9 . | 8 3 2
------+-------+------
. 6 . | . 5 . | 9 . .
1 9 4 | . 7 . | . 5 3
. . 8 | 9 . 4 | 2 . .
------+-------+------
. 1 3 | . . . | . . 9
8 . . | 5 3 . | 4 . 1
. . . | . 1 9 | . . .
```

Puzzle 717 - Medium

```
. . . | . 3 . | . . 9
3 6 9 | . 2 4 | . . 7
. . 2 | 9 . 8 | . 3 5
------+-------+------
. . . | . 5 . | . . .
2 7 . | 4 . . | 3 . .
. 3 8 | 7 6 . | . 4 2
------+-------+------
7 . . | 1 6 . | . . .
. 2 . | . . . | 7 6 .
8 4 . | . 9 . | . . .
```

Puzzle 718 - Medium

```
. 5 . | . 8 . | 3 . .
3 8 . | . . 6 | 5 9 .
. . . | 4 . 3 | 7 8 1
------+-------+------
4 6 . | . . 2 | 8 . .
. . . | . 1 8 | . . 9
9 . 8 | . . 4 | . 3 .
------+-------+------
8 . . | . . . | 1 7 3
6 . . | . 3 . | 9 4 .
5 . . | . 4 9 | 2 6 .
```

Puzzle 719 - Medium

```
2 . . | 7 . . | . . .
. . 5 | . 6 9 | . . 4
6 . 9 | . . 8 | . 1 .
------+-------+------
. . 7 | 6 . 4 | . 3 9
4 5 . | . . 3 | . . .
. . 1 | . 7 . | . . 5
------+-------+------
. 1 . | . 9 . | . 6 .
. 3 . | 4 . 6 | 7 5 1
7 6 . | 1 8 . | . 2 .
```

Puzzle 720 - Medium

```
8 . . | . . 2 | 4 6 5
2 5 . | . 3 4 | 9 1 8
6 . . | 8 . 1 | . . .
------+-------+------
. . . | . 6 7 | . 3 2
. 6 3 | . . . | . . .
. . . | . . 3 | 5 . .
------+-------+------
9 . . | . 1 . | 2 5 3
. . 6 | 2 . . | . 8 9
4 . . | 5 . . | . . .
```

Puzzle 721 - Medium

1	4					2	3	
3		8		7	5		6	
		9					5	4
9	3		1	8		6		5
			7	9	6			8
			3		4	7	2	
	1	9			7			3
			5	4			9	
4		2			1			6

Puzzle 722 - Medium

		8	3	5	9			7
3	5		8					
	9	7	6			8	3	5
			7	3	1	6	5	
						1	9	4
		5				7	2	3
4	1	2		8	3			
		6			7	3		2
			2				8	

Puzzle 723 - Medium

	9						8	
		8		3		4		5
4					8		9	1
	8	4	6		7	9	3	
1				8	3			7
	3	2	5		4		6	
8	4	6		2	1			
3			8	5		6		4
	2			7				

Puzzle 724 - Medium

9		7			5			
1						5	2	7
				3	6			
		2		9	1			
	7	5			8	3	1	
3	1			5	7			
5	9	4						2
			6	7		4		5
7	3		5	4	2	1	9	8

Puzzle 725 - Medium

		6	4					3
5		3	2			1		
9			6	3			8	
8		4			2	6	3	
3		1		8			5	2
6	7		5					
		5	8		4		9	
7				6			4	8
4			3		7	5		

Puzzle 726 - Medium

								9
9	6							2
4	7		3		1	5	6	
1	3			8		6		
							4	
2	9	5				7		1
	1		8		7		3	
		3		4				7
8		7	2		3			6

Puzzle 727 - Medium

5			8					
		2	4			5		8
					9	2		4
2		9				3		
		6			2	4		
8	1		7			9	2	
6				4				
	2		1	8	7			9
	8		2	6			4	3

Puzzle 728 - Medium

		7			6	2		8
	5		8				7	
		1			7			6
9	1			8		6		5
								1
		4	2			7		
		5	1			8	3	
		8	5	7		1	4	2
1	4		6			3		7

Puzzle 729 - Medium

				4	7	3		
		4						1
7		3	8			4	2	5
	7	8		9	4	5	1	6
6						8		4
	4		6	1		2		7
8			4					2
				8		1		
9	5		1	3		7		

Puzzle 730 - Medium

	9		5			2	4	8
				1			6	
		3				1		
		2	7					
3	8							5
5	4					6	9	
2	1			9	7	8		
	7		3	8	5	4		1
			6	2	1			7

Puzzle 731 - Medium

	5				3	9		
6	3	2	4			8	5	1
7		9	5	8	1		2	
1	7							
		5						
4	2		9			1	6	
3		8		6	9			7
						2		6
			7	1			9	

Puzzle 732 - Medium

	7		4	6	1	5		8
1	6			9			3	7
8	5			7	2		6	
				5	3	2	4	
	1			4		8		3
3					9	1		6
		8		3			1	
				8				4
	3	6			4			2

Puzzle 733 - Medium

	2			5				8
1	7	8		3			9	5
				4	1			
7	6	5			2			4
			7			5		
8			4			7	6	2
5			1	2	9	8		6
9	8			4		2		
4			6		8			7

Puzzle 734 - Medium

						8		4
5	4	3		1	7			
		9				1		7
	7	2		9	8			5
	5		3	6		9		1
		1						8
	9	6		8	1			
			2			6	5	
		5	7		9			6

Puzzle 735 - Medium

		6						
	3	2				7		4
1			2			5	8	
		9		2	1	6	7	
			8	3	7	1		
3			9	6			5	
	7			4				5
6	2	3	7	5		4		
			9	6	8			7

Puzzle 736 - Medium

	2		9	3		6		
9	6		7	1	8	3		5
7	3		2	6		4		1
	5							6
			8		6	5		2
	9	6			3			8
			3		9		7	
	7		1	5				3
				4				9

Puzzle 737 - Medium

		5	3			8		
			8			5		1
6	1							2
3			7	2			6	9
				1				3
	9	6		4		2	7	
	6	1	4					8
8	3		6		2			
	2		3	5	8	6		7

Puzzle 738 - Medium

	2	5						1
8	1	7						6
3	6				7		8	9
2		3		9	5			
						8	3	
				3	8	2		5
1		2	6	4		7		
5							1	3
					9	6	4	

Puzzle 739 - Medium

		7	9	6	3	4		8
	4	6			8		9	1
		8				6		5
		2			4	8		
7					5	3	4	6
8	3			9	1			
	1			3	6			4
6				4			7	3
		3	1			6		

Puzzle 740 - Medium

4	5		1		9		8	
	9		5			6		7
	1		2	8			4	
8		2					5	6
6		5				1	3	9
	3						7	2
			1					
5	6	4		9		7		1
1			6	5		3		4

Puzzle 741 - Medium

		1	8	2			5	7
	4	7		9		2		
2					7		9	8
		6		3	1	7	4	
9		4			6			2
			4			5	3	
3			7	4	2	9		
7		9	6	5		8		4
							7	

Puzzle 742 - Medium

						5	1	
5	4				1			9
3	6					2	4	
8			9	5				2
4	5	3		1	7			
	1							
6		7	9					1
		4		7	6			
	3	5		2		6	7	

Puzzle 743 - Medium

		1		2	5	4		
			4	9			2	7
				7		9		1
3	8	9	6	4				
	4	5	3			6		8
2	1	6	8			4	7	
	9		7					
4						3	8	
1	3	8	9		4			

Puzzle 744 - Medium

5	3			2				
1				3	4			
		6	9		1			3
3	9				2	6		
2			3		5			
	5		8	4			2	
	6	2	4	5			3	7
	7				3	9		4
	1		6			7	5	

Puzzle 745 - Medium

4		9	6		3		2	5
3	5	2	9					1
7	6					3	8	
		4		5	8			7
6		3		9			5	2
5			3		6		1	
			7	4	2			
8		7			5		4	6

Puzzle 746 - Medium

6	3		8				2	
8	5	4	6	1	2			
		1		3			6	
		9						
	8						4	
7		2		9		5	8	
2			1	6		4	7	
				2			1	6
		8			4	2		

Puzzle 747 - Medium

1	7	5		2	3	8	6	9
		4				1		
		6					5	
9			5		6		7	
6	4				8			
7		2		3	4		8	
	1		3			7	4	2
				4	1		3	
						5		

Puzzle 748 - Medium

	7					8		9
6	2	8	4		9		1	
		5		7	2	4	3	
	8			6	3			
4	9	6				3	5	
5					4			
			4			7	2	
	5	1	3	2			4	8
	4			1				

Puzzle 749 - Medium

	7	6	8		2		4	
		4		3				
5	3	8		6		9	2	
8		9	7		4			
	1	2	3		9		7	
			2	1				
					1	3	6	8
6					8		1	9
				2		4	5	

Puzzle 750 - Medium

	8			1		5		
	5		9		8	3		
	9	3		5				
	1			3	6			7
4				8			5	3
3	2		5	7		6	8	
		1				4		
7		2		9	1		3	
			7					6

Puzzle 751 - Medium

3	8	1		5				9
6	7	2			9			3
	4			7		1		
			1		7	2	4	
	2		9		3			7
1	6		5	2		3	9	
			8					2
4		9			6	7		
2			7	9	1			

Puzzle 752 - Medium

		4	7				2	
			3	5		9	6	
1						5		
	6	9		1				
	7					3	5	1
5		1	3		4			7
	5			7		4	1	6
7	1	3	4				8	
	4			2		7		5

Puzzle 753 - Medium

5	6	2	7	8	3		9	
		3	1		5			
4	2		3	7	6			8
				1				2
	9	8	4	5				7
	8		5		7			
		4	6		8		2	1
6		9		4		7		5

Puzzle 754 - Medium

	3				7	6		
			9	2			1	3
2				8		5		
	9		6	4	1			8
3			8	7				4
4	6				9			
	8	9	7	6		2		1
		7	5			8	4	6
	4		1					

Puzzle 755 - Medium

3	1		7	4			8	
4	2	8			5		6	3
7		6	8	2				
		9			1			
			9		2	4	7	
		7	3	5	4	1		
	3	4				6	2	
	8	1		6	9	3		7
						9		

Puzzle 756 - Medium

1		8		3	4			
4			9					
5						9		
	5		3					1
						5		9
6		4	5	1	9	2		8
			1	5		8		7
7				9	8			2
		1	7	2				3

Puzzle 757 - Hard

6	4		3	9			1	7
	3			2		5		4
			6	1				
		7			9	6		
	9		8		2	1	4	3
								9
8					7	4	2	
							3	
	6		2	4		8		1

Puzzle 758 - Hard

			8	5				7
5		9					8	
8		1		2	4			
2				8		4		
	8	4						5
7	1			9				3
	5			8		9	3	
1		6			3		5	
		8	7			6		

Puzzle 759 - Hard

4	8	6				2	3	
		3			4			9
1				5	6			8
	3						9	
2	4	5				3		
9					2	4	1	
	5			2				4
					5	2		
			6					

Puzzle 760 - Hard

	4					5		8
		5	7					
	3	1					9	2
3			5			2		
			9	6				
7	5			2		3		
			5	8		9	2	
5			4			3		
	2							4

Puzzle 761 - Hard

	8	6	3			9	4	
		2	4				1	6
7	5		6	9	1			
		1				8	3	
			6	8	5			
								4
			8	7		4	6	
4	9	7						
	2				4		7	

Puzzle 762 - Hard

		1						
	7		1	8	6			
6	3		7	9	4			
			5	2		1	6	
2			3			8	9	
	5							2
			6			3		
	9	7	2				8	
5		6				2		

Puzzle 763 - Hard

		1			9			5
	4		2	6	1			
			9		6		5	1
6		8		2				
	1	3		7				
	3			9	5	7		
			4		7	2	6	8
4	7	6				3		

Puzzle 764 - Hard

	8			6	1		9	
4				9		7		
		1						
		2			4	3		
3				5			8	
1	7						5	4
		6	3	7			4	5
5	1			2				7
	4			8	5			6

Puzzle 765 - Hard

9	2					5		
			2	3		7	1	
7			4			2		8
	7	3	2				5	6
				5				
	5		3			8	4	
5						1		2
	8	2						5
3		6	5					

Puzzle 766 - Hard

	9		1		8	5	6	
	1	6						8
8		5	6			3		
	3	2		5	9		1	8
		7		1				
							3	
			9		2			
		3	8	6			5	
2						4		

Puzzle 767 - Hard

4			5	6		1		
		6						
	2	8			1	5		6
	9		6				8	
6		2	1	7	9	4	5	
				2				
			2					
							3	7
			3	9	8	6		

Puzzle 768 - Hard

					7			2
	8	2				6		
					2	3	1	
	5						3	
	9	4	2			1		
	7		8	1				
		6			4	2	8	1
9		8	3		1			6
1	2			7		4		3

Puzzle 769 - Hard

	8				3		9	4
			2			3	1	6
				7				
	3					6		
	6	9				4	3	7
	7	3		8				1
	2				4			
3				2				8
	7					1		

Puzzle 770 - Hard

	1	9		3				
					1	9	8	4
8								5
			9					6
1			2	8		4		
3		5		1			2	8
9	3							2
							4	
			3	2	6		7	

Puzzle 771 - Hard

	9	2		1		4		
			6		8	3		2
2				4		9		
	7		9		6		2	
	6				1		3	
9			1	8				
1	5		3	6				
	2		5				7	3

Puzzle 772 - Hard

7	1		6					
	8				2	7		
6			7		8		4	
		5					8	3
		1	5	2	6	4	7	
4		7				5		
	7		9		1	8		4
				5				6
1	2							

Puzzle 773 - Hard

			1		2	4		
3		1	8	9				
						7	8	
8				6	7			
	6		7			2		
	9		2		5	6	3	
		7	4	8				1
4	1			3				2
		3	6	2				

Puzzle 774 - Hard

	2		3					1
9	3	7	2				4	
	1		9					2
	5	8	4					
				2	7		6	
					1			4
		3		9		8	2	5
1	6		5					3
		2					1	

Puzzle 775 - Hard

```
8 . . | . . 1 | . 7 .
1 . 2 | 9 . . | . . .
4 . . | 6 8 . | . 3 .
------+-------+------
. . . | . . . | . . 5
. . . | 5 7 . | . 2 3
. . 7 | 1 2 . | 9 . 6
------+-------+------
6 . 3 | . . . | . . .
7 . 9 | . . . | 3 4 .
. 4 . | . . . | . . .
```

Puzzle 776 - Hard

```
. 4 . | . 9 . | 6 . .
. 3 . | 7 . . | . 2 4
. 8 . | . . 2 | 7 9 .
------+-------+------
. . 2 | . . . | . 5 8
8 6 . | 2 . . | . 7 .
3 . . | . 5 . | 2 . .
------+-------+------
. . 8 | 6 . 5 | . . .
6 . . | 8 7 9 | 1 . .
. . 1 | . . . | . . 5
```

Puzzle 777 - Hard

```
9 . . | 7 . . | . 3 .
1 . 7 | . . 6 | . . .
. . . | . 8 . | 4 1 .
------+-------+------
. . . | 4 3 8 | . . 9
. 3 2 | 1 . . | . . .
5 . . | . 2 . | 1 3 .
------+-------+------
7 . . | 9 . . | 3 8 .
. . . | 3 6 . | 5 . .
. 5 . | . . . | . . 4
```

Puzzle 778 - Hard

```
. 3 . | . 8 . | 1 . .
. . . | . . . | . . .
. 6 1 | . . . | 3 8 5
------+-------+------
. . . | . 1 . | . 3 .
. 2 . | 9 . . | . . 6
8 . . | 2 7 . | 9 . .
------+-------+------
. 9 . | 8 1 7 | 6 . .
. . . | . 4 . | 8 7 9
7 . . | . . . | . . .
```

Puzzle 779 - Hard

```
. . . | . . . | . . .
. . . | 5 . . | . 7 .
. 6 . | . 4 1 | 8 3 .
------+-------+------
1 . . | 5 4 9 | 6 7 .
. . . | . . . | . . .
4 8 . | 6 . . | . . .
------+-------+------
3 . . | . 5 . | 4 8 .
9 . 1 | 2 . 6 | . . .
. . 2 | . 3 . | 1 . .
```

Puzzle 780 - Hard

```
. 5 4 | . . . | . 2 3
. . . | . 2 . | . 6 5
. . 3 | 4 9 . | . . .
------+-------+------
9 . 8 | . 5 . | 2 . 6
. 4 . | 6 . . | . . .
. 3 . | 9 1 . | . 7 4
------+-------+------
. . 2 | . 8 . | . . .
. . 9 | . . . | 1 5 .
5 . . | . . . | . . 8
```

Puzzle 781 - Hard

```
. . . | . . . | . . .
. 4 . | 2 . . | . 7 8
1 8 7 | . . 5 | 2 . .
------+-------+------
. 7 . | . . 8 | 3 . .
9 . . | . 7 . | 1 . .
. . 4 | . . . | 6 7 .
------+-------+------
7 . . | 9 . . | 2 . .
. 1 . | . 7 . | 9 . 3
6 . 5 | . . . | 7 . .
```

Puzzle 782 - Hard

```
6 . . | 9 . . | 4 3 .
3 . . | 7 . . | . 5 6
. 4 . | . . . | 1 . 7
------+-------+------
1 . 2 | 5 9 . | 7 . .
. 7 . | 4 . 1 | . . 5
. 9 . | . 8 . | . . .
------+-------+------
. . . | . . . | . . .
2 . 1 | . 4 . | 8 . .
. 9 . | 1 . . | . 4 2
```

Puzzle 783 - Hard

```
. . 2 | 9 4 . | . . 6
8 . . | . . . | . . .
3 4 . | 2 . . | . 9 .
------+-------+------
7 1 . | . . 6 | . 3 .
. . 5 | . . 9 | . . .
. 2 . | 1 . 3 | . . 5
------+-------+------
. . . | . 8 2 | 4 1 .
. . . | . 6 . | 7 . .
. 8 . | . . . | 2 . .
```

Puzzle 784 - Hard

```
1 . 3 | . . . | . . .
9 . . | 4 1 . | . 5 .
. 5 . | . 3 . | 8 . 6
------+-------+------
. 7 5 | . . . | . 3 .
. 1 . | 2 . . | . 9 4
. 9 . | . . . | . 2 .
------+-------+------
. . . | . 6 . | 5 . .
3 . 2 | 7 . 4 | 6 . .
. 4 5 | 1 . . | 3 . .
```

Puzzle 785 - Hard

```
. 7 . | 8 . . | . . .
2 9 1 | . . 3 | 8 . .
. . 5 | . . 9 | 4 . .
------+-------+------
1 . . | 8 5 9 | . . .
. . . | . . 3 | . 5 .
7 . . | . . . | . . .
------+-------+------
. . 6 | 9 . 5 | . 4 8
. 8 . | 1 . 4 | 3 . .
. . 7 | . . . | 5 6 .
```

Puzzle 786 - Hard

```
6 8 4 | . 1 . | . . .
7 . 2 | . . . | . 1 6
. 1 . | . . 7 | 8 2 .
------+-------+------
. 7 . | . . . | . . .
3 . . | 6 . . | . . 8
. 1 . | . . . | . 4 2
------+-------+------
. . . | 7 8 . | . 9 5
. 2 . | . . 1 | . . 4
. . . | . 9 . | 3 . .
```

Puzzle 787 - Hard

		7	3				8	6
1	6			7		2		
								7
			2				9	4
2					9	1		3
9				3	4	7		2
			4					
3	8		1			4	2	
6				8		5		

Puzzle 788 - Hard

		6	8	3				
9	3		5			4		
				7	6			
			9					
							4	8
	4		7		2	6	3	9
2		1	9				6	3
8	6				7		5	
	7		6			8		

Puzzle 789 - Hard

			5	3				
		3	6	7				
2			8					9
	2			9		8	4	
	9	4			1		7	
8		1		4				5
	1			3		6		
						9	1	2
						7		

Puzzle 790 - Hard

6			5				8	4
8			1	9			3	
4					6	7	9	2
	2		7	8				
	8	2				7		
	6							
	9		7			5		
	8	3			5	4	1	
						8	2	9

Puzzle 791 - Hard

5		9	7	1		2		
2			5	4	9		3	
3					6		9	5
	5		8			3		
	3		9		4		8	
			1	9				
	6					5		
	1	5	3		2			4

Puzzle 792 - Hard

		8	7			5	2	3
							4	
		5	8		3			
		2						1
			2	6				
7		3			6	4	8	
	7	1				3		
	9		4					
4		6				1		7

Puzzle 793 - Hard

	4		2		3		8	
		5			7			3
	3					7		
4							9	1
1		2		3	8		7	
		6	8	7	2		4	9
7		4		9				
			6			8		

Puzzle 794 - Hard

			2		7	8		
7		2		8				4
	9	8	5					
	3	4	7		2	6		8
2							9	
			9			3		5
								6
				8			5	
	8	5			3	2		

Puzzle 795 - Hard

		4		1	7			
1	9	2						
		3		9	6			
	5	1	7		2	4		
		6			5			
		7	8	3	9			6
			5		1	3	4	
7				8			6	2
	4			6				7

Puzzle 796 - Hard

	9		3		8	1	7	
1	7							
		2	1				5	
	6		8	3			4	
		7	6					8
8			2	7	5		9	
3							1	
				8	9			
		9			6	7		

Puzzle 797 - Hard

		5		9				7
	8			5				3
	7	2		6		5		
	2		7					
		8		5				
	6	1			3			2
		9		8			7	1
							9	
	1		2		7	4	8	

Puzzle 798 - Hard

3		2	5			1		9
1	5		3	9	4	2		
	8		2			3		6
6			4	2		8		1
8				3				4
		5		8				
				4				
				5	7		1	
	3					6		

Puzzle 799 - Hard

			4	9		1		
	6				1		3	4
				6		8		
9	8					2		
		5			2			9
			7	3				
	7	6			4			1
3						7	2	6
		9						5

Puzzle 800 - Hard

		8	9		7	2	1	
		9	3	5	2			
7	2		1	6				5
		4				9	2	
		5					4	
			8	4				7
		7				5	3	4
	4					6		
	3	6	4		9			

Puzzle 801 - Hard

	3	4			5		2	6
		5			9		4	
	1							7
				5		4	1	2
	5							
	6		1	7	2	9		
1						6		4
	9		2				5	3
	4	6					7	

Puzzle 802 - Hard

			5	2				6
7								4
						9		
1	5				9	4	7	
6	7				1	5		9
9	8							
		9		7	6			3
		6			5			
4				8		9		

Puzzle 803 - Hard

8				6				
1	2		3	4	5	6	8	9
			1			2		4
						5		
5		3						1
	1		8			9		
	3		6	7				
7	4	6		3				8
2	5			1				3

Puzzle 804 - Hard

	7				1		5	6
		1	9			8		4
	8		4	6				
	5	8				2	4	
					3			
	1		2	9		7	8	
		5	7	4	2	9	6	
		2		1		5		
			5					1

Puzzle 805 - Hard

		9	5		6	3		1
6			2		3		9	
				4		2		8
		5		6				
				7		9	2	
7	4				8			5
			8					
8	9		6		4	7	1	2
		7		3		6		

Puzzle 806 - Hard

				6		5		2
		9	7					
1	6					9	3	
6			4					1
				7				3
7	4	5		9		6		
2							7	
	7	3			8	1	6	
			7	3			8	

Puzzle 807 - Hard

1	3		5					6
		6	4	1				
	9			6		3	7	
			6	9		2		
	4		8	5		7		
			7					
4			7					
	6	9		2				
7				9				5

Puzzle 808 - Hard

	2		5			9		
5			1			6		
	4	1	6				7	3
							3	
7		9		1		2		6
	6		5		7	8		
	9		8	7				
	1						8	2
6	8		9			3		

Puzzle 809 - Hard

			8			4		2
4		9						
	6			7	4			
6		3		1				4
		8					1	7
	5		6	8				
	2					8		1
	1		3		8		7	
	4	5		2		3		9

Puzzle 810 - Hard

8				6	2	7	4	3
	3				7	2		
7		2		1				
			3	5			1	2
		1		2	9	6	3	5
9			2	7	3			
			9	4				
						4		6

Puzzle 811 - Hard

	4	6						
	3					2	4	
5					3	1	7	
	5		6	4	9			
2			1				5	7
9			2					1
		7				3	2	
3			6	9	7		1	
		9		5				

Puzzle 812 - Hard

	6	7		3	1	5		
	4	3			8			
1	9	5	4			3		
9						8		
		4		1			5	
	3		7	2				
3	7			4			2	
		9		3	4			
	1	2						8

Puzzle 813 - Hard

		3	5					4
9	6		4					8
		1			3			
			9		2			
7			4	8	6		5	
1					3	8	4	
	8			5		9		
3	9		6					1
4	7				8			5

Puzzle 814 - Hard

	9		1	2	7			
3					4		8	2
5					1			
			3	4	6			1
					5	3	6	
9						2		
1	5			9			7	6
	6		5					3

Puzzle 815 - Hard

	4	9	2		5		8	
	2		8		6	5		
		1		7		9		6
			6	5				
5	7	8		3		4		
4	1					9		
		5	9					7
		3		8	7	1	4	

Puzzle 816 - Hard

							2	4
	7	6	4		9			8
	3	8			7			
							2	
			8					7
7					1		8	9
		7		6			1	4
5			9		1			
	6				2			3

Puzzle 817 - Hard

					4	6	5	
			5				3	9
4			1			7		8
	1		8		2			7
5				9				
	2		6		1	8		
1			9	2			8	4
3		4					6	2
8	5							

Puzzle 818 - Hard

	1	2				5	9	
	6			4	3			1
	8			5				2
			8			4		
9			3				2	
				7				
2			4			8	7	
8				7	1		3	
	7							4

Puzzle 819 - Hard

	2							4
1		6		8				7
				5	1	2		
		3	4	7				
				2		4	3	
5	4		3				1	
	1	4			6	9		
	9	8						6
	6		7		9	3		

Puzzle 820 - Hard

3			8					
			5	7				6
9	7	6	3	2			5	
6	3	7					1	8
	9					2		
							4	7
5		9			7			
	4					1	7	
			2		1		9	

Puzzle 821 - Hard

	2			3	5			
	6		2				4	
7							5	
		7	5			2		
				9	1	5	6	
		5	3	6	2		7	4
						1		
		2				6		7
8	7	1				4		3

Puzzle 822 - Hard

		5						9
9	6	2			1			3
4			3	5		6		
	5							
				4	8	6		
		7	2	6				
	7					3		
6			9			5		
	2	4			5	7		6

Puzzle 823 - Hard

```
. . 9 | 7 2 . | . 6 .
. 7 . | . . 4 | . . .
6 . 2 | 9 . 8 | 7 . 3
------+-------+------
4 9 8 | . . 5 | . . .
7 2 6 | . . . | . . 8
1 . . | . . . | 6 . 7
------+-------+------
2 8 7 | . . . | . . .
9 3 . | . . 7 | . . .
. . . | . 1 . | 8 . .
```

Puzzle 824 - Hard

```
. . 4 | . 2 . | . . 8
. . 7 | . . . | . . .
. 5 . | 4 . 8 | . 9 3
------+-------+------
6 4 . | 7 . 1 | . . .
. 3 9 | . . . | 6 2 7
5 . . | . 2 . | . 1 .
------+-------+------
. . . | 1 . . | 3 . 6
. 9 . | 6 . . | 7 5 .
. 6 3 | . 7 . | . . .
```

Puzzle 825 - Hard

```
. . . | 6 8 7 | . 3 5
9 . . | . . 3 | . 8 1
. . . | . . 9 | . . .
------+-------+------
3 . . | . . 5 | . 2 7
. 6 . | . 2 4 | 3 . .
. . . | . . . | . . .
------+-------+------
. . 9 | 3 8 6 | . . .
. . 3 | 4 . . | . . .
1 4 8 | . 9 . | . . .
```

Puzzle 826 - Hard

```
3 8 . | 7 . 1 | . 4 9
5 . 7 | . 4 . | . 3 .
. . 4 | . . . | 7 . .
------+-------+------
. . . | . . . | . . 8
. . 2 | . . 5 | 4 9 3
. 3 . | . 8 . | 6 . 5
------+-------+------
1 9 . | 6 . . | . 8 .
. . 8 | . . . | 9 . .
. . . | 9 2 . | 5 . .
```

Puzzle 827 - Hard

```
. . . | 3 . . | . . .
. 3 9 | 2 . . | . 8 6
. . 4 | . 5 . | . . .
------+-------+------
3 . . | . . 9 | . 5 4
. 4 . | 8 2 . | 9 1 .
. 9 5 | . . . | 8 . .
------+-------+------
. 7 . | . 8 . | 4 . .
5 8 . | . 6 . | . . .
. . . | . . . | 2 7 .
```

Puzzle 828 - Hard

```
. 1 . | 2 4 . | . 8 5
. 9 . | . 3 . | . . 4
. 5 . | . . 6 | . . 3
------+-------+------
5 . 4 | . 9 . | 6 . .
. 8 . | 3 . . | . . .
7 6 . | . . 4 | . . .
------+-------+------
2 . . | . 5 . | 8 7 .
6 . . | . 1 3 | . . .
. . . | . . . | . . .
```

Puzzle 829 - Hard

```
. 1 6 | 9 . . | . . .
. . 5 | 1 2 . | 6 . .
8 2 . | . . . | 9 1 .
------+-------+------
. . . | . 6 . | . 7 .
. 8 . | 5 . . | 4 2 .
. . . | . . . | 8 6 3
------+-------+------
. 4 . | . . . | . . .
1 3 . | 4 9 8 | . . 6
. . . | 7 . 1 | 3 9 4
```

Puzzle 830 - Hard

```
4 1 . | . 7 5 | . 6 .
. . . | . 6 1 | . . 3
5 . 6 | . 9 . | 7 2 .
------+-------+------
2 . . | 1 3 . | . . 5
. . 4 | . . . | . . .
. 5 . | . 2 7 | 4 3 .
------+-------+------
. . . | 6 . 9 | . . .
. . . | 7 . . | . 8 .
6 2 . | 8 . . | 1 . 7
```

Puzzle 831 - Hard

```
. . . | . 3 . | 6 2 .
. . . | 7 6 . | . . .
. 8 . | 4 . . | . 9 .
------+-------+------
2 6 3 | 9 . . | . . 5
. . 4 | . . 1 | . 6 .
1 . . | 6 3 4 | . . .
------+-------+------
. . . | 3 7 . | 9 . 4
. . 2 | . . 9 | . . .
9 . . | . . 6 | 1 5 .
```

Puzzle 832 - Hard

```
. . 7 | 8 4 . | 9 . 6
. . 6 | . . . | . . .
8 . . | 2 . 9 | . . .
------+-------+------
1 9 . | . . . | . 6 3
. . . | 4 . 6 | . 7 .
. . . | 5 3 . | . . 9
------+-------+------
2 5 . | . . . | 3 4 9
. . . | . . . | . . .
. 4 8 | . . 2 | . 3 1
```

Puzzle 833 - Hard

```
. . . | . . . | 1 . .
3 5 . | . . . | . . 6
1 7 4 | . 2 . | 3 . .
------+-------+------
. . . | . . . | 9 . .
. 2 6 | 8 3 . | . . 7
. . 3 | . . . | . . .
------+-------+------
. . . | . 6 . | 5 . .
2 8 7 | 5 . . | 6 . 4
6 . . | . 4 . | 2 . .
```

Puzzle 834 - Hard

```
. . . | . 6 . | . . .
. 8 . | 9 . 5 | 4 . .
. . 1 | . 3 . | . . 5
------+-------+------
. . . | 2 . . | . 3 .
. . . | . 7 . | . . .
. . . | 4 1 . | 5 6 7
------+-------+------
6 2 5 | 1 . . | . . 9
. 4 . | 5 9 2 | . . .
8 3 . | 7 . . | . . .
```

Puzzle 835 - Hard

```
1 9 . | 4 . . | . 2 .
3 . . | . . 2 | 7 9 .
. . 7 | . . . | 6 3 .
------+-------+------
. . . | 7 . . | . . .
5 . . | . 2 . | . . 4
. 4 . | 1 5 9 | . . .
------+-------+------
. . . | . . . | . . 2
. . . | . . 4 | . 9 .
9 7 . | 3 . 1 | 5 . .
```

Puzzle 836 - Hard

```
. . 6 | . . . | . 7 5
3 4 . | . 5 9 | 1 . .
9 . . | 1 . 6 | 4 . 2
------+-------+------
4 9 1 | . . . | 8 . .
5 . . | . . . | . . .
6 . . | 5 9 2 | 3 . .
------+-------+------
2 . 3 | . 4 . | . . 7
8 . . | 7 1 5 | . . .
. . . | . 9 . | . . .
```

Puzzle 837 - Hard

```
. 1 . | 3 . . | . 7 .
. . 8 | . 1 9 | . 5 .
. . 4 | . . . | 8 3 .
------+-------+------
. . . | . 7 . | . 8 .
3 6 . | . . 1 | 7 9 4
. . . | . . . | . . .
------+-------+------
. 5 . | 1 . 7 | . . 8
9 . . | 4 . 2 | . . .
4 7 . | 8 . . | 9 6 .
```

Puzzle 838 - Hard

```
7 . . | 8 9 . | . . .
8 . . | . 7 2 | 3 4 9
. . 2 | 1 4 . | . . .
------+-------+------
5 . . | . . 4 | . 3 2
. 2 . | . . . | 1 . .
. . 9 | . . . | 8 . .
------+-------+------
. . 6 | . 5 . | . . .
2 7 . | . . . | . 5 .
. 5 . | . 1 7 | . 6 .
```

Puzzle 839 - Hard

```
. . 5 | 6 . . | . . .
. 2 . | 7 . . | . . 3
9 . 6 | . 2 . | . 5 8
------+-------+------
1 4 . | . 7 6 | 5 . .
. . . | 9 . 3 | 1 . .
. 5 . | 2 . . | 8 . 6
------+-------+------
6 . . | . . 2 | 9 . 1
5 . 2 | . . 7 | . . .
. . 4 | . . . | . . .
```

Puzzle 840 - Hard

```
. . . | 8 4 . | 1 6 7
4 . . | 3 . . | . . .
. . 7 | 9 . . | . . .
------+-------+------
. . . | 2 . . | 3 5 .
. 8 . | 3 . . | . . .
. 6 . | 5 1 . | 2 . .
------+-------+------
. 2 . | 7 . . | 4 . .
. 5 . | 1 . . | . 9 .
. 9 . | 4 . . | . 8 .
```

Puzzle 841 - Hard

3		5	2	8		9	1	
4		7			5		3	
9	8		4			7		
	5				8	3		
8	4		3		2			9
			5	7		4		
	9						8	1
	2					6		
						2	5	

Puzzle 842 - Hard

6						5	9	
	8			3				2
3			5				6	4
	3				4	1		
			1	5	3			
					2	3	5	
		3	9					7
2			4		6			
8				1			2	

Puzzle 843 - Hard

			5	1				
6				2			8	
						7	2	
3			2		7			1
		2			1			
1	6		3	9	5			
4		1	6				7	9
		8		6				3
		6	9					

Puzzle 844 - Hard

		4		7		9	8	
		8			3			
	6	5		8	9			
				2			4	6
6		1	3		5	8		
				6				
	8				4		1	2
9	4				6		7	
		7			2	4		

Puzzle 845 - Hard

	1					4	6	
2	8		1		5	7		
4		3			7		5	
6		2		1		3		
3	7							
			7				8	4
				6				
		4	5			9	6	
		7	9	8				3

Puzzle 846 - Hard

		9				4	7	
		5					3	6
			8	4	7		5	
4								8
					3	2	1	5
	2	6		9	8	3		
	7	3	6				2	1
		8	7			9		4
		4		1				

Puzzle 847 - Hard

```
. 3 . | . . 5 | . . .
8 . 5 | . . 1 | 7 9 .
. . . | . 9 . | . . .
------+-------+------
3 . . | 7 8 . | 6 . .
. . 1 | . . . | . . 8
. . 6 | . 3 . | . . .
------+-------+------
4 1 . | . . . | 5 . 9
2 . 9 | . 6 . | . 3 1
. . 3 | . 1 . | 8 2 .
```

Puzzle 848 - Hard

```
3 8 . | . . . | 4 . 6
. . 9 | . . 8 | . . 3
1 . 7 | . 4 . | . . 9
------+-------+------
4 . 5 | 2 . . | 7 . .
. . . | . 6 . | 4 . .
. . . | 8 . . | . . 2
------+-------+------
. . 8 | . 1 5 | . . .
. . 6 | . 7 . | . . .
. 7 . | . . . | . 3 1
```

Puzzle 849 - Hard

```
. 9 . | 6 . 4 | . 7 3
. . 8 | . 7 . | 1 . .
. . 3 | . . . | . 2 9
------+-------+------
. . 6 | 7 . 5 | . . 8
3 . . | 2 6 8 | . 9 .
. . 2 | . 4 . | 3 6 .
------+-------+------
. . 9 | . . 6 | . . 2
. 7 . | 5 . . | . 3 .
. . 4 | . . 7 | . . .
```

Puzzle 850 - Hard

```
. 4 5 | . 6 . | 1 . 9
1 . . | . . . | 7 2 .
. . . | 1 9 . | . . .
------+-------+------
. . . | 9 7 . | . . .
7 . 6 | . . 4 | . . .
. 3 . | . 8 1 | . . .
------+-------+------
4 7 . | . . . | . 6 .
. 5 . | . 1 . | 9 3 .
. . . | 5 3 . | . . 7
```

Puzzle 851 - Hard

```
9 . . | . 2 . | . . 1
. . 4 | . . 9 | . . .
. . . | . 1 . | 2 4 .
------+-------+------
7 . 9 | 1 3 . | . . .
. 1 . | 2 . 4 | 7 . .
5 . . | 7 . . | . . 8
------+-------+------
2 3 . | . . . | 4 . .
4 9 . | 8 . . | . 1 .
8 . . | 3 . 9 | . 6 .
```

Puzzle 852 - Hard

```
2 . 7 | . . . | . . .
8 9 . | 2 5 . | 1 . 7
. . . | 4 . . | 8 6 .
------+-------+------
. . 9 | . . 5 | . . .
. . . | . . . | 7 1 4
. . 3 | 6 . . | 2 9 5
------+-------+------
. . . | 3 1 . | . . .
9 6 . | 7 8 . | 4 . .
7 . 8 | . 9 . | . . .
```

Puzzle 853 - Hard

	2							
	1	5	9	3		6	7	
			7		1			
3			7					
	6	1	3		9		4	
9	8			6	2		5	
				4	5	9		
5				2		8		
		2			1			6

Puzzle 854 - Hard

7		3	6	8				
		5	2					
		9				6	5	
4		5			3	8	6	9
9	6					3	1	2
			9					
6			4			9	3	5
		3			8	4		6
						2		

Puzzle 855 - Hard

3			2	7				4
	9	7			8	6		
		4						
	7		9					8
		7			5			1
		4		5				
	1	5	3	2				
	2				8			6
	9		7	6		1		

Puzzle 856 - Hard

				8		8		
1			8				9	
	8		6	1		7	3	
	7	2		6			8	3
3		1				5	7	
	9	8			5			4
	5		1					
	2	3	5			4	1	
7			3					9

Puzzle 857 - Hard

								6
	4	3	2	7	6		9	1
		8			1	2	3	
		6					2	5
		1		2	3			4
				7	9	1	8	
			1	2		7		
1			3				8	
3					4			

Puzzle 858 - Hard

1			7	3	8			
2		3		5			4	
	5							
9	3		8					
		7		6		5		
6					3			8
		1					8	
	6		3					4
8			2		6		3	

Puzzle 859 - Hard

```
2 . . | 8 6 . | . . 7
. . . | . 4 . | 6 8 .
1 . . | . . 6 | 5 7 8
------+-------+------
7 . . | 2 . . | . . 3
. . 6 | 5 . . | 4 1 .
. 7 . | . . . | . 9 .
------+-------+------
. . . | . 2 . | . . .
. . . | . . . | . . .
6 9 2 | . . . | 7 4 .
```

Puzzle 860 - Hard

```
4 6 . | . . 8 | . . .
5 . 8 | . 6 . | . 9 2
. 7 . | 9 . . | 6 8 4
------+-------+------
2 . . | . . . | . 3 .
3 . . | . 8 . | . 2 .
8 . . | . 9 . | 7 . 6
------+-------+------
. . . | . 2 7 | 8 . .
. . . | . 1 5 | 2 . .
. . . | . . . | . . .
```

Puzzle 861 - Hard

```
3 . . | . . . | . 6 1
. . 2 | 7 . . | . . 5
. 5 . | . . 6 | 7 2 .
------+-------+------
5 6 . | 9 . 3 | . 7 .
. . 8 | . . . | 9 . .
. . . | . . . | 5 3 .
------+-------+------
. 1 . | . . 2 | . . .
2 . 6 | . . . | . . 9
7 . . | 6 . . | 3 . .
```

Puzzle 862 - Hard

```
. . . | . . . | 3 . .
. 5 . | . 3 . | . 7 4
. 3 . | . . . | . 1 .
------+-------+------
. 8 7 | . . . | 9 . .
5 . 1 | . . 6 | . . .
9 . . | . 5 8 | . . .
------+-------+------
. . . | 4 . . | 2 3 .
3 . 9 | . 6 . | 4 . .
4 . . | 3 . . | 7 . 8
```

Puzzle 863 - Hard

```
. . . | 2 . 5 | 7 4 .
. . 6 | . . . | 3 . .
. 7 . | . 6 . | 5 9 .
------+-------+------
9 3 . | . . 7 | 6 . .
. . 1 | 9 2 . | . . .
. . . | 5 . . | 1 . .
------+-------+------
4 9 . | 8 . . | 6 . 7
. . . | 6 3 . | . . .
. 6 . | . . . | 2 9 .
```

Puzzle 864 - Hard

```
. 9 . | 1 7 . | . . .
2 1 . | 4 3 . | . . 5
. . . | 2 6 . | . . .
------+-------+------
9 2 . | 5 . . | . . 6
6 8 . | . . 2 | 5 . .
. . 1 | 3 . . | 2 7 .
------+-------+------
. 6 . | 9 4 . | 5 . 8
. 8 . | . . . | . . .
5 . . | . . . | 3 . .
```

Puzzle 865 - Hard

```
. 8 . | . . 2 | . . 6
2 . . | 8 . . | . . .
. . . | 7 9 1 | . . .
------+-------+------
. 6 . | . 8 9 | 5 . .
. . 2 | . . 3 | . . .
5 4 3 | 1 . . | . . 8
------+-------+------
. . 5 | 9 . . | 8 . 1
6 1 . | 5 . . | . . .
. 3 9 | 6 1 . | 2 4 .
```

Puzzle 866 - Hard

```
. . . | . . 6 | 2 . .
. . 2 | 9 . . | 1 . 3
. 9 . | 7 1 . | 6 . .
------+-------+------
. . . | 5 . . | . 2 6
. . . | 9 . . | 7 5 8
2 5 4 | 6 7 . | . . .
------+-------+------
. 6 . | . . . | . . 9
. . 5 | . 4 . | . . 7
. 4 3 | . . . | . . .
```

Puzzle 867 - Hard

```
8 . . | . . 7 | . . .
. 2 5 | . . . | 9 . .
6 7 . | 2 3 . | 5 4 .
------+-------+------
. . . | . . . | . . 9
. . . | 7 . 8 | 5 . .
. . 8 | . 6 4 | . 3 7
------+-------+------
. . 7 | . 1 5 | . 4 6
. . . | . . 3 | . . .
3 1 . | . 8 2 | . . .
```

Puzzle 868 - Hard

```
. 8 6 | 5 . 7 | . . .
7 5 2 | . 4 . | . 6 .
. . 4 | 6 3 8 | 7 . .
------+-------+------
6 . . | . . . | . 9 .
. 1 9 | . 7 4 | . . .
. . 7 | 2 9 . | . . .
------+-------+------
. . . | 1 . . | 9 . .
. . 3 | . . . | 6 . 5
8 2 . | . . . | . . .
```

Puzzle 869 - Hard

```
7 . . | 6 . . | 5 . 3
. . 3 | . . . | . . .
. . . | 8 . 7 | . . .
------+-------+------
. 1 8 | . . 2 | . 9 6
2 . . | 6 9 . | . . .
. 9 . | . 1 . | 8 . .
------+-------+------
. 4 . | 2 . . | . . 9
. . . | . 6 . | 5 4 .
9 . . | . 7 . | 2 . .
```

Puzzle 870 - Hard

```
5 6 . | 2 . . | . . 7
. . . | . . . | 9 5 3
. . 9 | 7 . . | 8 . 6
------+-------+------
. . 4 | . . 1 | . . 2
6 . . | 3 2 . | . 7 .
. 8 5 | . . . | 6 . .
------+-------+------
9 . . | 5 6 8 | . 1 .
. . . | . 3 . | . . .
8 1 3 | . . 7 | . . .
```

Puzzle 871 - Hard

3		6		1				7
					2			1
			5	9		3	6	
	5		3				1	
	9		5		6		8	4
	2			9				
	7			6	3			
6	4		9	8		2	7	
9				4				

Puzzle 872 - Hard

				9			1	
6			2			9	3	5
1			4	6				8
3	7		2		6			
2						7		1
5		6						
9	2	8		3	7	1		
							9	
				5				

Puzzle 873 - Hard

		2				5	7	
		7	1				3	9
		8		2	9			4
	7	5		4				
3				8				
						9	4	2
				6				
	8	3	4			5		
6	5				1	3		

Puzzle 874 - Hard

				9		3		
	3					9		5
	9	6	3			8	4	
		1			8			
	6					7	5	9
7		9	6			2		
6			4	5	3			
9				7		5	8	
4				1				7

Puzzle 875 - Hard

9	3	2					4	7
4				3		8		
						6		
	4						5	
8			3					
6		7		5	2		9	
						7		
7	8	4	2	6				
				5	9		2	

Puzzle 876 - Hard

				2		9		6
2		3	4				7	
	6	5	8		9			
	2			7			1	9
	8					5		
	3	9				6		
				8	1			
		8	3	5		4		
		1						

Puzzle 877 - Hard

```
. . 8 | . 3 . | 6 . .
4 . . | 7 . . | 9 . .
. 7 . | . . . | 5 . .
------+-------+------
. . . | 5 . 6 | 2 . .
9 8 . | 1 . 3 | . . 7
. 6 . | 2 . . | 4 3 .
------+-------+------
. 4 . | . . 7 | . . .
1 . . | . 8 . | . . 5
. . . | . . . | 7 . 4
```

Puzzle 878 - Hard

```
. 5 7 | 9 . . | . 3 8
. . . | 7 8 . | 4 . .
. . 8 | . . . | . 1 .
------+-------+------
3 9 . | . . . | . . .
. . 6 | . . 2 | 3 9 7
7 4 5 | . 1 . | . . .
------+-------+------
. 2 . | 1 . 3 | . . .
. . 9 | . . . | . . 3
. . . | 6 4 . | . . .
```

Puzzle 879 - Hard

```
3 . . | . . . | 2 . .
9 . . | . . 6 | 7 . 4
. . . | 2 4 5 | . . 9
------+-------+------
2 8 . | . 9 . | . . 6
. 4 . | . 8 . | . . .
. . . | . 7 . | 8 . .
------+-------+------
. 9 . | . . . | 3 4 .
. 6 . | 9 . . | . 1 .
5 . 2 | 4 . . | . . .
```

Puzzle 880 - Hard

```
3 . 7 | . . . | 6 . 1
. . 8 | 5 2 . | . . .
2 . . | . 7 . | . . 5
------+-------+------
4 . 1 | . . . | 8 . .
. . . | 2 . . | 3 6 .
. . . | . . . | 1 . .
------+-------+------
1 7 2 | 8 . . | . . 6
. . 9 | . 5 . | . . 3
. . . | 9 . . | 7 . .
```

Puzzle 881 - Hard

```
. 2 . | . . . | . . 4
8 6 . | 4 3 . | 1 . .
4 . . | . 8 . | . . .
------+-------+------
. . . | 5 . . | 2 4 1
. . . | 3 . . | . 7 6
. 5 4 | . . 2 | . . 8
------+-------+------
3 4 . | . 2 . | . . 5
5 . . | 6 . 7 | . 2 .
. . 6 | . . . | 9 . .
```

Puzzle 882 - Hard

```
9 . . | . 8 5 | 4 . .
. . 2 | 9 1 . | . 6 .
4 1 . | . . . | . . 5
------+-------+------
3 2 . | . 9 . | . . 6
. . . | 4 . 1 | 2 . .
6 . . | . 2 . | 7 8 1
------+-------+------
. . . | . 7 . | . . 3
. 3 . | 5 . . | . 9 4
8 9 . | . . . | . . .
```

Puzzle 883 - Hard

				5		9		
			8			4		
	3	8		6			2	5
	1	7	9		5		4	8
5			3			1		2
				4				
1		3	5			2		
	6		4		8			
			7	1			6	

Puzzle 884 - Hard

	3						1	
	1							4
	7	4				3	5	
					1	2	3	5
8		1	6					
	9	5	7	3				
7				8	6			
	5		1			7	8	
1		2	5	4			6	3

Puzzle 885 - Hard

			7	3	4	6		2
4			5		2			3
7				8		5		
								9
	5		2					
8			4	7	5			
6	4	7			9			
1	9	8					2	
5	3							6

Puzzle 886 - Hard

4				1		8	9	
						3		6
9	7	3			2			
			9	1		6		
	9	4				5		7
			5			2		
2		9	3		8		5	1
3	5				7			
	8		1		5		4	

Puzzle 887 - Hard

7	3	8					5	
						2		8
	9		6					
1		6	5	3				4
9	7		8					
	5	4		1		7	2	
	8	9						
		5	4	9		8	7	
		7			8	4		9

Puzzle 888 - Hard

7			4		9		5	
5	6	2				1	9	
		1		5				3
		4				5		
		8						7
2	5		1	8				9
4	3						2	6
8				6			3	
		7	3					5

Puzzle 889 - Hard

			3			7	6	9
	9	7			4			
						2		
	3	5		2	1	8	9	
7						4		6
					5	3		
	5	3				6		
		6	5			9	8	
		9		6				

Puzzle 890 - Hard

		9	3					
			6	2				5
8					5			
7		6	5	9				
		5		2	3	8		
2		4		7				9
6				5		3		1
								4
	9					7	8	

Puzzle 891 - Hard

8		7		2		1	9	
		1						2
		6		9	7	5	8	
5			7			6		
1		9					5	
7	8	3			5		2	
			9		3			
	3	2		7			6	
	1				2			7

Puzzle 892 - Hard

			8	5		1		
	1	5	4	6				7
	9						4	
			7			3	6	
9			6					
5	7		9		4	2		
		1		2		7		
4					3		1	6

Puzzle 893 - Hard

	9	7			8	4		
			4	7	9			5
	3	8						
	5					3		4
				4				
6	2				3		7	
			1	3	4	5		9
9				5	7	1		8
		6		9	2		4	

Puzzle 894 - Hard

6					4			7
5	1		9		7			
		7	4				8	1
		4		8		2		5
8		3	7			6		
		9						
		5	1			8	7	9
	8			4	3			
		1	5	7		3		

Puzzle 895 - Hard

1		5						
			9				1	4
		4			5	3		8
	2	3		6	4			5
	1		2					9
4		9				8	2	3
			6			4		
7				2				
9	5		4				8	1

Puzzle 896 - Hard

	6				5	3	2	
		5		3	8		9	
	8					1	6	
			6					
4						9	7	3
	7		9			2		
7			3	9		5		
	2				4			
	1		2	7		8	9	

Puzzle 897 - Hard

1	6	9			5			
8				2	4			
			1			5		
		7		8		4		
		3						
6			7		9		8	
		8			6			
9		6	5	3		1		
	5		9			3	6	

Puzzle 898 - Hard

9	5	3					1	2
2		6						
		8		1				
			4	7		3		
			8				4	1
			5			6	7	
	8	2			1		9	5
			9	5		8		
		7	2				3	

Puzzle 899 - Hard

1		2		9	6			
				2			5	1
	7	9		3	6	2		
			2	7				
3		7				4		
2					3	1	7	8
9		3		2				
			4			8		
		6		8	5	7		9

Puzzle 900 - Hard

		9			3			5
		3	2			7		
	4	5			8			3
		7		5	4		6	
6	8			7	9	1		
				2				
						7		1
3			1	9				
4			8	6			7	

			2	5		8	7	
		1		7				
8		9	1				2	6
9			3	8	6			
	3	2		1		7	6	
	8	5				3		9
					1			
	4					6	8	
	9	8						4

7				6				
	9		1	8			5	
	5	8						2
			2	6		4		
1						2		
				7		9	8	
	7			1	9			
9	2		6			8	5	7
		6			2			4

	6		2					
8	9			1	6			
		4		5			1	
			3		1		9	
				8				
		8		2	9	4	6	
7	5	3				6		
4			5			7		3
		6	1				4	9

			7		8			
				1				
	2	9					4	8
		2			1			4
		8		6				
6	5	1		9				
		8				1	2	6
					4	5		
2		3			6	4	8	

4								3
	2					5	8	
	1	8						4
		6		9			4	1
9				2				
					5	2		9
1			8		3	7		
6		5				3	1	
						4		6

8		5	2					
7		2		3	9	6		
9						7		
	7	6		2		3	1	
			7		4			
		9	6	5		8		
							4	
6	2			8	7			
						5	8	6

Puzzle 907 - Hard

2	7	8				4		
			3	2		9		
3		1			5			
4	5							
8				5			9	6
				7	8			2
			9					
9		6	8	1				
		2				3	7	9

Puzzle 908 - Hard

						6		
			9				7	8
	3	8	7	6	5		4	
8		5		3	6			4
					1		3	2
3		1		4	7	8		6
	4					1		
	1	2	3					
7								5

Puzzle 909 - Hard

			1	3			4	
						2	1	3
	4		5			7		
				1	5			
5	3	4	6		8			
8			9		2			5
	9				3		2	
	6		4				5	
3								

Puzzle 910 - Hard

		6	1	2				7
8			5		7	6		
	3							2
	5			8	1			
			6		3		5	
3				1				
			7	1		8		
				4			9	
7			9		5		6	

Puzzle 911 - Hard

					6		8	
4				8			1	2
		5	9			6	3	7
			4			8		
			1					3
8		9		7	2			
	6	8	2	3		1	5	
7			8				2	9
				1	9			8

Puzzle 912 - Hard

	1			5	6	2		
				1	8	4		
5	7			2			8	
2		7						8
							5	
	3		6			7		
7	2		9					4
				7			1	
	9	4		8		6	2	

Puzzle 913 - Hard

			1		8		5	
	1			6	2	9		
		9		2	4	8		
		4		2	9			
	6	7						
1				6		7		
		6	2					4
2				5		3	7	
	4		6		7	2		

Puzzle 914 - Hard

			5	6			8	9
								6
5	6	1		8		2		
			9	3				
2			1		6		9	
9		4	7					3
						4		
3	2		4	7		9		
	1		6				7	8

Puzzle 915 - Hard

			5	8		1		3
		5	4					
		7			1			
	5			4	2			
2		1	8					
9	3				5		1	2
5	8		7			2		
	2	9					5	6
	1	6	2				9	4

Puzzle 916 - Hard

		7	8			5	6	4
	9		2	4	5			
1				6	3	9		
			6	8			9	5
								2
		4		1			7	
5				3		4		
3	6						1	
								9

Puzzle 917 - Hard

8		9			1	2		
	4		8		6	1		9
			5		8			
		4		3				
5				6		3		2
			7			5	8	4
	7	8			2			5
	9		1		4			6
		6		7	3			

Puzzle 918 - Hard

	4	9			6			
	6			7	8		4	
	8			4	9			
		8	3					1
1	3			2				7
	5		9					
	7		2				6	
8								
			8	4	3	7		9

Puzzle 919 - Hard

```
. . . | . . . | 4 . .
. . . | 7 1 . | . . .
. . . | 4 8 . | 5 1 .
------+-------+------
5 4 . | . 2 1 | 9 . .
. 6 . | 9 . . | . . .
2 . . | 6 . 3 | 1 7 .
------+-------+------
1 . 2 | . 6 8 | 4 . 9
8 . 4 | . 7 . | . . 3
. 9 . | 1 . . | 2 . .
```

Puzzle 920 - Hard

```
. . 5 | . 3 . | 9 . 4
. 3 5 | . 6 . | . 7 .
1 6 . | 2 4 7 | . . .
------+-------+------
5 . . | . . 1 | 8 3 6
. 1 . | 7 . . | . . 9
8 . 2 | . . . | 4 . .
------+-------+------
. 1 . | . . . | . . 5
. 4 . | 6 5 9 | 3 . .
. . . | . . . | . . .
```

Puzzle 921 - Hard

```
. 7 . | 9 8 2 | . . 4
6 . . | . . . | . . 2
2 . 9 | . . 7 | . . .
------+-------+------
. . 6 | . . 5 | 9 3 .
8 . . | 6 9 . | . . .
5 . . | 2 . . | . . 6
------+-------+------
. 5 . | . . 3 | . . 1
. . 8 | 7 . . | 2 . .
7 . 4 | . . . | . . .
```

Puzzle 922 - Hard

```
2 . . | 3 7 6 | . . 1
. . . | . . . | . . 6
. . 3 | 5 . . | 7 4 .
------+-------+------
4 . 5 | 9 . . | . 8 .
. 8 . | . 5 7 | . . .
. . . | . 4 . | . 6 3
------+-------+------
7 . 2 | 8 . 9 | . . .
. 5 . | 7 . . | . 3 .
. 4 . | . . . | 6 . .
```

Puzzle 923 - Hard

```
6 7 . | . . 2 | . . .
. 3 . | 6 8 1 | . . .
. 8 1 | 4 5 . | . . .
------+-------+------
. . . | . 1 4 | . . 5
. . 5 | . . . | 4 9 .
. 1 . | . . 5 | 3 . .
------+-------+------
. 9 8 | 5 3 . | . . .
3 . 7 | . . . | . . 6
1 . 6 | . . 9 | . 5 3
```

Puzzle 924 - Hard

```
7 . 6 | 8 2 . | 4 . .
. . . | . . 5 | . . .
. . 9 | . 1 7 | 6 . .
------+-------+------
. . 5 | . . 3 | . 4 .
. 1 . | 7 . . | . . 8
. . . | . . 8 | . 9 6
------+-------+------
4 5 7 | 2 . . | . . .
. 2 . | . . . | . . .
9 . . | . . . | 8 . 2
```

Puzzle 925 - Hard

5		2	4					6
3				6	2	1		
					5			
			6	7		8		
		5		8			4	
	1		9	5	4	3	6	2
	4	7		3	9			
						2		
	3					5	7	

Puzzle 926 - Hard

				8		6		
8						3	5	
3				9		2		
6						9		
9	5		4			8	7	
	7							6
			5					
	5		9	6		4		
1	2		3			5		9

Puzzle 927 - Hard

					3			
2			4	1				
6		7						
1		3	7		9	2	5	
7				4	9	3		
		5	3			8	7	
			1					2
	6		8		5			
8		1		4	5			9

Puzzle 928 - Hard

			2	1				
8		1			6	5		3
					3	7		
			9			8		5
5		7						
	2	9		3		1	4	
		6					7	9
		8		4		2		
	7						5	

Puzzle 929 - Hard

	3			1				8
	8	7			2			4
6				4				
5			2		9			3
3			8					
	9	8						
	1			6		3		
		3			5		7	
	2	9		8				1

Puzzle 930 - Hard

4	5	1						
			9	8		2		
9						3		
						7		
8	4	9			2	1		
7	3			4	1			8
5	9	3					8	
		4	7					
	1	7	9	8				5

Puzzle 931 - Hard

					9	5		
	7				5		4	
8						1		
1					4	8		
3	2	8				4	7	
			2		3			5
4				2				1
			1	4	8	6		
2				9				

Puzzle 932 - Hard

9	2		5					
				9		3		
			1			2	6	
				6	8			
	8							3
	9	6		7	5		8	1
2			6			7	8	
7	4		8					
		1			4			

Puzzle 933 - Hard

6	7		2		1	9		
4		2	5				3	
	5							
	1	6		9		4	2	
		9						
	2	8						1
	3	4	6		7	2	1	
			8				5	4
				1			6	

Puzzle 934 - Hard

						2	3	
	2	1		4	5	6	7	9
					7			5
5			6			9		
4							1	3
	8	7	9					
	6	4	2		9			
				6	1		2	
			7	5			9	

Puzzle 935 - Hard

		1				7		3
	2	9		8				6
					1			
	9		2					
2				3	1			
	8	7	6					4
		8			9		5	
9	1	6						7
5			1	6		9		

Puzzle 936 - Hard

9				7	1		3	
				9	5	1		6
				8	6	2		9
				4		7		
2	7			3	6	4		
		3		8	5		2	
3	5	7						
8								4
	4	6		2		5		

Puzzle 937 - Hard

1		7		8				
			9			2		
	4					7	1	
2		6	7	5				
7	9	4	1	2	8			
				6				
6		3				5		8
						1		2
		2	5	4	3			9

Puzzle 938 - Hard

	6		9					
			1			3	5	
3		5					4	1
	5			3				9
	7		8	4				2
6								
								7
7		9	2		1		8	4
5		1	4					

Puzzle 939 - Hard

	1	7				5		
8		6				3		
		7		6				2
	8		9			4		
	7		3					
	1		5	7		9	2	3
	2	3		4		8		9
					7		4	

Puzzle 940 - Hard

3		1	2		8			7
		4		7	9	3		8
	8	7		3			2	
			6	7				5
2			8					6
				5	2			9
	1		7				5	
		6			4		8	
			3	8		4		

Puzzle 941 - Hard

				9		4		
8			5		7			9
9			3	8		1		
		1				6		2
			6		9	5		
		6		3	2	8		1
1					6	7		8
5	3			7	8			
		7		2			5	

Puzzle 942 - Hard

		1	8					
	4			6	7	1		5
	5				3		9	
9			7			8	5	1
		2						3
5		6		8		2		
	2			3	5			8
		8	2					
3			1	7		9		

Puzzle 943 - Hard

	8					2	9	1
		6						
	3	9		5	2	6	8	
				8				
	4				9		1	6
			6	1		8		3
	2	3			4		5	
9		8						
6		4			1			9

Puzzle 944 - Hard

8	5	4	2					3
9		1		5				
	7							
							2	9
			8	9			6	
		9			5			
	3	2				7		
		5		8	1		9	2
1					2	3		5

Puzzle 945 - Hard

	2		1	4	7			9
			3			7	1	2
				9			4	6
	1	4			2			
7			5					
		6	4			2		8
3		7			6			
		8	9		4			7
2				3	5	1		

Puzzle 946 - Hard

6				8	1			
						6	4	3
		4		2				
	4	1	7		8	5		
3	2	6	9			1	8	
5				1				
		8						
	5		1	9			2	
								4

Puzzle 947 - Hard

		5			4	7		
	9					4		
4			5	9	1			
				1		3	2	
		7		2	5		1	
1			4				7	6
		3						
6	1		7	5		8		2
9		2	8		6			7

Puzzle 948 - Hard

6		2	9	5	7			
5	9							
			6					9
8		9	4			2		
			5		2			
3		4				6	5	
2	8		9		4	1		
	3		7		5			
	1		2			7	4	

Puzzle 949 - Hard

1			7	4	6			
							7	3
			9		5		2	
	9	4		8	2			
	8	6	1					7
					9			2
	2					7	3	
								8
3		9			8		1	

Puzzle 950 - Hard

	6			3	9	1		
	9	8	5	6	7		3	4
7			1					
						3		
6	3	7				9		
		1	9	5	3	6		7
	9					7		
		2				8		3
	2						1	

Puzzle 951 - Hard

	4		3	6				
2		1				3		
		3	7		2	4		9
6					7	9		8
						1	3	4
3		5		9				
			4			5		3
	3	6						2
4	2			8		6		

Puzzle 952 - Hard

5					6		9	
								8
				4		3		2
			6	2	4		8	
	5							4
						9	7	2
		6	5	3	8	2		
				7			3	6
7	1							9

Puzzle 953 - Hard

	5	1						8
	9							
3				5		6		
			5		8	2	7	
			2				8	6
7	2			6			3	4
4				9			6	
		3	6	8				
		5	7	3		4		

Puzzle 954 - Hard

					5		9	
				7	8	1	5	6
	5			9		8		7
		2		8				
		5				7		1
4		8		1				
2		9	1		6			3
				3			6	8
6				5			1	

Puzzle 955 - Hard

7		9	3		1	6		
	8	2		7		1		
		3			2	5		
3	2	4		8				
8			6	1	3			
1			2			8		7
	5		4				1	
2					8			
4			5					8

Puzzle 956 - Hard

	3			6			9	1
					8		6	
2		6					4	
	5	1		2	7			
	2	4					8	5
	8		7	5		4		
		5		8	2	6		9
				6				

Puzzle 957 - Hard

		4				2		
3					6			8
	9		1		7			
7						9	5	3
8	2	9				1	7	4
	5				1			
	3							
		5	6	9	2			
2	8			4	3	6	1	9

Puzzle 958 - Hard

8						6		
			6	4	1			
2					8			
					4	9	2	
1	9			7				4
			8		5	1		7
		6					3	
	8	4	5			2		
		7		6	3			1

Puzzle 959 - Hard

		3			5			1
6				9				
	5							
4		6	1	7	5			9
	3	8			9			
	1		6	5	3			2
		5	1				9	
	6	1		7				
		9	5	4		8	1	

Puzzle 960 - Hard

		4	3	6			2	
							9	
		3		4			5	
7								
3	4	9	1					
2		1	9	3				8
4			8	6				5
			2		4			1
	9		5		7			

Puzzle 961 - Hard

	1							
		6						9
	9	2			8		3	7
6				1	5			
		1	7	2	6			
5			8	9	3	4		1
				5				
				8			9	
7	3				9	5		

Puzzle 962 - Hard

3	6		2	1			9	8
					8	4	3	
			3			5		2
			5					4
	8		7					
		6		3			7	
		7						6
1	9		8	6	5	3	4	
				7	3	9		

Puzzle 963 - Hard

4				3				2
5			1					
	1	8		6	9			
		4			8	6		1
6								
	9	1			3			
	5			4		2	9	6
	6	3		8			5	7
								3

Puzzle 964 - Hard

	4	1	6	8				
						6		4
7			4	5			8	
	3	9		6	4		1	
	1			3				
		4				2		9
			9	5				
	8			2		7		1
	3	7	1			9		8

Puzzle 965 - Hard

					6	1		
6			4			7	5	3
			7		3			
				5	6			4
	7		1			9		
8			7			3		
		7						8
5	4		6	8		2		
2				4				9

Puzzle 966 - Hard

2				3	4	9		
	3	9	8					5
			6	5		3		
		3				1		
8	5				9			
9					3			4
5	9	7						
		1		6		5		
						4		3

Puzzle 967 - Hard

2					1			
	7							6
3							9	7
7		8			2			4
	4		8				1	2
	6							3
8	5			1	7	4		
	1	4	5	2		3		
				3		5		

Puzzle 968 - Hard

				4	1		5	
			2	9		7		
4			7	5			2	
8		4		2		9		7
1		9				4		
7							6	5
3	4	8			2			
	2						8	3
5							9	

Puzzle 969 - Hard

5	1	3				9		
		6		4				5
	4	2	6			3		
7			8	5		9		
						3	8	
	8			7				1
6			9					2
	8		2					
	2	4	7	5	8			

Puzzle 970 - Hard

				1		8		7
	6		4					3
8	7					5	4	
			3	8		6		
	8						3	4
1		4	7			9	8	
			8	2		1		5
2				7				9
	1					3		

Puzzle 971 - Hard

					1		7	
	1					3		
7	4					8		2
1		9		6		4		8
2		6				7		
		8	5				2	
				5				7
5			9			6		
6		4			2		8	

Puzzle 972 - Hard

8		3	7				5	4
				9		3	8	
1	2		4					
	6	8				5	4	
					8			
			2	1	4		3	6
	1	4		6			7	
7	5				2			
		2	1			9		5

Puzzle 973 - Hard

	6		4		7		9	
			3	9				
3	9					4		2
	7						4	
8			7	6	9			1
1	3		5			8		
		7	1	5	2			9
			2		4			8
	1					8	6	

Puzzle 974 - Hard

5		8			4	1		
	4	7				6		
			2					
9			4	3		7		
				9				2
4	8		7					5
		9	1					
1			8			5		
	7	6				9	4	3

Puzzle 975 - Hard

	6		5	7	9	1		
5								6
4		7	2					
	8	4	6	3			5	
								8
	5	9	1	4			7	
				9			2	1
	2			6	3			
	4	5				9		7

Puzzle 976 - Hard

			9	8	6		5	
4				5		3		
5	6						1	9
	8			1	4			
1		4		2		6		7
	3		8		9			2
	5		7	4				
				9				
	2		3		5			1

Puzzle 977 - Hard

			5	4				3
6			3	8				1
			1			6		9
		3	4	7	9			
	4	2	8			3		5
		6						
			9			7		
5	6		1			7	9	
8				4		5		

Puzzle 978 - Hard

7		3	2	9				
								6
5	2				7			
		4	7		8		5	3
				5			2	9
	1			3				
1			4	8		3		2
	9		5		2			
								8

Puzzle 979 - Hard

	7	4		5		2	6	
				7	2			
	9				6		4	7
	3				9		8	
				4				
9		5	3	6			7	
	2	3				5	1	
							3	
7	4	8						

Puzzle 980 - Hard

		5		7	2	4		
2	3	6						5
		4					9	
						5	2	6
	5				3			8
		2	4		6			9
			1	4	7			
		7						4
5			1	6	3			

Puzzle 981 - Hard

1	6					8	7	2
	7				1	9	3	
			8					
	9	1		5				
	5	6	9		8			
3		4		1				
	8	9						6
	1			2				5
5		2			9			

Puzzle 982 - Hard

	8		9				5	
		6	8	5				
	7	3		1	6	8		
			6	3			1	8
9	5			2				
6						9	4	
					2		7	4
		4				5		
	1		3	9			8	

Puzzle 983 - Hard

	4							
		3	9	6		2		
2	7				5			
4	9			8				6
		5				4		2
		8					9	
9	2				1	3	7	
	3	8			7	5		
		5			4	9		

Puzzle 984 - Hard

9			5					
		6				8	3	2
		8						
	5		6	4			1	
						6	2	
		9	7		8			
2				6	7			
8		7		3		2	5	1
3				2	5	7		4

Puzzle 985 - Hard

		4	5					
7		8				3		1
3						8	4	
1		5	3		2			
		7	1	4	5	2	6	3
			7			1		
5	6				9	4		
			2				1	
			3				8	9

Puzzle 986 - Hard

5	9				2		1	8
	2				1			
1	4		9					2
	5	1		4	3		7	6
		4			9	3		
6			1		7	5		
				6			2	
3	8							
		9						3

Puzzle 987 - Hard

1	9					3		
	7	8						
					1	8	9	5
	8	5			4	7	2	
7			2		8			6
		3						8
		6	9		5			3
				8		6	5	9
	5	1						

Puzzle 988 - Hard

	6				1			3
			4			1		
				9	3			
	9						7	4
2	1						3	
7	8		2		6			9
		5	3					
				1	4	3		
8			6		7		9	

Puzzle 989 - Hard

				7	9			
	9	8	6	5		7		
		3	2					6
			7	1	8			
			6	5	2	1		
6		7	4					
	4					7		
3	8	6				2		
7					1	6	5	

Puzzle 990 - Hard

8		9	7			2		
4			5	3				
3	1	6	9					
		5						6
2								3
	7	3	1					9
								5
9	4						6	8
5				9	1			2

Puzzle 991 - Hard

.	2	.	.	9	.	.	.	1
1	8	.	7	.	.	2	.	.
.	.	.	.	5	.	.	.	9
.	4	.	8	7	.	3	5	.
3	.	5	.	.	.	.	.	.
6	1	.	.	4	2	.	.	.
1	.	.	3	.	9	6	8	.
.	4	9	.	.	.	.	.	2
.	.	3	7	.	.	.	.	4

Puzzle 992 - Hard

.	.	3	1	.	6	.	7	.
.	.	5	.	8	7	.	3	.
.	1	.	.	3	.	.	.	.
1	.	.	.	7	9	.	.	.
.	.	2	.	.	.	.	.	.
.	.	.	.	4	.	7	1	.
5	8	.	6	.	.	.	.	7
.	.	4	.	2	.	.	5	.
.	3	7	.	1	4	8	.	6

Puzzle 993 - Hard

6	.	.	.	.	.	.	.	2
.	.	.	1	4	.	3	.	.
.	4	3	8	9	.	1	.	.
.	7	5	.	.	6	.	.	.
.	.	.	.	9	.	2	.	.
.	2	.	7	.	6	9	1	.
.	5	.	1	.	.	.	4	.
.	.	.	4	.	.	.	6	9
4	.	6	9	3	5	.	.	.

Puzzle 994 - Hard

.	.	.	3	4	.	9	.	5
6	.	4	.	1	9	.	.	.
.	9	7	.	5	.	4	1	8
.	.	5	1	.	.	2	.	.
7	.	.	.	9	5	.	.	.
9	4	.	.	.	.	.	.	.
5	.	9	.	6	.	.	7	.
1	.	.	.	.	.	.	.	.
.	.	.	.	2	.	6	5	.

Puzzle 995 - Hard

.	.	5	.	1	.	.	.	9
.	.	3	.	.	5	.	.	.
4	7	.	5	6	9	8	.	.
3	8	.	.	.	1	9	.	5
.	2	.	.	.	3	.	.	.
.	.	9	8	.	.	1	.	7
5	1	.	6	.	4	.	9	.
.	.	.	7	.	.	.	.	.
.	.	6	.	.	8	.	.	.

Puzzle 996 - Hard

.	.	8	.	7	.	.	.	.
4	9	.	.	.	.	.	1	.
.	.	2	.	.	3	.	9	.
1	.	5	.	.	9	.	.	2
.	3	.	.	2	.	5	.	.
.	.	.	.	.	6	1	9	4
3	.	.	.	7	5	.	.	.
7	1	6	2	.	.	.	3	9
.	2	.	3	.	.	4	8	.

Puzzle 997 - Hard

		2	4	8		9		
			1	5	6			
4				9		5		1
2		9	3		4			
		5				1		2
								6
	9		7		8		5	4
		4	6	2	1			
3	2			4				8

Puzzle 998 - Hard

			7					
1	2		3		5			
9		5		2	1			4
7	1	3			2			
			7			6		
	4			3		5	2	
2		1			6		3	
		6	4	1		2		
4				3			6	

Puzzle 999 - Hard

1		6		3				2
	7		5				3	8
8	3			4	2		7	
		3			7	5		
								6
5	1		4		6			
4	8		9			3		6
			1	8		2		
	9					8	5	

Puzzle 1000 - Hard

8			7			4		
	3		2	8				
	1		9			8		7
3						6		5
		4	8	3				2
		1		6	5			3
			9			7	6	
			1	7				
			6	5	8		9	

Puzzle 1001 - Hard

2	1	6		8		5	4	
		9	3		1		6	
3								
		3				2	1	
4	6		9		5		8	
		7		3		4	5	
				6	4	8	9	2
	9							
			8				7	

Puzzle 1002 - Hard

				4				
	2	7				8		
8					1			
		9					5	2
5				1				3
	3	2			4		9	8
	7	8	4		5	9		1
	1	5		3	9			
9		3				6		

Puzzle 1003 - Hard

		5			8			3
6					2			
1	3				7	9		
			4		1		6	
	2	6		8				
		4		2	5	8		
5						7	9	
	6		3					1
				5	9		3	

Puzzle 1004 - Hard

	8	2				7	3	
							2	
		1		7	8			
	3				1		7	8
1	4		5					9
		9			8	5		
	7		3	9				
	3						9	
	1						8	5

Puzzle 1005 - Hard

	2					5		8
	4			8				
5	7		3		6			
						7	9	
4		2	1	7		6		
7	9			6	5			
	4				2			6
5	7		4			8		
1						4		

Puzzle 1006 - Hard

				2				
		1	4		7			2
		2	6	5		3		1
	9	5	2			4		
	8	3					7	
4	7				9		2	
			7				6	
	2					5		4
6		4	3			7		

Puzzle 1007 - Hard

		6	8	7	1			
	3							1
9			3			5		
2	1				9			
		8	2		3			
			5	6				
		4	6					9
8	6	9		2		3	4	
							8	

Puzzle 1008 - Hard

		2	8					
4								
8	7				4	2	9	
			7			6		
7				6				9
1	6	9		3		4	8	
2		7					6	8
	4		1	8			2	
		8	2			1		4

ANSWERS

Puzzle 1

3	1	9	8	2	4	7	6	5
5	8	4	7	3	6	1	9	2
2	6	7	1	5	9	4	3	8
4	9	2	3	8	5	6	1	7
8	5	3	6	7	1	9	2	4
1	7	6	4	9	2	5	8	3
6	4	8	5	1	3	2	7	9
7	2	1	9	4	8	3	5	6
9	3	5	2	6	7	8	4	1

Puzzle 2

7	9	2	4	6	5	1	3	8
1	8	6	7	2	3	4	5	9
4	3	5	8	1	9	2	7	6
5	7	9	3	4	2	6	8	1
3	1	4	6	9	8	5	2	7
2	6	8	1	5	7	3	9	4
9	2	1	5	8	6	7	4	3
6	5	3	9	7	4	8	1	2
8	4	7	2	3	1	9	6	5

Puzzle 3

1	3	9	4	2	6	8	7	5
8	4	7	1	5	9	6	3	2
2	6	5	8	3	7	9	4	1
4	8	2	6	9	3	1	5	7
7	5	6	2	8	1	3	9	4
9	1	3	5	7	4	2	8	6
3	9	4	7	1	2	5	6	8
5	7	1	9	6	8	4	2	3
6	2	8	3	4	5	7	1	9

Puzzle 4

2	1	3	7	6	9	4	5	8
7	5	9	4	3	8	6	1	2
4	6	8	5	2	1	7	9	3
1	3	7	6	4	5	8	2	9
6	2	4	8	9	3	1	7	5
8	9	5	2	1	7	3	4	6
5	8	1	3	7	2	9	6	4
9	4	2	1	8	6	5	3	7
3	7	6	9	5	4	2	8	1

Puzzle 5

1	6	2	7	3	9	8	5	4
8	3	9	5	2	4	1	6	7
7	4	5	1	8	6	2	3	9
4	7	6	3	5	1	9	8	2
5	2	8	4	9	7	3	1	6
9	1	3	2	6	8	7	4	5
6	8	4	9	7	3	5	2	1
3	5	7	6	1	2	4	9	8
2	9	1	8	4	5	6	7	3

Puzzle 6

5	9	8	4	6	7	1	3	2
2	3	1	9	5	8	4	6	7
6	7	4	1	3	2	5	8	9
3	1	9	7	8	5	6	2	4
4	6	5	3	2	9	8	7	1
7	8	2	6	4	1	9	5	3
8	4	3	2	9	6	7	1	5
1	2	6	5	7	4	3	9	8
9	5	7	8	1	3	2	4	6

Puzzle 7

8	1	9	6	3	5	7	2	4
5	7	2	8	1	4	9	3	6
4	6	3	2	9	7	5	1	8
7	9	1	5	6	3	8	4	2
6	2	8	4	7	9	1	5	3
3	4	5	1	2	8	6	7	9
1	5	6	9	4	2	3	8	7
9	3	4	7	8	1	2	6	5
2	8	7	3	5	6	4	9	1

Puzzle 8

6	3	7	9	8	4	1	2	5
8	4	2	1	7	5	9	3	6
1	9	5	6	3	2	8	7	4
9	1	6	2	4	7	5	8	3
5	8	4	3	9	6	2	1	7
2	7	3	8	5	1	4	6	9
3	2	9	5	6	8	7	4	1
4	5	8	7	1	3	6	9	2
7	6	1	4	2	9	3	5	8

Puzzle 9

2	5	3	4	6	9	7	8	1
7	8	9	3	5	1	2	4	6
6	4	1	8	7	2	5	9	3
5	9	2	7	8	6	1	3	4
1	3	4	2	9	5	8	6	7
8	7	6	1	4	3	9	2	5
4	6	7	5	2	8	3	1	9
9	1	8	6	3	7	4	5	2
3	2	5	9	1	4	6	7	8

Puzzle 10

2	6	3	7	8	4	1	9	5
9	1	8	3	5	6	4	7	2
5	4	7	9	2	1	6	3	8
1	5	2	6	9	7	3	8	4
4	3	9	5	1	8	2	6	7
8	7	6	4	3	2	5	1	9
6	8	4	1	7	5	9	2	3
3	2	1	8	4	9	7	5	6
7	9	5	2	6	3	8	4	1

Puzzle 11

6	2	5	9	8	4	7	3	1
3	4	1	7	5	6	8	9	2
9	7	8	3	2	1	6	4	5
8	5	9	4	6	3	1	2	7
1	3	4	8	7	2	5	6	9
7	6	2	1	9	5	4	8	3
5	9	7	2	4	8	3	1	6
4	1	6	5	3	9	2	7	8
2	8	3	6	1	7	9	5	4

Puzzle 12

1	6	8	2	3	7	5	4	9
5	3	7	4	1	9	6	8	2
4	2	9	8	6	5	1	3	7
7	5	3	6	2	1	4	9	8
2	8	1	7	9	4	3	6	5
6	9	4	3	5	8	2	7	1
3	4	5	9	8	2	7	1	6
8	7	2	1	4	6	9	5	3
9	1	6	5	7	3	8	2	4

Puzzle 13

6	3	2	5	1	7	8	9	4
8	4	9	2	3	6	1	5	7
7	5	1	4	9	8	6	3	2
2	8	4	6	5	1	9	7	3
5	6	3	8	7	9	4	2	1
9	1	7	3	4	2	5	8	6
4	2	5	1	8	3	7	6	9
1	7	6	9	2	5	3	4	8
3	9	8	7	6	4	2	1	5

Puzzle 14

5	7	6	3	1	4	8	9	2
3	4	9	2	8	7	1	5	6
2	1	8	9	5	6	7	3	4
6	2	4	1	9	5	3	8	7
1	5	3	4	7	8	2	6	9
8	9	7	6	2	3	4	1	5
9	3	2	5	4	1	6	7	8
7	6	5	8	3	2	9	4	1
4	8	1	7	6	9	5	2	3

Puzzle 15

6	3	9	2	8	7	5	1	4
7	4	8	3	1	5	2	6	9
2	5	1	6	9	4	3	8	7
5	6	7	1	2	9	8	4	3
3	8	4	5	7	6	1	9	2
1	9	2	8	4	3	7	5	6
4	7	3	9	5	8	6	2	1
9	2	5	7	6	1	4	3	8
8	1	6	4	3	2	9	7	5

Puzzle 16

9	2	7	8	4	3	1	5	6
8	3	6	2	1	5	4	7	9
4	5	1	6	9	7	8	3	2
2	6	5	3	8	4	7	9	1
7	8	4	9	6	1	3	2	5
1	9	3	7	5	2	6	4	8
6	4	8	5	3	9	2	1	7
5	1	2	4	7	8	9	6	3
3	7	9	1	2	6	5	8	4

Puzzle 17

8	3	4	1	7	9	5	2	6
5	1	7	8	2	6	9	3	4
6	2	9	5	3	4	1	8	7
7	5	3	4	8	1	2	6	9
9	8	2	3	6	7	4	5	1
1	4	6	2	9	5	3	7	8
2	9	8	6	4	3	7	1	5
3	7	1	9	5	8	6	4	2
4	6	5	7	1	2	8	9	3

Puzzle 18

1	3	7	5	8	4	2	9	6
9	2	8	1	3	6	5	7	4
4	5	6	9	7	2	8	3	1
6	1	5	4	2	9	3	8	7
2	8	9	3	1	7	6	4	5
7	4	3	8	6	5	1	2	9
5	6	4	2	9	8	7	1	3
3	7	2	6	4	1	9	5	8
8	9	1	7	5	3	4	6	2

Puzzle 19

2	3	7	6	8	5	4	9	1
6	1	9	2	4	7	8	3	5
5	4	8	3	9	1	6	7	2
8	9	2	1	7	4	5	6	3
1	5	6	9	2	3	7	8	4
4	7	3	5	6	8	2	1	9
9	8	5	4	1	6	3	2	7
3	6	1	7	5	2	9	4	8
7	2	4	8	3	9	1	5	6

Puzzle 20

9	1	6	3	8	4	2	7	5
4	5	8	6	2	7	3	9	1
3	2	7	5	9	1	8	6	4
8	9	4	1	7	3	5	2	6
6	7	5	8	4	2	1	3	9
1	3	2	9	5	6	7	4	8
2	6	9	7	1	5	4	8	3
5	4	3	2	6	8	9	1	7
7	8	1	4	3	9	6	5	2

Puzzle 21

8	6	3	5	2	9	1	7	4
4	5	7	3	1	8	2	6	9
9	1	2	4	7	6	8	5	3
6	8	9	7	5	1	3	4	2
1	2	4	6	8	3	7	9	5
7	3	5	9	4	2	6	8	1
3	7	8	2	9	4	5	1	6
5	4	6	1	3	7	9	2	8
2	9	1	8	6	5	4	3	7

Puzzle 22

2	9	8	1	5	7	3	6	4
4	6	7	2	8	3	1	5	9
5	3	1	4	6	9	8	2	7
7	1	5	9	2	4	6	3	8
9	2	3	8	7	6	4	1	5
6	8	4	3	1	5	7	9	2
1	4	2	6	9	8	5	7	3
3	7	9	5	4	1	2	8	6
8	5	6	7	3	2	9	4	1

Puzzle 23

2	8	3	4	9	5	1	6	7
5	6	1	7	2	3	9	4	8
7	4	9	8	1	6	5	3	2
4	2	6	5	8	1	7	9	3
9	7	8	6	3	4	2	5	1
3	1	5	9	7	2	4	8	6
1	9	7	3	5	8	6	2	4
6	3	2	1	4	9	8	7	5
8	5	4	2	6	7	3	1	9

Puzzle 24

3	1	2	7	9	4	5	6	8
7	8	9	5	3	6	1	2	4
5	4	6	2	8	1	9	3	7
8	5	1	3	2	7	6	4	9
2	9	7	6	4	5	8	1	3
4	6	3	9	1	8	7	5	2
9	3	5	1	7	2	4	8	6
1	2	4	8	6	9	3	7	5
6	7	8	4	5	3	2	9	1

Puzzle 25

6	4	9	8	3	5	1	2	7
1	3	8	2	6	7	9	4	5
2	5	7	1	4	9	6	3	8
8	1	2	5	7	6	3	9	4
5	9	4	3	2	8	7	1	6
7	6	3	9	1	4	8	5	2
4	2	1	7	8	3	5	6	9
3	8	5	6	9	2	4	7	1
9	7	6	4	5	1	2	8	3

Puzzle 26

2	3	7	9	6	1	5	4	8
8	6	5	3	4	2	7	9	1
1	9	4	8	5	7	6	2	3
3	8	2	7	9	6	1	5	4
6	7	9	5	1	4	3	8	2
4	5	1	2	8	3	9	7	6
5	2	3	1	7	8	4	6	9
9	4	8	6	3	5	2	1	7
7	1	6	4	2	9	8	3	5

Puzzle 27

6	3	5	8	1	7	9	4	2
1	4	2	5	6	9	7	8	3
7	8	9	3	4	2	1	6	5
3	1	6	9	7	5	8	2	4
8	2	7	6	3	4	5	1	9
5	9	4	1	2	8	6	3	7
2	7	8	4	5	1	3	9	6
4	6	1	7	9	3	2	5	8
9	5	3	2	8	6	4	7	1

Puzzle 28

5	2	6	4	3	1	7	8	9
8	4	7	9	6	2	5	1	3
9	3	1	8	5	7	4	6	2
2	5	8	3	1	4	9	7	6
1	9	3	5	7	6	2	4	8
6	7	4	2	9	8	3	5	1
3	1	2	7	8	5	6	9	4
4	6	5	1	2	9	8	3	7
7	8	9	6	4	3	1	2	5

Puzzle 29

1	5	8	6	4	9	7	2	3
2	4	6	3	8	7	9	1	5
3	7	9	5	1	2	8	4	6
5	8	4	2	9	6	3	7	1
7	9	1	4	3	8	5	6	2
6	3	2	1	7	5	4	8	9
9	6	3	8	2	4	1	5	7
8	1	5	7	6	3	2	9	4
4	2	7	9	5	1	6	3	8

Puzzle 30

1	6	8	5	3	9	2	7	4
9	4	7	8	2	1	5	3	6
2	3	5	4	6	7	1	9	8
3	8	4	9	7	5	6	1	2
6	1	2	3	8	4	7	5	9
7	5	9	2	1	6	8	4	3
8	7	3	1	9	2	4	6	5
4	9	1	6	5	8	3	2	7
5	2	6	7	4	3	9	8	1

Puzzle 31

8	3	2	7	6	5	4	1	9
6	5	7	1	9	4	8	3	2
4	1	9	8	2	3	5	6	7
7	4	8	2	1	9	6	5	3
5	9	1	6	3	7	2	8	4
2	6	3	4	5	8	7	9	1
1	8	6	3	4	2	9	7	5
3	2	5	9	7	6	1	4	8
9	7	4	5	8	1	3	2	6

Puzzle 32

4	2	8	1	5	9	3	7	6
9	5	7	4	3	6	8	1	2
3	6	1	2	7	8	9	5	4
1	9	3	6	4	7	2	8	5
5	4	2	9	8	3	1	6	7
8	7	6	5	1	2	4	3	9
6	8	4	7	2	1	5	9	3
2	1	9	3	6	5	7	4	8
7	3	5	8	9	4	6	2	1

Puzzle 33

9	3	1	2	7	4	8	6	5
2	7	5	6	1	8	9	3	4
4	6	8	9	3	5	7	1	2
7	9	2	3	5	6	1	4	8
8	4	3	7	2	1	5	9	6
5	1	6	4	8	9	3	2	7
3	8	4	1	6	7	2	5	9
1	5	9	8	4	2	6	7	3
6	2	7	5	9	3	4	8	1

Puzzle 34

7	3	5	8	9	1	4	2	6
4	2	8	7	6	5	9	3	1
6	1	9	2	3	4	5	7	8
8	9	6	5	1	3	7	4	2
1	4	2	6	7	9	8	5	3
5	7	3	4	8	2	6	1	9
2	8	1	9	4	7	3	6	5
3	6	7	1	5	8	2	9	4
9	5	4	3	2	6	1	8	7

Puzzle 35

4	1	5	2	3	8	7	9	6
3	8	6	7	4	9	1	2	5
2	9	7	6	1	5	8	3	4
7	5	3	8	2	6	4	1	9
6	4	1	9	5	3	2	7	8
8	2	9	1	7	4	5	6	3
9	7	4	5	6	2	3	8	1
5	6	2	3	8	1	9	4	7
1	3	8	4	9	7	6	5	2

Puzzle 36

7	4	9	5	6	1	8	2	3
5	2	6	9	8	3	4	1	7
1	3	8	7	2	4	9	5	6
6	5	4	2	7	9	1	3	8
3	1	7	8	4	6	2	9	5
8	9	2	1	3	5	6	7	4
9	7	5	6	1	8	3	4	2
4	8	1	3	5	2	7	6	9
2	6	3	4	9	7	5	8	1

Puzzle 37

8	6	4	3	7	5	1	9	2
1	2	5	9	8	6	3	4	7
3	9	7	4	1	2	5	6	8
4	3	2	6	5	9	7	8	1
5	1	9	7	4	8	6	2	3
6	7	8	2	3	1	9	5	4
7	5	3	8	6	4	2	1	9
9	4	6	1	2	3	8	7	5
2	8	1	5	9	7	4	3	6

Puzzle 38

8	2	9	6	7	5	3	4	1
7	1	6	4	9	3	2	8	5
3	4	5	1	2	8	9	6	7
5	9	4	8	1	7	6	2	3
1	7	8	3	6	2	4	5	9
6	3	2	9	5	4	7	1	8
2	8	7	5	4	9	1	3	6
4	5	1	7	3	6	8	9	2
9	6	3	2	8	1	5	7	4

Puzzle 39

2	9	8	5	1	3	6	4	7
1	4	3	8	6	7	5	9	2
7	6	5	2	4	9	8	1	3
3	2	4	7	9	5	1	8	6
6	7	1	4	8	2	3	5	9
8	5	9	1	3	6	7	2	4
9	8	7	3	5	4	2	6	1
5	3	6	9	2	1	4	7	8
4	1	2	6	7	8	9	3	5

Puzzle 40

7	3	9	5	1	6	2	4	8
4	5	1	8	3	2	6	7	9
8	6	2	7	9	4	1	5	3
6	8	5	4	7	3	9	2	1
2	4	7	1	6	9	3	8	5
9	1	3	2	5	8	7	6	4
5	9	6	3	8	7	4	1	2
3	2	8	6	4	1	5	9	7
1	7	4	9	2	5	8	3	6

Puzzle 41

8	1	4	6	7	2	3	9	5
9	7	2	4	3	5	1	8	6
5	6	3	1	8	9	2	7	4
2	5	1	7	9	8	6	4	3
7	9	6	3	4	1	5	2	8
3	4	8	2	5	6	9	1	7
4	2	5	8	1	3	7	6	9
6	8	9	5	2	7	4	3	1
1	3	7	9	6	4	8	5	2

Puzzle 42

9	1	8	7	4	3	5	2	6
5	6	7	8	9	2	4	3	1
3	4	2	5	6	1	7	9	8
7	3	1	9	5	4	8	6	2
4	2	5	1	8	6	3	7	9
6	8	9	3	2	7	1	5	4
1	5	4	2	3	9	6	8	7
8	9	6	4	7	5	2	1	3
2	7	3	6	1	8	9	4	5

Puzzle 43

8	9	6	1	3	2	5	4	7
2	3	7	4	9	5	6	8	1
4	1	5	6	7	8	3	2	9
1	2	8	9	4	6	7	5	3
7	5	9	8	1	3	2	6	4
3	6	4	5	2	7	1	9	8
9	7	2	3	5	4	8	1	6
6	4	3	2	8	1	9	7	5
5	8	1	7	6	9	4	3	2

Puzzle 44

2	8	5	7	6	9	4	1	3
1	6	3	4	5	2	8	7	9
4	9	7	1	3	8	5	2	6
3	2	4	5	8	6	1	9	7
8	1	9	3	7	4	2	6	5
7	5	6	2	9	1	3	8	4
6	4	2	9	1	5	7	3	8
5	7	8	6	2	3	9	4	1
9	3	1	8	4	7	6	5	2

Puzzle 45

2	4	7	6	8	3	9	1	5
1	9	5	4	2	7	8	3	6
6	8	3	1	5	9	2	4	7
3	6	2	9	7	4	5	8	1
5	7	9	2	1	8	4	6	3
4	1	8	3	6	5	7	2	9
8	2	1	5	9	6	3	7	4
7	5	4	8	3	1	6	9	2
9	3	6	7	4	2	1	5	8

Puzzle 46

2	1	8	4	3	7	9	5	6
3	4	5	8	6	9	2	1	7
6	9	7	5	1	2	4	8	3
1	8	9	3	5	6	7	2	4
4	7	2	9	8	1	6	3	5
5	3	6	7	2	4	8	9	1
7	6	1	2	9	5	3	4	8
8	2	4	1	7	3	5	6	9
9	5	3	6	4	8	1	7	2

Puzzle 47

6	2	4	8	5	9	1	3	7
7	3	5	1	6	4	2	9	8
8	1	9	7	2	3	6	4	5
9	7	2	4	3	6	5	8	1
4	5	8	9	7	1	3	2	6
1	6	3	5	8	2	9	7	4
5	8	6	3	9	7	4	1	2
2	9	1	6	4	8	7	5	3
3	4	7	2	1	5	8	6	9

Puzzle 48

5	6	9	7	3	1	2	8	4
4	2	1	6	5	8	7	3	9
7	3	8	2	9	4	6	1	5
2	5	6	4	7	3	8	9	1
8	4	3	9	1	2	5	7	6
9	1	7	5	8	6	4	2	3
3	8	4	1	2	5	9	6	7
1	9	5	8	6	7	3	4	2
6	7	2	3	4	9	1	5	8

Puzzle 49

3	9	7	6	8	1	2	4	5
2	4	1	5	3	9	8	6	7
6	8	5	4	7	2	1	9	3
5	6	9	1	4	7	3	2	8
7	2	8	3	9	6	4	5	1
1	3	4	8	2	5	9	7	6
4	1	2	7	6	8	5	3	9
8	7	3	9	5	4	6	1	2
9	5	6	2	1	3	7	8	4

Puzzle 50

4	6	9	1	2	3	5	7	8
7	1	3	9	8	5	6	4	2
2	5	8	6	4	7	3	9	1
9	7	4	2	5	8	1	6	3
6	2	1	4	3	9	8	5	7
3	8	5	7	1	6	9	2	4
1	4	6	3	9	2	7	8	5
8	9	2	5	7	1	4	3	6
5	3	7	8	6	4	2	1	9

Puzzle 51

1	5	3	6	8	9	4	7	2
8	6	4	7	5	2	9	3	1
7	2	9	4	3	1	6	8	5
5	7	6	9	1	3	2	4	8
9	8	1	5	2	4	3	6	7
3	4	2	8	7	6	5	1	9
6	9	7	1	4	5	8	2	3
2	1	5	3	6	8	7	9	4
4	3	8	2	9	7	1	5	6

Puzzle 52

8	5	2	9	1	7	4	6	3
7	6	1	3	8	4	2	5	9
9	4	3	6	2	5	8	1	7
3	1	9	4	6	8	7	2	5
5	2	8	1	7	9	3	4	6
6	7	4	2	5	3	9	8	1
2	3	7	5	4	1	6	9	8
1	8	6	7	9	2	5	3	4
4	9	5	8	3	6	1	7	2

Puzzle 53

9	8	5	6	2	4	7	1	3
7	4	6	8	1	3	5	2	9
2	3	1	7	5	9	6	4	8
6	2	8	4	7	5	9	3	1
3	5	7	1	9	8	4	6	2
1	9	4	2	3	6	8	7	5
5	6	9	3	4	2	1	8	7
8	1	2	9	6	7	3	5	4
4	7	3	5	8	1	2	9	6

Puzzle 54

3	9	2	1	7	6	8	5	4
5	6	4	2	3	8	1	9	7
1	7	8	5	9	4	2	6	3
6	5	7	8	2	3	4	1	9
8	4	1	9	5	7	6	3	2
9	2	3	6	4	1	5	7	8
2	8	5	7	6	9	3	4	1
7	3	6	4	1	2	9	8	5
4	1	9	3	8	5	7	2	6

Puzzle 55

2	5	1	9	8	4	3	6	7
3	4	8	7	6	1	5	2	9
6	9	7	3	2	5	1	4	8
4	3	9	8	5	7	2	1	6
7	8	5	2	1	6	9	3	4
1	6	2	4	3	9	8	7	5
9	2	3	6	4	8	7	5	1
5	7	4	1	9	3	6	8	2
8	1	6	5	7	2	4	9	3

Puzzle 56

1	2	7	9	5	3	4	6	8
3	5	9	4	6	8	7	2	1
4	6	8	7	1	2	5	3	9
9	8	4	6	2	5	3	1	7
7	3	5	1	8	4	2	9	6
6	1	2	3	7	9	8	4	5
5	4	1	8	3	6	9	7	2
2	9	6	5	4	7	1	8	3
8	7	3	2	9	1	6	5	4

Puzzle 57

8	4	5	3	7	6	9	1	2
3	6	2	4	9	1	7	5	8
7	9	1	8	5	2	3	6	4
4	2	9	1	8	5	6	3	7
6	5	3	9	4	7	2	8	1
1	8	7	6	2	3	4	9	5
5	3	6	2	1	4	8	7	9
9	7	4	5	3	8	1	2	6
2	1	8	7	6	9	5	4	3

Puzzle 58

8	5	7	4	3	9	6	2	1
6	2	3	5	7	1	4	8	9
9	4	1	2	8	6	5	3	7
4	9	6	8	2	5	1	7	3
3	8	2	1	6	7	9	5	4
7	1	5	9	4	3	8	6	2
2	6	4	3	1	8	7	9	5
1	7	9	6	5	2	3	4	8
5	3	8	7	9	4	2	1	6

Puzzle 59

7	9	8	1	4	6	3	5	2
1	5	4	3	9	2	7	6	8
6	2	3	7	5	8	9	1	4
9	1	5	2	8	7	4	3	6
4	6	7	5	1	3	8	2	9
8	3	2	9	6	4	1	7	5
2	8	9	6	7	1	5	4	3
5	7	6	4	3	9	2	8	1
3	4	1	8	2	5	6	9	7

Puzzle 60

3	5	9	2	8	7	6	4	1
1	7	4	9	5	6	8	2	3
8	6	2	3	4	1	7	9	5
9	8	3	7	6	2	5	1	4
5	2	7	4	1	8	3	6	9
6	4	1	5	9	3	2	7	8
4	9	6	8	2	5	1	3	7
7	1	5	6	3	9	4	8	2
2	3	8	1	7	4	9	5	6

Puzzle 61

2	8	4	7	1	6	5	3	9
7	6	3	8	9	5	1	2	4
9	1	5	2	3	4	7	8	6
3	5	2	4	7	8	9	6	1
8	7	9	3	6	1	4	5	2
1	4	6	9	5	2	8	7	3
4	3	7	6	8	9	2	1	5
6	9	1	5	2	7	3	4	8
5	2	8	1	4	3	6	9	7

Puzzle 62

5	9	2	4	3	7	1	8	6
6	8	7	2	9	1	5	3	4
3	4	1	5	8	6	9	7	2
9	7	3	1	4	2	6	5	8
2	6	8	9	7	5	4	1	3
4	1	5	8	6	3	7	2	9
7	3	9	6	1	8	2	4	5
8	2	4	7	5	9	3	6	1
1	5	6	3	2	4	8	9	7

Puzzle 63

3	5	2	8	6	1	9	4	7
4	6	7	5	3	9	1	8	2
9	1	8	2	7	4	5	6	3
5	8	4	7	9	3	6	2	1
7	9	1	6	2	8	4	3	5
2	3	6	1	4	5	8	7	9
6	4	9	3	5	2	7	1	8
8	7	3	9	1	6	2	5	4
1	2	5	4	8	7	3	9	6

Puzzle 64

9	8	2	5	7	6	3	4	1
6	1	5	4	9	3	7	2	8
4	7	3	1	8	2	5	6	9
7	9	1	8	6	4	2	3	5
2	4	8	3	1	5	6	9	7
5	3	6	9	2	7	1	8	4
1	6	4	2	5	9	8	7	3
3	5	7	6	4	8	9	1	2
8	2	9	7	3	1	4	5	6

Puzzle 65

9	1	8	7	5	6	2	4	3
2	5	4	1	3	9	7	6	8
3	6	7	8	4	2	5	9	1
7	2	3	6	1	4	8	5	9
8	4	1	3	9	5	6	7	2
6	9	5	2	8	7	3	1	4
1	7	6	4	2	3	9	8	5
5	8	2	9	7	1	4	3	6
4	3	9	5	6	8	1	2	7

Puzzle 66

3	4	9	5	8	6	2	7	1
2	1	5	9	3	7	6	4	8
6	8	7	1	2	4	5	9	3
9	5	4	2	7	3	1	8	6
1	3	6	8	5	9	7	2	4
8	7	2	6	4	1	3	5	9
4	6	8	7	1	5	9	3	2
7	2	1	3	9	8	4	6	5
5	9	3	4	6	2	8	1	7

Puzzle 67

6	4	7	2	9	3	5	1	8
3	2	1	6	8	5	9	4	7
9	5	8	4	7	1	2	6	3
2	9	3	8	4	7	6	5	1
1	8	4	9	5	6	3	7	2
7	6	5	1	3	2	8	9	4
8	3	6	5	1	4	7	2	9
5	1	9	7	2	8	4	3	6
4	7	2	3	6	9	1	8	5

Puzzle 68

7	6	3	5	9	1	2	4	8
4	1	2	7	6	8	9	5	3
9	5	8	2	4	3	7	6	1
2	3	9	1	8	5	6	7	4
8	4	5	9	7	6	3	1	2
6	7	1	3	2	4	5	8	9
3	9	4	6	1	7	8	2	5
1	2	7	8	5	9	4	3	6
5	8	6	4	3	2	1	9	7

Puzzle 69

7	5	8	1	6	3	9	2	4
3	1	2	7	9	4	8	5	6
6	4	9	2	8	5	3	7	1
4	3	6	8	1	2	7	9	5
8	7	1	4	5	9	2	6	3
2	9	5	3	7	6	4	1	8
1	8	3	6	2	7	5	4	9
5	2	4	9	3	1	6	8	7
9	6	7	5	4	8	1	3	2

Puzzle 70

9	1	6	5	7	8	4	2	3
5	3	7	1	2	4	8	6	9
8	4	2	3	9	6	5	1	7
1	2	3	9	5	7	6	8	4
4	9	5	6	8	3	2	7	1
7	6	8	2	4	1	3	9	5
6	5	1	8	3	9	7	4	2
2	7	9	4	6	5	1	3	8
3	8	4	7	1	2	9	5	6

Puzzle 71

2	8	6	9	3	1	7	4	5
7	1	9	5	4	6	3	8	2
5	4	3	7	2	8	6	1	9
4	9	5	8	7	3	2	6	1
3	2	8	6	1	9	4	5	7
1	6	7	2	5	4	8	9	3
8	5	1	3	6	2	9	7	4
6	7	2	4	9	5	1	3	8
9	3	4	1	8	7	5	2	6

Puzzle 72

1	3	9	7	4	5	6	8	2
7	6	8	1	9	2	4	5	3
5	4	2	6	3	8	9	7	1
6	8	4	3	1	7	2	9	5
3	5	7	2	6	9	1	4	8
9	2	1	5	8	4	7	3	6
4	9	3	8	2	6	5	1	7
2	1	5	4	7	3	8	6	9
8	7	6	9	5	1	3	2	4

Puzzle 73

8	3	1	6	7	2	5	4	9
7	6	2	9	4	5	3	1	8
5	4	9	3	1	8	6	7	2
9	1	6	5	8	3	4	2	7
4	5	7	1	2	9	8	3	6
2	8	3	4	6	7	9	5	1
6	7	4	8	5	1	2	9	3
3	2	5	7	9	6	1	8	4
1	9	8	2	3	4	7	6	5

Puzzle 74

3	8	7	6	2	5	1	9	4
5	1	9	7	4	3	8	2	6
6	2	4	9	1	8	3	7	5
9	5	6	1	3	4	7	8	2
1	3	2	8	7	6	5	4	9
4	7	8	2	5	9	6	1	3
2	6	3	4	8	1	9	5	7
7	9	1	5	6	2	4	3	8
8	4	5	3	9	7	2	6	1

Puzzle 75

5	4	1	6	8	2	3	9	7
3	8	6	4	7	9	5	1	2
7	9	2	1	5	3	4	6	8
4	2	8	9	3	5	1	7	6
6	3	7	2	1	4	8	5	9
9	1	5	8	6	7	2	3	4
8	5	3	7	2	6	9	4	1
2	6	4	5	9	1	7	8	3
1	7	9	3	4	8	6	2	5

Puzzle 76

4	6	1	7	8	5	3	9	2
9	5	2	3	6	1	4	7	8
3	7	8	2	4	9	1	5	6
7	1	3	5	9	2	8	6	4
8	2	4	6	3	7	5	1	9
6	9	5	8	1	4	2	3	7
1	3	9	4	7	8	6	2	5
2	4	7	1	5	6	9	8	3
5	8	6	9	2	3	7	4	1

Puzzle 77

4	3	6	7	9	8	1	5	2
9	2	7	6	1	5	8	3	4
5	1	8	2	4	3	6	7	9
2	5	3	8	6	7	4	9	1
1	6	4	9	5	2	7	8	3
7	8	9	4	3	1	2	6	5
6	4	2	5	8	9	3	1	7
8	9	1	3	7	4	5	2	6
3	7	5	1	2	6	9	4	8

Puzzle 78

9	3	7	6	1	5	4	8	2
1	4	6	3	8	2	5	7	9
8	2	5	7	9	4	6	1	3
3	5	1	8	4	9	2	6	7
6	8	4	1	2	7	3	9	5
2	7	9	5	6	3	8	4	1
7	1	2	4	3	6	9	5	8
5	6	3	9	7	8	1	2	4
4	9	8	2	5	1	7	3	6

Puzzle 79

7	2	4	5	9	3	1	6	8
9	8	6	2	4	1	3	5	7
5	1	3	8	7	6	9	4	2
6	9	1	4	3	8	2	7	5
8	4	2	9	5	7	6	1	3
3	7	5	1	6	2	8	9	4
4	5	8	6	2	9	7	3	1
1	6	7	3	8	4	5	2	9
2	3	9	7	1	5	4	8	6

Puzzle 80

4	9	7	2	1	8	5	3	6
6	3	5	9	4	7	1	8	2
2	8	1	6	3	5	7	9	4
5	4	2	1	8	3	6	7	9
7	6	8	5	2	9	3	4	1
9	1	3	4	7	6	8	2	5
1	7	6	3	9	4	2	5	8
3	5	4	8	6	2	9	1	7
8	2	9	7	5	1	4	6	3

Puzzle 81

1	7	6	4	5	8	2	3	9
5	8	3	2	7	9	4	1	6
9	4	2	1	3	6	8	7	5
3	9	7	5	1	4	6	8	2
4	2	5	6	8	7	3	9	1
8	6	1	9	2	3	7	5	4
7	1	4	3	6	5	9	2	8
6	5	8	7	9	2	1	4	3
2	3	9	8	4	1	5	6	7

Puzzle 82

4	6	1	2	5	3	7	9	8
2	5	9	7	4	8	1	3	6
7	8	3	6	1	9	2	4	5
9	7	6	1	2	5	4	8	3
8	2	5	9	3	4	6	1	7
3	1	4	8	6	7	9	5	2
1	3	7	4	8	2	5	6	9
6	9	8	5	7	1	3	2	4
5	4	2	3	9	6	8	7	1

Puzzle 83

8	2	4	5	6	3	7	9	1
6	1	9	2	4	7	8	3	5
3	7	5	8	1	9	2	4	6
2	9	8	6	5	1	3	7	4
5	4	3	9	7	8	6	1	2
1	6	7	3	2	4	5	8	9
4	5	1	7	8	2	9	6	3
9	8	6	1	3	5	4	2	7
7	3	2	4	9	6	1	5	8

Puzzle 84

7	8	5	6	2	9	4	3	1
9	2	6	3	4	1	8	5	7
4	3	1	5	8	7	6	9	2
2	4	3	1	9	8	5	7	6
8	6	9	7	3	5	2	1	4
1	5	7	4	6	2	9	8	3
5	1	8	2	7	6	3	4	9
6	7	4	9	5	3	1	2	8
3	9	2	8	1	4	7	6	5

Puzzle 85

6	3	7	1	9	8	2	5	4
2	4	5	3	7	6	8	9	1
1	9	8	5	2	4	7	6	3
5	2	9	6	8	3	4	1	7
8	7	6	9	4	1	5	3	2
3	1	4	7	5	2	9	8	6
4	8	3	2	6	5	1	7	9
9	5	1	4	3	7	6	2	8
7	6	2	8	1	9	3	4	5

Puzzle 86

5	2	8	9	1	6	3	4	7
4	1	9	3	8	7	2	6	5
3	6	7	4	2	5	9	8	1
1	8	4	5	3	2	7	9	6
7	3	2	8	6	9	1	5	4
6	9	5	7	4	1	8	3	2
9	4	1	6	7	3	5	2	8
8	7	3	2	5	4	6	1	9
2	5	6	1	9	8	4	7	3

Puzzle 87

5	2	1	9	7	3	4	6	8
7	8	4	5	6	1	2	9	3
6	3	9	4	8	2	5	7	1
4	6	8	2	9	7	3	1	5
9	5	7	1	3	8	6	2	4
2	1	3	6	5	4	9	8	7
8	9	5	7	4	6	1	3	2
1	7	6	3	2	5	8	4	9
3	4	2	8	1	9	7	5	6

Puzzle 88

8	6	9	2	5	1	4	3	7
1	5	4	7	6	3	2	9	8
2	7	3	8	4	9	6	5	1
7	3	5	1	9	2	8	4	6
6	4	1	5	3	8	9	7	2
9	8	2	4	7	6	5	1	3
3	1	6	9	2	4	7	8	5
5	9	8	6	1	7	3	2	4
4	2	7	3	8	5	1	6	9

Puzzle 89

3	9	2	8	5	1	7	4	6
7	4	8	2	3	6	5	9	1
1	5	6	4	7	9	3	2	8
2	7	9	1	8	4	6	5	3
6	8	1	5	9	3	4	7	2
5	3	4	6	2	7	8	1	9
4	6	5	3	1	2	9	8	7
8	1	7	9	6	5	2	3	4
9	2	3	7	4	8	1	6	5

Puzzle 90

1	8	6	2	4	9	3	5	7
5	7	4	6	8	3	1	2	9
2	9	3	7	1	5	4	8	6
3	2	7	5	6	8	9	1	4
4	6	1	3	9	2	5	7	8
8	5	9	4	7	1	6	3	2
7	1	5	9	2	6	8	4	3
9	4	8	1	3	7	2	6	5
6	3	2	8	5	4	7	9	1

Puzzle 91

5	2	6	4	3	9	8	1	7
3	8	1	2	5	7	4	6	9
4	9	7	1	8	6	3	5	2
1	6	8	3	9	5	7	2	4
2	7	5	8	1	4	6	9	3
9	3	4	7	6	2	5	8	1
7	1	2	5	4	8	9	3	6
8	4	9	6	2	3	1	7	5
6	5	3	9	7	1	2	4	8

Puzzle 92

1	8	9	7	5	4	6	2	3
5	4	6	1	3	2	8	9	7
7	3	2	6	8	9	4	5	1
6	5	7	2	4	8	1	3	9
9	1	3	5	7	6	2	4	8
4	2	8	9	1	3	5	7	6
2	6	4	3	9	1	7	8	5
8	9	5	4	6	7	3	1	2
3	7	1	8	2	5	9	6	4

Puzzle 93

8	5	4	1	6	3	7	2	9
6	2	3	7	5	9	8	4	1
1	9	7	2	4	8	5	3	6
2	1	5	9	7	6	4	8	3
7	4	8	5	3	1	6	9	2
3	6	9	4	8	2	1	5	7
9	7	1	8	2	5	3	6	4
4	8	6	3	9	7	2	1	5
5	3	2	6	1	4	9	7	8

Puzzle 94

4	7	9	2	5	3	1	8	6
2	5	3	6	1	8	4	9	7
6	1	8	9	4	7	5	2	3
8	4	6	1	7	5	9	3	2
5	9	2	3	8	6	7	1	4
7	3	1	4	2	9	8	6	5
1	2	5	8	6	4	3	7	9
3	6	7	5	9	1	2	4	8
9	8	4	7	3	2	6	5	1

Puzzle 95

8	2	9	7	3	4	1	5	6
4	1	3	6	5	2	9	7	8
5	6	7	8	9	1	2	3	4
3	8	2	9	4	7	6	1	5
1	4	5	2	8	6	7	9	3
7	9	6	3	1	5	8	4	2
2	5	8	1	7	3	4	6	9
6	7	4	5	2	9	3	8	1
9	3	1	4	6	8	5	2	7

Puzzle 96

7	2	6	8	4	5	9	3	1
9	3	8	7	2	1	5	6	4
5	4	1	9	3	6	8	2	7
8	5	9	6	7	3	1	4	2
4	7	3	5	1	2	6	9	8
6	1	2	4	9	8	7	5	3
1	8	5	2	6	4	3	7	9
2	6	7	3	8	9	4	1	5
3	9	4	1	5	7	2	8	6

Puzzle 97

6	1	7	3	4	8	2	5	9
9	8	5	6	1	2	4	3	7
4	2	3	7	9	5	8	6	1
5	4	6	1	3	7	9	8	2
3	7	1	2	8	9	5	4	6
2	9	8	4	5	6	1	7	3
1	5	4	9	7	3	6	2	8
7	6	9	8	2	4	3	1	5
8	3	2	5	6	1	7	9	4

Puzzle 98

3	7	5	4	8	2	9	6	1
8	9	2	6	5	1	3	7	4
6	4	1	7	3	9	5	2	8
9	6	4	2	7	8	1	5	3
7	1	8	3	6	5	4	9	2
5	2	3	9	1	4	6	8	7
1	3	9	8	2	6	7	4	5
2	5	6	1	4	7	8	3	9
4	8	7	5	9	3	2	1	6

Puzzle 99

4	5	2	1	8	3	6	9	7
8	7	6	5	4	9	2	3	1
9	3	1	6	2	7	5	8	4
5	8	3	7	6	2	4	1	9
2	6	7	4	9	1	3	5	8
1	4	9	8	3	5	7	6	2
7	1	4	9	5	6	8	2	3
6	2	8	3	1	4	9	7	5
3	9	5	2	7	8	1	4	6

Puzzle 100

6	5	9	4	2	8	1	7	3
3	8	4	6	1	7	2	5	9
7	1	2	5	9	3	6	4	8
5	3	8	7	6	1	4	9	2
1	4	7	9	3	2	8	6	5
9	2	6	8	5	4	3	1	7
2	9	3	1	7	6	5	8	4
8	7	1	3	4	5	9	2	6
4	6	5	2	8	9	7	3	1

Puzzle 101

7	6	1	5	8	4	3	9	2
2	9	5	1	3	6	4	7	8
4	3	8	7	2	9	6	5	1
8	2	6	9	4	3	7	1	5
9	4	7	8	5	1	2	3	6
5	1	3	2	6	7	9	8	4
1	5	9	4	7	2	8	6	3
3	8	2	6	9	5	1	4	7
6	7	4	3	1	8	5	2	9

Puzzle 102

9	5	4	8	3	2	6	7	1
6	8	2	1	4	7	9	5	3
1	7	3	5	6	9	2	4	8
7	4	9	2	8	5	1	3	6
2	1	5	6	9	3	4	8	7
3	6	8	7	1	4	5	9	2
4	2	1	9	7	8	3	6	5
5	9	7	3	2	6	8	1	4
8	3	6	4	5	1	7	2	9

Puzzle 103

2	1	7	4	8	5	6	3	9
8	6	9	1	3	7	4	5	2
5	4	3	6	2	9	7	8	1
1	8	5	2	4	3	9	7	6
3	9	2	5	7	6	1	4	8
4	7	6	9	1	8	3	2	5
9	2	8	7	6	4	5	1	3
6	3	4	8	5	1	2	9	7
7	5	1	3	9	2	8	6	4

Puzzle 104

9	4	8	5	2	1	7	6	3
1	5	7	3	8	6	2	4	9
2	3	6	9	4	7	1	5	8
3	9	2	6	5	4	8	1	7
8	1	5	2	7	3	4	9	6
6	7	4	8	1	9	5	3	2
5	2	9	4	3	8	6	7	1
4	6	1	7	9	2	3	8	5
7	8	3	1	6	5	9	2	4

Puzzle 105

9	3	7	4	2	6	8	1	5
4	5	8	9	7	1	6	2	3
2	6	1	8	5	3	7	4	9
5	4	3	6	1	7	9	8	2
7	1	2	5	9	8	3	6	4
8	9	6	2	3	4	5	7	1
1	7	4	3	8	5	2	9	6
6	2	5	7	4	9	1	3	8
3	8	9	1	6	2	4	5	7

Puzzle 106

3	7	8	2	1	6	4	5	9
2	5	1	9	3	4	7	6	8
6	9	4	5	8	7	1	3	2
8	1	5	3	9	2	6	7	4
4	2	6	7	5	1	9	8	3
9	3	7	6	4	8	5	2	1
7	6	3	1	2	9	8	4	5
5	8	9	4	7	3	2	1	6
1	4	2	8	6	5	3	9	7

Puzzle 107

6	3	8	5	4	7	1	9	2
4	2	7	1	8	9	6	5	3
5	9	1	6	3	2	7	8	4
9	6	4	2	5	3	8	7	1
7	1	3	8	6	4	9	2	5
8	5	2	9	7	1	4	3	6
1	7	9	4	2	5	3	6	8
3	8	5	7	1	6	2	4	9
2	4	6	3	9	8	5	1	7

Puzzle 108

4	9	6	3	7	5	8	1	2
7	5	8	2	1	4	3	6	9
2	3	1	9	8	6	5	4	7
3	2	4	7	9	8	6	5	1
1	7	9	5	6	3	4	2	8
6	8	5	4	2	1	9	7	3
8	4	7	6	3	2	1	9	5
5	1	2	8	4	9	7	3	6
9	6	3	1	5	7	2	8	4

Puzzle 109

8	2	6	5	9	4	1	3	7
3	1	4	8	7	2	6	9	5
7	9	5	1	6	3	2	8	4
4	6	1	9	8	5	7	2	3
2	5	7	4	3	1	9	6	8
9	8	3	6	2	7	4	5	1
1	4	2	3	5	6	8	7	9
6	3	9	7	4	8	5	1	2
5	7	8	2	1	9	3	4	6

Puzzle 110

6	8	4	3	7	1	5	9	2
2	3	9	4	5	8	6	7	1
1	5	7	6	9	2	3	8	4
4	1	3	8	2	7	9	5	6
9	7	8	5	4	6	2	1	3
5	2	6	1	3	9	7	4	8
7	9	1	2	6	4	8	3	5
8	6	5	7	1	3	4	2	9
3	4	2	9	8	5	1	6	7

Puzzle 111

3	4	7	5	2	9	1	8	6
1	9	5	6	7	8	2	4	3
8	6	2	1	4	3	5	7	9
2	3	1	4	8	5	6	9	7
4	8	6	2	9	7	3	5	1
7	5	9	3	1	6	8	2	4
5	1	4	9	6	2	7	3	8
9	7	3	8	5	1	4	6	2
6	2	8	7	3	4	9	1	5

Puzzle 112

7	1	2	9	4	5	3	8	6
5	9	4	6	3	8	1	7	2
3	6	8	7	1	2	9	4	5
6	4	5	8	7	1	2	9	3
1	3	7	5	2	9	4	6	8
8	2	9	3	6	4	7	5	1
4	8	3	1	5	7	6	2	9
2	5	1	4	9	6	8	3	7
9	7	6	2	8	3	5	1	4

Puzzle 113

2	7	4	1	9	3	5	6	8
9	3	5	7	8	6	4	2	1
6	8	1	5	4	2	9	7	3
3	9	2	6	1	8	7	4	5
1	6	8	4	5	7	3	9	2
5	4	7	2	3	9	8	1	6
8	5	6	9	7	1	2	3	4
7	2	3	8	6	4	1	5	9
4	1	9	3	2	5	6	8	7

Puzzle 114

5	4	6	7	9	3	8	1	2
2	3	7	4	8	1	5	6	9
1	8	9	5	6	2	7	4	3
8	7	1	3	4	5	2	9	6
6	2	4	1	7	9	3	5	8
3	9	5	6	2	8	1	7	4
9	1	3	2	5	4	6	8	7
4	6	2	8	1	7	9	3	5
7	5	8	9	3	6	4	2	1

Puzzle 115

3	7	8	4	5	9	2	6	1
6	2	5	1	7	3	9	4	8
1	9	4	6	8	2	7	5	3
5	1	3	2	9	6	8	7	4
4	8	9	7	3	1	6	2	5
7	6	2	5	4	8	1	3	9
8	5	7	9	2	4	3	1	6
2	3	1	8	6	5	4	9	7
9	4	6	3	1	7	5	8	2

Puzzle 116

1	8	9	4	5	6	2	7	3
5	3	6	9	2	7	1	4	8
4	7	2	8	1	3	9	6	5
6	4	5	2	9	8	7	3	1
9	1	7	5	3	4	6	8	2
8	2	3	6	7	1	5	9	4
3	5	4	1	6	9	8	2	7
2	6	8	7	4	5	3	1	9
7	9	1	3	8	2	4	5	6

Puzzle 117

4	7	6	2	5	3	9	1	8
1	5	8	9	4	6	3	2	7
9	3	2	8	1	7	6	4	5
3	6	7	5	8	4	2	9	1
8	2	9	3	7	1	4	5	6
5	1	4	6	2	9	8	7	3
2	8	1	4	6	5	7	3	9
6	9	5	7	3	2	1	8	4
7	4	3	1	9	8	5	6	2

Puzzle 118

4	9	5	7	1	2	8	3	6
7	1	8	3	9	6	5	4	2
3	2	6	4	8	5	7	9	1
6	3	7	2	5	1	4	8	9
2	4	1	8	7	9	3	6	5
5	8	9	6	4	3	1	2	7
8	7	2	5	6	4	9	1	3
9	5	3	1	2	8	6	7	4
1	6	4	9	3	7	2	5	8

Puzzle 119

3	8	1	2	7	9	6	4	5
7	6	4	1	3	5	9	8	2
5	9	2	8	6	4	3	1	7
8	7	5	4	1	6	2	3	9
9	2	3	7	5	8	1	6	4
1	4	6	3	9	2	5	7	8
2	3	7	5	8	1	4	9	6
4	1	9	6	2	7	8	5	3
6	5	8	9	4	3	7	2	1

Puzzle 120

8	6	7	2	1	9	5	3	4
1	4	5	8	3	7	9	2	6
9	3	2	5	6	4	7	8	1
2	8	1	7	4	3	6	5	9
7	5	4	6	9	2	3	1	8
3	9	6	1	5	8	4	7	2
6	7	3	9	2	1	8	4	5
4	2	9	3	8	5	1	6	7
5	1	8	4	7	6	2	9	3

Puzzle 121

9	4	2	5	6	7	1	3	8
3	8	1	4	2	9	6	7	5
7	6	5	3	1	8	2	9	4
8	5	9	7	3	1	4	6	2
4	2	3	6	9	5	8	1	7
1	7	6	8	4	2	9	5	3
6	3	8	9	5	4	7	2	1
2	9	4	1	7	3	5	8	6
5	1	7	2	8	6	3	4	9

Puzzle 122

2	3	5	6	8	4	9	1	7
1	9	4	3	2	7	8	5	6
7	8	6	1	9	5	3	4	2
6	7	1	5	4	9	2	8	3
3	5	9	2	1	8	6	7	4
8	4	2	7	3	6	1	9	5
5	6	8	9	7	2	4	3	1
4	2	3	8	5	1	7	6	9
9	1	7	4	6	3	5	2	8

Puzzle 123

7	2	8	6	4	3	1	9	5
6	5	1	9	8	2	7	4	3
3	4	9	7	1	5	2	6	8
5	7	6	1	3	9	4	8	2
9	8	4	2	5	6	3	1	7
2	1	3	8	7	4	9	5	6
4	9	2	3	6	8	5	7	1
8	3	7	5	9	1	6	2	4
1	6	5	4	2	7	8	3	9

Puzzle 124

4	5	2	3	1	6	9	7	8
8	9	3	5	2	7	4	1	6
1	6	7	4	9	8	3	5	2
6	7	5	8	4	3	1	2	9
3	1	9	6	7	2	8	4	5
2	8	4	9	5	1	7	6	3
9	3	1	2	6	4	5	8	7
5	4	6	7	8	9	2	3	1
7	2	8	1	3	5	6	9	4

Puzzle 125

7	9	3	6	8	4	2	1	5
8	4	1	9	2	5	3	7	6
5	2	6	7	3	1	4	8	9
9	6	2	3	5	8	7	4	1
3	5	8	1	4	7	6	9	2
1	7	4	2	9	6	8	5	3
4	3	7	5	1	2	9	6	8
6	1	9	8	7	3	5	2	4
2	8	5	4	6	9	1	3	7

Puzzle 126

5	3	6	2	1	9	7	8	4
9	4	2	5	8	7	6	1	3
1	7	8	4	6	3	9	2	5
6	1	7	8	5	4	2	3	9
3	9	4	1	7	2	8	5	6
8	2	5	3	9	6	1	4	7
2	5	9	7	3	8	4	6	1
4	6	1	9	2	5	3	7	8
7	8	3	6	4	1	5	9	2

Puzzle 127

8	9	3	1	6	4	2	7	5
7	6	5	3	9	2	4	1	8
1	4	2	8	5	7	6	9	3
6	1	9	4	7	3	8	5	2
2	5	4	9	8	6	7	3	1
3	8	7	5	2	1	9	4	6
4	7	6	2	3	5	1	8	9
5	2	8	7	1	9	3	6	4
9	3	1	6	4	8	5	2	7

Puzzle 128

3	6	9	7	4	8	1	5	2
7	2	5	3	1	6	8	4	9
1	4	8	9	5	2	6	7	3
9	1	6	4	7	3	2	8	5
2	5	4	1	8	9	3	6	7
8	7	3	2	6	5	9	1	4
4	3	1	8	9	7	5	2	6
5	8	2	6	3	4	7	9	1
6	9	7	5	2	1	4	3	8

Puzzle 129

2	9	5	6	1	8	3	4	7
1	6	3	9	7	4	5	2	8
7	4	8	5	2	3	1	6	9
9	2	4	3	6	1	8	7	5
6	8	1	2	5	7	9	3	4
3	5	7	8	4	9	6	1	2
8	1	6	7	9	2	4	5	3
5	3	2	4	8	6	7	9	1
4	7	9	1	3	5	2	8	6

Puzzle 130

3	5	8	1	4	2	9	7	6
9	2	6	7	5	3	4	8	1
7	4	1	9	8	6	2	3	5
8	1	7	2	9	4	5	6	3
6	3	2	8	1	5	7	4	9
5	9	4	3	6	7	8	1	2
4	6	3	5	2	8	1	9	7
1	7	5	4	3	9	6	2	8
2	8	9	6	7	1	3	5	4

Puzzle 131

7	4	1	8	5	3	9	6	2
8	2	3	1	9	6	7	5	4
5	9	6	7	2	4	1	8	3
4	5	2	9	8	7	3	1	6
1	8	9	3	6	2	4	7	5
6	3	7	5	4	1	2	9	8
3	6	5	4	1	9	8	2	7
2	1	4	6	7	8	5	3	9
9	7	8	2	3	5	6	4	1

Puzzle 132

7	8	1	2	3	6	9	4	5
9	2	6	7	4	5	1	8	3
4	5	3	8	9	1	7	6	2
3	1	2	9	7	4	8	5	6
8	6	7	3	5	2	4	9	1
5	4	9	6	1	8	2	3	7
2	7	5	4	6	9	3	1	8
1	3	4	5	8	7	6	2	9
6	9	8	1	2	3	5	7	4

Puzzle 133

2	9	5	7	6	3	4	8	1
1	7	8	9	2	4	5	3	6
3	6	4	8	1	5	7	2	9
8	2	7	1	9	6	3	4	5
5	3	6	2	4	7	1	9	8
4	1	9	5	3	8	2	6	7
7	5	2	4	8	9	6	1	3
9	4	3	6	5	1	8	7	2
6	8	1	3	7	2	9	5	4

Puzzle 134

6	3	8	9	4	1	7	5	2
7	4	5	6	8	2	3	1	9
2	1	9	3	5	7	8	4	6
1	2	6	5	3	9	4	8	7
8	7	3	2	6	4	5	9	1
5	9	4	1	7	8	2	6	3
9	5	2	4	1	3	6	7	8
4	8	1	7	2	6	9	3	5
3	6	7	8	9	5	1	2	4

Puzzle 135

2	1	9	3	4	7	6	5	8
5	4	3	2	6	8	9	7	1
6	7	8	5	1	9	3	2	4
1	6	5	4	2	3	8	9	7
7	9	4	8	5	6	2	1	3
3	8	2	7	9	1	4	6	5
9	3	6	1	7	4	5	8	2
8	2	7	9	3	5	1	4	6
4	5	1	6	8	2	7	3	9

Puzzle 136

4	7	3	5	6	9	8	2	1
8	1	9	3	4	2	7	5	6
6	5	2	8	7	1	4	9	3
5	3	7	4	1	8	9	6	2
9	6	1	2	5	7	3	4	8
2	4	8	9	3	6	5	1	7
1	8	4	7	2	5	6	3	9
3	9	6	1	8	4	2	7	5
7	2	5	6	9	3	1	8	4

Puzzle 137

1	8	2	6	7	5	4	9	3
9	7	3	8	4	2	1	5	6
6	5	4	1	9	3	2	7	8
3	6	9	4	5	7	8	2	1
7	2	5	3	1	8	6	4	9
4	1	8	9	2	6	7	3	5
8	3	7	2	6	9	5	1	4
2	9	1	5	8	4	3	6	7
5	4	6	7	3	1	9	8	2

Puzzle 138

1	9	2	4	7	3	6	8	5
7	4	3	5	6	8	2	9	1
5	8	6	1	9	2	4	7	3
8	2	9	6	5	1	3	4	7
3	5	7	9	2	4	8	1	6
6	1	4	8	3	7	9	5	2
4	6	1	3	8	5	7	2	9
9	7	8	2	1	6	5	3	4
2	3	5	7	4	9	1	6	8

Puzzle 139

3	4	2	6	9	1	7	5	8
7	1	9	3	5	8	6	4	2
6	5	8	7	2	4	3	9	1
1	9	4	8	6	7	2	3	5
5	2	7	9	4	3	1	8	6
8	3	6	5	1	2	9	7	4
9	8	1	4	3	6	5	2	7
2	7	5	1	8	9	4	6	3
4	6	3	2	7	5	8	1	9

Puzzle 140

6	2	7	9	5	8	1	4	3
9	4	3	7	2	1	6	8	5
1	5	8	6	4	3	7	2	9
2	7	9	8	1	6	5	3	4
5	8	1	3	9	4	2	7	6
4	3	6	5	7	2	8	9	1
8	1	2	4	6	9	3	5	7
3	9	5	1	8	7	4	6	2
7	6	4	2	3	5	9	1	8

Puzzle 141

1	5	8	2	3	9	7	4	6
6	2	4	7	8	5	3	9	1
9	3	7	4	1	6	5	8	2
5	9	3	6	2	8	4	1	7
7	4	6	9	5	1	8	2	3
8	1	2	3	4	7	6	5	9
2	7	1	5	6	4	9	3	8
4	8	9	1	7	3	2	6	5
3	6	5	8	9	2	1	7	4

Puzzle 142

7	5	9	2	4	6	1	8	3
3	8	4	5	1	9	7	2	6
6	1	2	8	7	3	4	5	9
9	7	6	1	8	5	3	4	2
8	3	1	6	2	4	5	9	7
4	2	5	9	3	7	6	1	8
5	4	8	7	6	2	9	3	1
2	9	7	3	5	1	8	6	4
1	6	3	4	9	8	2	7	5

Puzzle 143

1	6	2	3	5	7	4	9	8
9	4	8	1	6	2	5	3	7
3	5	7	4	8	9	2	6	1
5	9	1	2	3	4	7	8	6
8	3	6	7	9	5	1	2	4
2	7	4	8	1	6	3	5	9
6	1	5	9	4	3	8	7	2
7	8	3	6	2	1	9	4	5
4	2	9	5	7	8	6	1	3

Puzzle 144

5	8	2	1	7	9	6	3	4
6	1	7	5	3	4	9	8	2
9	4	3	2	8	6	7	1	5
7	3	1	6	9	2	5	4	8
8	2	5	7	4	1	3	9	6
4	9	6	8	5	3	1	2	7
3	6	8	4	1	7	2	5	9
1	7	4	9	2	5	8	6	3
2	5	9	3	6	8	4	7	1

Puzzle 145

4	5	6	3	8	2	7	1	9
2	3	8	9	7	1	4	5	6
9	1	7	5	4	6	8	2	3
3	8	4	7	2	9	1	6	5
1	6	2	4	5	8	9	3	7
5	7	9	6	1	3	2	8	4
6	2	5	1	9	4	3	7	8
7	4	1	8	3	5	6	9	2
8	9	3	2	6	7	5	4	1

Puzzle 146

2	3	6	4	1	9	5	8	7
1	5	7	3	8	2	9	6	4
9	8	4	5	6	7	2	3	1
6	7	5	8	3	1	4	2	9
3	1	8	9	2	4	7	5	6
4	2	9	7	5	6	8	1	3
8	9	3	1	7	5	6	4	2
5	4	2	6	9	3	1	7	8
7	6	1	2	4	8	3	9	5

Puzzle 147

4	5	6	9	8	2	1	7	3
8	7	9	1	5	3	6	2	4
3	2	1	7	4	6	8	9	5
2	9	7	5	1	8	3	4	6
6	4	5	3	7	9	2	1	8
1	8	3	6	2	4	7	5	9
5	1	4	8	3	7	9	6	2
9	3	2	4	6	1	5	8	7
7	6	8	2	9	5	4	3	1

Puzzle 148

3	4	5	1	8	9	2	6	7
6	7	9	2	3	4	8	5	1
1	2	8	5	6	7	9	4	3
4	9	6	3	7	8	1	2	5
5	1	3	6	9	2	4	7	8
2	8	7	4	5	1	3	9	6
7	5	1	9	4	3	6	8	2
8	3	4	7	2	6	5	1	9
9	6	2	8	1	5	7	3	4

Puzzle 149

5	4	9	3	8	6	7	2	1
8	6	7	1	4	2	9	3	5
3	1	2	5	7	9	6	4	8
9	2	8	4	1	5	3	7	6
7	3	6	9	2	8	1	5	4
1	5	4	6	3	7	2	8	9
4	7	5	2	9	1	8	6	3
6	8	1	7	5	3	4	9	2
2	9	3	8	6	4	5	1	7

Puzzle 150

2	9	4	6	8	3	5	7	1
6	8	7	9	5	1	2	3	4
5	1	3	4	7	2	9	6	8
3	2	9	5	1	7	4	8	6
8	4	5	3	9	6	7	1	2
7	6	1	2	4	8	3	9	5
9	5	8	7	6	4	1	2	3
4	3	6	1	2	9	8	5	7
1	7	2	8	3	5	6	4	9

Puzzle 151

1	3	4	7	9	2	6	8	5
8	9	2	6	5	1	4	3	7
6	5	7	3	4	8	2	1	9
9	6	3	4	7	5	1	2	8
4	8	1	2	6	9	5	7	3
2	7	5	8	1	3	9	6	4
7	4	9	1	8	6	3	5	2
3	1	8	5	2	4	7	9	6
5	2	6	9	3	7	8	4	1

Puzzle 152

9	8	4	7	5	3	2	6	1
6	7	3	2	4	1	5	8	9
2	1	5	6	8	9	7	3	4
5	6	1	8	7	4	3	9	2
3	4	2	5	9	6	1	7	8
8	9	7	1	3	2	4	5	6
7	5	6	4	2	8	9	1	3
4	3	8	9	1	7	6	2	5
1	2	9	3	6	5	8	4	7

Puzzle 153

1	2	7	8	4	6	5	9	3
6	8	3	2	5	9	1	4	7
4	9	5	1	3	7	2	6	8
8	5	2	7	6	3	4	1	9
3	7	6	9	1	4	8	2	5
9	1	4	5	8	2	7	3	6
2	4	9	3	7	5	6	8	1
7	6	1	4	9	8	3	5	2
5	3	8	6	2	1	9	7	4

Puzzle 154

8	4	2	5	3	6	1	9	7
3	6	9	4	7	1	5	2	8
1	7	5	8	9	2	3	4	6
7	8	6	9	2	3	4	5	1
4	9	3	6	1	5	8	7	2
5	2	1	7	4	8	6	3	9
9	1	8	2	5	4	7	6	3
6	5	7	3	8	9	2	1	4
2	3	4	1	6	7	9	8	5

Puzzle 155

4	5	2	9	8	1	3	6	7
1	9	6	4	3	7	2	8	5
3	8	7	5	6	2	4	9	1
8	6	4	2	9	5	7	1	3
5	1	9	6	7	3	8	2	4
2	7	3	1	4	8	6	5	9
7	3	5	8	2	9	1	4	6
9	4	8	3	1	6	5	7	2
6	2	1	7	5	4	9	3	8

Puzzle 156

5	3	6	4	9	7	2	1	8
8	9	4	6	2	1	3	7	5
7	2	1	3	5	8	4	6	9
1	8	9	7	4	6	5	2	3
2	6	5	1	3	9	8	4	7
4	7	3	2	8	5	6	9	1
3	4	8	9	1	2	7	5	6
6	1	2	5	7	3	9	8	4
9	5	7	8	6	4	1	3	2

Puzzle 157

8	9	6	3	1	7	4	5	2
5	1	4	6	8	2	7	9	3
2	7	3	9	4	5	8	6	1
4	5	7	1	6	3	9	2	8
1	3	9	7	2	8	5	4	6
6	2	8	4	5	9	1	3	7
3	4	1	8	9	6	2	7	5
9	6	5	2	7	1	3	8	4
7	8	2	5	3	4	6	1	9

Puzzle 158

8	5	6	2	4	9	1	7	3
1	3	2	5	7	6	9	4	8
4	9	7	1	8	3	5	6	2
7	4	1	3	2	8	6	9	5
5	2	8	9	6	4	7	3	1
3	6	9	7	1	5	8	2	4
9	8	3	6	5	2	4	1	7
6	7	5	4	3	1	2	8	9
2	1	4	8	9	7	3	5	6

Puzzle 159

7	3	5	9	6	4	1	8	2
8	9	1	2	5	3	6	7	4
2	4	6	8	1	7	9	3	5
3	6	9	7	2	1	4	5	8
5	2	8	6	4	9	7	1	3
1	7	4	3	8	5	2	9	6
4	8	7	1	3	2	5	6	9
6	1	2	5	9	8	3	4	7
9	5	3	4	7	6	8	2	1

Puzzle 160

6	9	2	1	5	7	4	8	3
5	4	1	6	3	8	2	7	9
3	7	8	4	9	2	1	6	5
8	6	4	2	1	5	9	3	7
1	3	5	7	4	9	8	2	6
9	2	7	3	8	6	5	1	4
7	5	9	8	6	1	3	4	2
2	1	3	5	7	4	6	9	8
4	8	6	9	2	3	7	5	1

Puzzle 161

9	4	3	5	6	7	1	8	2
5	7	8	3	2	1	6	4	9
2	1	6	4	8	9	5	7	3
1	3	2	8	9	6	4	5	7
4	9	5	1	7	3	8	2	6
6	8	7	2	5	4	9	3	1
7	2	9	6	4	5	3	1	8
3	6	4	7	1	8	2	9	5
8	5	1	9	3	2	7	6	4

Puzzle 162

9	7	1	2	6	5	3	4	8
2	8	3	9	1	4	6	5	7
5	4	6	8	3	7	2	1	9
4	9	8	1	5	6	7	2	3
6	1	7	3	8	2	4	9	5
3	5	2	7	4	9	1	8	6
7	2	5	4	9	3	8	6	1
8	3	9	6	2	1	5	7	4
1	6	4	5	7	8	9	3	2

Puzzle 163

3	8	2	9	6	5	4	7	1
4	9	7	3	8	1	6	5	2
5	6	1	7	4	2	8	9	3
9	7	8	5	1	4	3	2	6
2	1	3	8	9	6	7	4	5
6	5	4	2	3	7	1	8	9
7	4	9	1	2	3	5	6	8
1	2	5	6	7	8	9	3	4
8	3	6	4	5	9	2	1	7

Puzzle 164

8	6	7	5	1	4	3	9	2
5	1	4	3	2	9	8	7	6
9	2	3	6	8	7	5	4	1
3	5	1	7	6	8	4	2	9
4	8	6	2	9	1	7	3	5
2	7	9	4	3	5	1	6	8
1	9	2	8	7	3	6	5	4
7	4	8	9	5	6	2	1	3
6	3	5	1	4	2	9	8	7

Puzzle 165

3	6	2	9	7	4	1	5	8
8	7	4	1	3	5	9	6	2
5	9	1	8	6	2	3	7	4
7	2	3	4	9	8	5	1	6
6	5	8	2	1	3	4	9	7
4	1	9	6	5	7	2	8	3
9	4	6	7	2	1	8	3	5
2	3	7	5	8	9	6	4	1
1	8	5	3	4	6	7	2	9

Puzzle 166

1	2	3	7	9	8	6	5	4
7	5	8	6	3	4	9	1	2
9	4	6	2	1	5	8	3	7
3	8	4	1	5	7	2	6	9
2	6	9	8	4	3	5	7	1
5	1	7	9	6	2	3	4	8
4	7	2	5	8	6	1	9	3
6	3	1	4	2	9	7	8	5
8	9	5	3	7	1	4	2	6

Puzzle 167

7	2	1	6	5	3	4	8	9
8	3	4	9	7	1	6	5	2
5	6	9	2	8	4	7	1	3
6	8	2	4	1	7	9	3	5
4	1	5	8	3	9	2	6	7
9	7	3	5	6	2	1	4	8
3	4	6	7	9	8	5	2	1
1	5	7	3	2	6	8	9	4
2	9	8	1	4	5	3	7	6

Puzzle 168

5	4	9	7	1	8	2	6	3
3	2	1	6	4	5	8	7	9
8	6	7	2	9	3	5	4	1
1	3	6	4	8	2	7	9	5
9	5	8	3	7	6	4	1	2
2	7	4	1	5	9	3	8	6
4	9	3	8	2	1	6	5	7
7	1	2	5	6	4	9	3	8
6	8	5	9	3	7	1	2	4

Puzzle 169

4	5	7	9	2	1	6	8	3
3	8	1	7	5	6	4	2	9
2	9	6	3	4	8	5	1	7
8	7	4	2	6	5	9	3	1
5	3	2	1	8	9	7	6	4
6	1	9	4	3	7	8	5	2
9	2	5	8	7	3	1	4	6
7	6	3	5	1	4	2	9	8
1	4	8	6	9	2	3	7	5

Puzzle 170

4	1	9	6	8	3	5	2	7
5	6	7	2	1	9	8	3	4
3	8	2	7	4	5	6	1	9
8	3	6	5	2	7	4	9	1
2	4	1	3	9	8	7	6	5
7	9	5	1	6	4	2	8	3
1	7	3	8	5	6	9	4	2
6	5	4	9	3	2	1	7	8
9	2	8	4	7	1	3	5	6

Puzzle 171

1	9	3	6	2	8	7	5	4
8	4	2	3	7	5	6	9	1
7	6	5	4	1	9	2	8	3
9	8	6	2	4	3	5	1	7
3	5	7	1	9	6	4	2	8
4	2	1	5	8	7	9	3	6
5	7	4	8	3	2	1	6	9
2	3	9	7	6	1	8	4	5
6	1	8	9	5	4	3	7	2

Puzzle 172

2	4	1	7	6	3	9	8	5
3	7	6	5	8	9	2	4	1
5	8	9	2	4	1	3	7	6
7	6	5	3	1	4	8	9	2
1	2	4	8	9	7	5	6	3
9	3	8	6	2	5	7	1	4
8	9	3	1	5	6	4	2	7
4	1	7	9	3	2	6	5	8
6	5	2	4	7	8	1	3	9

Puzzle 173

5	1	2	9	3	6	7	4	8
4	9	6	7	1	8	5	3	2
3	8	7	2	4	5	1	9	6
2	3	1	4	5	9	6	8	7
9	7	5	8	6	1	4	2	3
6	4	8	3	7	2	9	1	5
1	2	3	5	9	7	8	6	4
8	5	9	6	2	4	3	7	1
7	6	4	1	8	3	2	5	9

Puzzle 174

6	9	1	8	3	2	5	4	7
7	8	3	5	6	4	9	2	1
2	4	5	7	1	9	6	3	8
4	5	8	1	2	3	7	6	9
9	3	2	6	7	5	8	1	4
1	6	7	4	9	8	3	5	2
3	2	4	9	5	7	1	8	6
8	7	6	3	4	1	2	9	5
5	1	9	2	8	6	4	7	3

Puzzle 175

3	8	9	4	6	2	1	5	7
2	7	4	9	1	5	8	3	6
5	1	6	7	8	3	4	9	2
6	5	2	3	7	4	9	8	1
7	3	1	6	9	8	2	4	5
9	4	8	5	2	1	7	6	3
4	2	7	8	5	6	3	1	9
1	6	3	2	4	9	5	7	8
8	9	5	1	3	7	6	2	4

Puzzle 176

3	2	7	5	9	8	1	6	4
6	1	9	3	2	4	5	8	7
4	8	5	6	1	7	2	9	3
9	6	4	1	7	2	8	3	5
2	5	3	8	6	9	4	7	1
8	7	1	4	3	5	6	2	9
5	9	6	7	8	1	3	4	2
7	4	8	2	5	3	9	1	6
1	3	2	9	4	6	7	5	8

Puzzle 177

9	7	3	6	2	4	8	5	1
2	5	6	8	1	7	4	3	9
4	1	8	3	9	5	7	2	6
7	2	9	4	5	3	6	1	8
3	8	4	2	6	1	9	7	5
5	6	1	7	8	9	3	4	2
6	9	7	1	3	2	5	8	4
8	3	2	5	4	6	1	9	7
1	4	5	9	7	8	2	6	3

Puzzle 178

8	3	5	9	6	7	1	4	2
2	1	4	3	8	5	7	9	6
9	6	7	2	4	1	8	5	3
3	8	6	4	9	2	5	1	7
5	2	1	7	3	8	4	6	9
7	4	9	5	1	6	3	2	8
1	7	2	8	5	9	6	3	4
4	5	8	6	2	3	9	7	1
6	9	3	1	7	4	2	8	5

Puzzle 179

4	9	3	7	5	6	2	8	1
6	7	1	4	2	8	5	3	9
8	2	5	9	1	3	4	6	7
9	1	8	2	3	7	6	5	4
2	3	4	5	6	1	7	9	8
5	6	7	8	9	4	3	1	2
7	5	2	6	8	9	1	4	3
3	8	6	1	4	2	9	7	5
1	4	9	3	7	5	8	2	6

Puzzle 180

2	1	3	7	5	4	8	9	6
4	7	5	8	6	9	3	2	1
6	9	8	1	3	2	5	7	4
1	8	4	3	9	7	2	6	5
5	6	7	4	2	8	1	3	9
3	2	9	5	1	6	4	8	7
9	3	2	6	4	1	7	5	8
7	4	6	2	8	5	9	1	3
8	5	1	9	7	3	6	4	2

Puzzle 181

1	2	3	5	6	9	4	8	7
5	4	8	3	2	7	9	6	1
7	9	6	1	4	8	5	2	3
6	7	4	9	5	3	2	1	8
3	8	9	7	1	2	6	4	5
2	1	5	6	8	4	3	7	9
8	3	7	2	9	6	1	5	4
9	5	2	4	7	1	8	3	6
4	6	1	8	3	5	7	9	2

Puzzle 182

5	2	6	8	1	9	3	7	4
9	3	7	4	5	2	1	8	6
4	8	1	7	6	3	2	5	9
2	6	4	5	7	1	8	9	3
7	9	8	6	3	4	5	1	2
1	5	3	9	2	8	4	6	7
3	1	9	2	8	7	6	4	5
8	7	5	3	4	6	9	2	1
6	4	2	1	9	5	7	3	8

Puzzle 183

1	7	5	8	9	2	4	3	6
8	9	6	7	4	3	2	1	5
3	2	4	1	5	6	9	7	8
9	4	1	3	8	5	6	2	7
6	3	2	4	1	7	8	5	9
5	8	7	2	6	9	1	4	3
2	6	8	5	3	1	7	9	4
7	5	9	6	2	4	3	8	1
4	1	3	9	7	8	5	6	2

Puzzle 184

4	3	1	9	2	6	5	7	8
7	8	5	1	3	4	6	9	2
6	9	2	5	8	7	1	3	4
2	5	3	7	9	8	4	6	1
9	4	6	3	1	5	8	2	7
8	1	7	4	6	2	9	5	3
3	2	9	6	4	1	7	8	5
1	7	8	2	5	9	3	4	6
5	6	4	8	7	3	2	1	9

Puzzle 185

1	4	7	6	9	2	8	3	5
9	8	5	3	4	1	6	7	2
3	2	6	5	8	7	4	9	1
8	9	3	1	7	6	5	2	4
4	6	2	8	3	5	9	1	7
5	7	1	9	2	4	3	8	6
7	5	8	2	6	9	1	4	3
2	1	9	4	5	3	7	6	8
6	3	4	7	1	8	2	5	9

Puzzle 186

1	8	4	5	2	3	6	9	7
3	5	7	1	9	6	8	2	4
2	6	9	8	4	7	5	3	1
4	3	6	7	5	8	9	1	2
5	7	2	6	1	9	3	4	8
9	1	8	4	3	2	7	6	5
8	4	5	3	6	1	2	7	9
7	2	3	9	8	4	1	5	6
6	9	1	2	7	5	4	8	3

Puzzle 187

4	8	1	3	5	9	6	2	7
6	9	7	4	2	8	5	3	1
5	3	2	6	1	7	8	9	4
1	7	5	9	3	2	4	8	6
2	4	3	7	8	6	9	1	5
8	6	9	1	4	5	2	7	3
9	5	4	8	7	1	3	6	2
7	2	6	5	9	3	1	4	8
3	1	8	2	6	4	7	5	9

Puzzle 188

9	3	1	4	8	6	5	2	7
4	6	5	3	7	2	1	8	9
2	8	7	5	9	1	6	4	3
7	5	9	2	3	8	4	1	6
8	2	6	1	4	7	3	9	5
3	1	4	9	6	5	2	7	8
5	7	2	8	1	3	9	6	4
6	4	3	7	2	9	8	5	1
1	9	8	6	5	4	7	3	2

Puzzle 189

5	3	4	2	1	7	8	9	6
7	6	2	4	9	8	3	1	5
1	8	9	5	3	6	4	2	7
6	2	8	7	5	4	9	3	1
3	1	7	8	6	9	2	5	4
4	9	5	3	2	1	6	7	8
9	7	6	1	4	3	5	8	2
8	5	3	6	7	2	1	4	9
2	4	1	9	8	5	7	6	3

Puzzle 190

8	6	5	2	3	7	1	4	9
4	7	3	8	1	9	5	6	2
9	2	1	4	6	5	3	7	8
7	9	4	3	2	1	6	8	5
3	5	2	9	8	6	7	1	4
1	8	6	7	5	4	2	9	3
6	4	9	5	7	3	8	2	1
5	1	8	6	9	2	4	3	7
2	3	7	1	4	8	9	5	6

Puzzle 191

4	9	7	3	8	5	2	6	1
6	3	1	7	4	2	5	8	9
8	2	5	1	9	6	4	3	7
3	8	4	6	2	7	9	1	5
2	1	9	5	3	8	7	4	6
5	7	6	4	1	9	3	2	8
7	4	3	8	5	1	6	9	2
9	6	8	2	7	4	1	5	3
1	5	2	9	6	3	8	7	4

Puzzle 192

7	4	9	3	6	5	1	8	2
1	2	3	9	4	8	7	5	6
5	6	8	1	2	7	4	3	9
8	5	2	7	1	3	9	6	4
4	3	7	5	9	6	2	1	8
9	1	6	4	8	2	3	7	5
6	9	4	8	3	1	5	2	7
2	7	1	6	5	4	8	9	3
3	8	5	2	7	9	6	4	1

Puzzle 193

2	6	1	7	8	4	5	9	3
5	7	4	1	3	9	2	6	8
8	9	3	5	2	6	7	4	1
4	3	7	6	1	2	8	5	9
6	1	5	4	9	8	3	7	2
9	2	8	3	5	7	4	1	6
7	5	9	2	6	3	1	8	4
1	8	2	9	4	5	6	3	7
3	4	6	8	7	1	9	2	5

Puzzle 194

3	2	6	9	1	4	8	7	5
7	1	8	2	3	5	4	6	9
5	9	4	8	6	7	2	3	1
9	4	2	7	5	3	6	1	8
8	6	5	4	9	1	3	2	7
1	7	3	6	2	8	9	5	4
6	3	1	5	8	9	7	4	2
2	8	7	1	4	6	5	9	3
4	5	9	3	7	2	1	8	6

Puzzle 195

5	3	4	1	8	7	9	2	6
9	7	1	5	2	6	8	4	3
8	6	2	9	3	4	7	1	5
4	8	7	6	1	9	5	3	2
3	2	5	7	4	8	1	6	9
1	9	6	2	5	3	4	7	8
2	1	8	4	6	5	3	9	7
7	4	3	8	9	2	6	5	1
6	5	9	3	7	1	2	8	4

Puzzle 196

6	1	8	7	3	2	5	4	9
2	4	3	1	5	9	6	7	8
5	9	7	8	4	6	2	3	1
1	5	4	3	8	7	9	2	6
7	6	9	5	2	1	3	8	4
8	3	2	6	9	4	1	5	7
4	8	6	2	1	5	7	9	3
3	7	5	9	6	8	4	1	2
9	2	1	4	7	3	8	6	5

Puzzle 197

8	3	7	4	5	6	1	9	2
5	1	4	2	9	3	7	8	6
6	2	9	1	8	7	3	4	5
7	8	5	3	2	4	6	1	9
4	6	1	9	7	8	2	5	3
2	9	3	5	6	1	8	7	4
3	5	6	8	1	9	4	2	7
9	4	8	7	3	2	5	6	1
1	7	2	6	4	5	9	3	8

Puzzle 198

1	3	8	9	4	5	2	7	6
6	2	4	7	1	3	9	5	8
5	7	9	2	8	6	3	4	1
2	5	6	1	3	7	8	9	4
4	1	3	5	9	8	6	2	7
8	9	7	4	6	2	5	1	3
9	6	5	3	7	4	1	8	2
7	8	1	6	2	9	4	3	5
3	4	2	8	5	1	7	6	9

Puzzle 199

1	8	9	6	5	3	2	4	7
4	5	2	8	1	7	6	9	3
7	6	3	2	4	9	5	1	8
3	1	8	7	9	5	4	2	6
5	7	4	3	2	6	9	8	1
9	2	6	1	8	4	3	7	5
6	4	1	5	7	2	8	3	9
2	3	7	9	6	8	1	5	4
8	9	5	4	3	1	7	6	2

Puzzle 200

9	1	6	3	8	7	4	5	2
5	4	3	1	6	2	7	9	8
8	7	2	4	5	9	3	6	1
1	2	9	6	7	3	5	8	4
6	8	5	9	4	1	2	3	7
7	3	4	5	2	8	6	1	9
3	9	7	2	1	5	8	4	6
2	6	1	8	3	4	9	7	5
4	5	8	7	9	6	1	2	3

Puzzle 201

3	2	4	9	5	1	7	6	8
8	7	6	4	2	3	5	1	9
9	5	1	7	6	8	4	2	3
7	4	9	6	3	2	1	8	5
1	6	3	5	8	7	9	4	2
5	8	2	1	9	4	3	7	6
4	3	5	2	1	6	8	9	7
2	1	8	3	7	9	6	5	4
6	9	7	8	4	5	2	3	1

Puzzle 202

5	4	9	1	2	6	7	3	8
6	8	1	7	4	3	2	5	9
3	7	2	5	9	8	4	1	6
4	9	7	2	6	5	1	8	3
2	5	8	4	3	1	9	6	7
1	3	6	9	8	7	5	2	4
9	2	3	8	1	4	6	7	5
8	1	5	6	7	9	3	4	2
7	6	4	3	5	2	8	9	1

Puzzle 203

5	4	9	1	3	2	6	8	7
7	1	3	5	6	8	9	4	2
2	6	8	4	7	9	5	3	1
8	5	2	6	9	1	4	7	3
9	7	1	3	8	4	2	5	6
6	3	4	7	2	5	8	1	9
1	2	7	8	4	6	3	9	5
4	9	5	2	1	3	7	6	8
3	8	6	9	5	7	1	2	4

Puzzle 204

1	7	2	3	9	5	6	4	8
5	3	9	8	6	4	1	2	7
4	8	6	7	1	2	9	5	3
2	9	4	6	7	3	5	8	1
8	1	7	5	4	9	2	3	6
3	6	5	1	2	8	7	9	4
6	4	3	9	5	1	8	7	2
9	2	1	4	8	7	3	6	5
7	5	8	2	3	6	4	1	9

Puzzle 205

6	9	7	1	4	2	8	3	5
3	4	8	5	9	7	1	2	6
2	1	5	8	3	6	4	9	7
9	7	2	3	6	8	5	1	4
8	5	6	2	1	4	3	7	9
4	3	1	9	7	5	6	8	2
7	8	3	6	5	9	2	4	1
5	2	4	7	8	1	9	6	3
1	6	9	4	2	3	7	5	8

Puzzle 206

3	4	7	1	9	2	8	5	6
9	8	5	6	3	7	1	2	4
6	1	2	8	4	5	3	7	9
5	2	1	7	8	9	4	6	3
4	9	8	3	5	6	7	1	2
7	3	6	2	1	4	5	9	8
2	6	3	4	7	1	9	8	5
1	5	4	9	6	8	2	3	7
8	7	9	5	2	3	6	4	1

Puzzle 207

9	2	7	8	1	4	3	6	5
3	1	6	9	5	2	4	8	7
8	4	5	6	3	7	2	9	1
2	9	3	7	4	1	6	5	8
4	6	1	5	8	3	9	7	2
5	7	8	2	6	9	1	4	3
6	8	4	3	2	5	7	1	9
1	3	9	4	7	8	5	2	6
7	5	2	1	9	6	8	3	4

Puzzle 208

7	6	8	5	3	1	2	9	4
9	1	2	6	4	7	5	3	8
4	5	3	2	9	8	7	1	6
2	7	9	8	5	3	4	6	1
6	4	5	7	1	9	3	8	2
8	3	1	4	2	6	9	7	5
3	2	6	9	8	5	1	4	7
5	9	7	1	6	4	8	2	3
1	8	4	3	7	2	6	5	9

Puzzle 209

4	7	2	1	3	6	9	5	8
9	8	6	4	5	2	3	1	7
3	1	5	9	7	8	6	2	4
1	3	4	2	8	5	7	9	6
8	6	9	7	1	3	2	4	5
5	2	7	6	9	4	8	3	1
7	4	3	5	6	9	1	8	2
6	5	8	3	2	1	4	7	9
2	9	1	8	4	7	5	6	3

Puzzle 210

6	4	9	8	1	7	2	3	5
1	3	8	6	5	2	4	7	9
7	5	2	9	3	4	1	8	6
4	1	7	2	9	6	8	5	3
2	8	5	4	7	3	6	9	1
9	6	3	1	8	5	7	2	4
8	2	4	5	6	9	3	1	7
5	7	1	3	4	8	9	6	2
3	9	6	7	2	1	5	4	8

Puzzle 211

8	9	1	2	3	6	7	5	4
4	2	3	7	8	5	6	1	9
6	7	5	1	9	4	3	8	2
3	8	6	5	7	2	9	4	1
1	4	9	8	6	3	2	7	5
2	5	7	4	1	9	8	3	6
7	6	2	3	5	1	4	9	8
5	3	4	9	2	8	1	6	7
9	1	8	6	4	7	5	2	3

Puzzle 212

1	3	7	6	9	4	2	8	5
9	8	5	2	3	1	7	6	4
4	2	6	7	5	8	9	3	1
7	1	4	9	8	3	6	5	2
3	9	8	5	2	6	4	1	7
6	5	2	1	4	7	3	9	8
8	7	3	4	6	5	1	2	9
2	6	1	8	7	9	5	4	3
5	4	9	3	1	2	8	7	6

Puzzle 213

8	6	9	2	4	1	3	5	7
3	7	5	8	9	6	1	4	2
1	4	2	7	3	5	6	9	8
5	3	6	4	1	8	7	2	9
2	9	8	5	7	3	4	1	6
7	1	4	6	2	9	5	8	3
4	5	3	9	6	2	8	7	1
9	8	1	3	5	7	2	6	4
6	2	7	1	8	4	9	3	5

Puzzle 214

6	7	8	2	1	4	9	5	3
1	5	2	3	7	9	6	4	8
3	4	9	8	6	5	2	7	1
5	8	6	1	9	3	4	2	7
9	3	1	7	4	2	8	6	5
7	2	4	5	8	6	1	3	9
4	9	5	6	3	8	7	1	2
8	1	3	4	2	7	5	9	6
2	6	7	9	5	1	3	8	4

Puzzle 215

9	1	6	7	3	2	4	8	5
2	4	7	6	5	8	9	3	1
3	5	8	9	1	4	6	2	7
6	3	1	2	4	9	7	5	8
8	9	4	5	7	1	3	6	2
5	7	2	3	8	6	1	9	4
4	2	5	1	6	3	8	7	9
1	6	9	8	2	7	5	4	3
7	8	3	4	9	5	2	1	6

Puzzle 216

8	7	4	2	1	5	3	6	9
9	3	5	4	6	7	8	1	2
1	2	6	9	8	3	4	7	5
2	1	9	8	7	4	6	5	3
4	6	3	1	5	2	7	9	8
7	5	8	6	3	9	2	4	1
6	9	2	5	4	8	1	3	7
3	8	1	7	9	6	5	2	4
5	4	7	3	2	1	9	8	6

Puzzle 217

3	7	8	6	1	4	9	2	5
9	6	4	3	2	5	7	1	8
2	5	1	9	7	8	6	3	4
6	9	7	1	4	3	8	5	2
5	1	2	7	8	9	4	6	3
4	8	3	2	5	6	1	9	7
7	4	6	5	3	1	2	8	9
1	2	5	8	9	7	3	4	6
8	3	9	4	6	2	5	7	1

Puzzle 218

7	6	3	5	1	2	9	4	8
4	2	9	7	6	8	3	1	5
8	5	1	4	3	9	2	7	6
1	8	7	6	4	3	5	2	9
5	9	6	1	2	7	8	3	4
2	3	4	9	8	5	7	6	1
6	7	8	3	9	4	1	5	2
3	4	2	8	5	1	6	9	7
9	1	5	2	7	6	4	8	3

Puzzle 219

3	8	9	5	2	4	7	6	1
6	1	2	3	8	7	5	4	9
5	4	7	6	1	9	2	3	8
9	2	4	1	5	6	8	7	3
7	3	8	9	4	2	1	5	6
1	5	6	8	7	3	9	2	4
4	7	3	2	9	8	6	1	5
8	6	5	7	3	1	4	9	2
2	9	1	4	6	5	3	8	7

Puzzle 220

9	6	4	5	3	1	7	8	2
5	3	7	4	8	2	1	6	9
2	1	8	9	6	7	5	4	3
7	8	5	2	9	3	6	1	4
4	2	1	8	7	6	9	3	5
3	9	6	1	5	4	8	2	7
8	5	3	6	2	9	4	7	1
1	7	9	3	4	8	2	5	6
6	4	2	7	1	5	3	9	8

Puzzle 221

2	6	1	8	9	5	3	7	4
3	9	7	1	4	6	2	8	5
8	5	4	2	7	3	9	1	6
5	2	8	9	6	1	7	4	3
7	4	6	5	3	8	1	2	9
1	3	9	4	2	7	5	6	8
6	1	2	3	8	9	4	5	7
4	8	3	7	5	2	6	9	1
9	7	5	6	1	4	8	3	2

Puzzle 222

4	2	5	3	1	6	8	9	7
1	7	6	4	8	9	2	5	3
8	3	9	2	5	7	1	6	4
6	8	4	7	2	5	3	1	9
3	5	7	1	9	4	6	8	2
9	1	2	6	3	8	7	4	5
5	4	3	8	6	2	9	7	1
2	9	8	5	7	1	4	3	6
7	6	1	9	4	3	5	2	8

Puzzle 223

8	9	7	2	6	3	4	1	5
4	5	1	8	7	9	6	3	2
3	2	6	5	1	4	8	7	9
2	8	9	6	5	7	1	4	3
1	3	4	9	2	8	5	6	7
6	7	5	4	3	1	9	2	8
5	4	2	7	9	6	3	8	1
7	6	3	1	8	5	2	9	4
9	1	8	3	4	2	7	5	6

Puzzle 224

2	5	1	8	7	9	4	6	3
6	9	7	4	1	3	8	2	5
4	8	3	2	6	5	7	9	1
9	2	8	3	5	4	1	7	6
7	3	5	1	8	6	2	4	9
1	6	4	9	2	7	5	3	8
5	1	9	7	3	2	6	8	4
3	7	6	5	4	8	9	1	2
8	4	2	6	9	1	3	5	7

Puzzle 225

4	9	6	3	2	5	8	7	1
1	3	5	6	8	7	2	4	9
2	7	8	4	1	9	3	6	5
5	2	1	8	9	4	7	3	6
3	8	4	1	7	6	9	5	2
7	6	9	5	3	2	1	8	4
6	1	3	9	4	8	5	2	7
8	4	7	2	5	1	6	9	3
9	5	2	7	6	3	4	1	8

Puzzle 226

4	6	9	3	5	1	7	8	2
3	1	7	8	4	2	9	6	5
5	2	8	7	9	6	4	1	3
1	9	4	6	8	5	3	2	7
6	7	3	9	2	4	1	5	8
8	5	2	1	7	3	6	4	9
7	3	5	4	1	8	2	9	6
2	4	6	5	3	9	8	7	1
9	8	1	2	6	7	5	3	4

Puzzle 227

8	4	2	5	7	6	3	9	1
5	7	9	1	8	3	6	2	4
3	6	1	2	9	4	5	7	8
4	1	6	7	2	8	9	5	3
9	8	7	4	3	5	2	1	6
2	3	5	6	1	9	4	8	7
6	5	8	9	4	1	7	3	2
1	2	4	3	5	7	8	6	9
7	9	3	8	6	2	1	4	5

Puzzle 228

3	6	5	2	8	1	4	9	7
7	9	4	5	6	3	8	1	2
2	1	8	9	4	7	6	3	5
6	8	3	7	5	2	9	4	1
1	2	9	4	3	6	5	7	8
4	5	7	8	1	9	2	6	3
9	3	1	6	2	8	7	5	4
8	4	6	1	7	5	3	2	9
5	7	2	3	9	4	1	8	6

Puzzle 229

4	9	5	2	7	6	3	8	1
8	7	1	3	4	5	2	9	6
3	6	2	9	1	8	5	7	4
1	8	7	4	6	2	9	3	5
2	4	9	7	5	3	6	1	8
5	3	6	1	8	9	7	4	2
7	1	3	5	2	4	8	6	9
6	2	4	8	9	7	1	5	3
9	5	8	6	3	1	4	2	7

Puzzle 230

6	7	4	5	1	2	8	3	9
9	2	1	8	7	3	4	6	5
3	8	5	4	9	6	1	2	7
7	9	6	3	2	8	5	1	4
2	4	3	9	5	1	6	7	8
1	5	8	6	4	7	3	9	2
5	6	2	7	3	4	9	8	1
8	1	9	2	6	5	7	4	3
4	3	7	1	8	9	2	5	6

Puzzle 231

6	9	1	7	2	3	4	8	5
5	8	4	9	6	1	7	2	3
3	2	7	4	8	5	1	9	6
2	1	8	3	9	6	5	4	7
4	5	3	1	7	8	9	6	2
7	6	9	2	5	4	8	3	1
1	4	2	8	3	7	6	5	9
8	3	6	5	1	9	2	7	4
9	7	5	6	4	2	3	1	8

Puzzle 232

3	9	2	8	5	6	1	4	7
5	1	6	4	2	7	8	9	3
4	7	8	9	1	3	6	5	2
6	5	3	7	4	8	9	2	1
2	4	7	6	9	1	3	8	5
1	8	9	2	3	5	7	6	4
9	6	5	1	7	2	4	3	8
8	2	1	3	6	4	5	7	9
7	3	4	5	8	9	2	1	6

Puzzle 233

8	3	9	5	6	7	1	2	4
6	5	7	2	4	1	8	3	9
1	2	4	3	9	8	7	5	6
2	7	8	4	5	6	9	1	3
3	6	1	7	8	9	2	4	5
4	9	5	1	3	2	6	8	7
5	8	2	6	7	3	4	9	1
9	4	6	8	1	5	3	7	2
7	1	3	9	2	4	5	6	8

Puzzle 234

7	2	4	6	5	8	9	3	1
5	9	1	2	7	3	6	8	4
6	3	8	9	1	4	5	2	7
1	4	5	3	6	9	8	7	2
2	6	3	4	8	7	1	9	5
9	8	7	1	2	5	4	6	3
3	5	6	7	9	1	2	4	8
4	1	9	8	3	2	7	5	6
8	7	2	5	4	6	3	1	9

Puzzle 235

9	6	4	1	7	2	5	3	8
2	8	7	5	3	6	1	9	4
1	3	5	8	9	4	6	2	7
8	4	9	3	6	1	7	5	2
5	2	6	7	4	9	3	8	1
7	1	3	2	5	8	4	6	9
3	5	8	4	2	7	9	1	6
6	7	1	9	8	3	2	4	5
4	9	2	6	1	5	8	7	3

Puzzle 236

8	3	1	2	9	6	7	4	5
7	4	5	1	8	3	9	6	2
2	9	6	7	4	5	8	1	3
6	7	3	5	1	2	4	8	9
5	2	4	9	6	8	3	7	1
9	1	8	4	3	7	2	5	6
3	5	7	6	2	4	1	9	8
4	8	9	3	5	1	6	2	7
1	6	2	8	7	9	5	3	4

Puzzle 237

7	8	1	3	5	2	6	9	4
9	2	3	8	4	6	7	1	5
6	5	4	7	9	1	8	2	3
3	1	2	6	7	8	4	5	9
5	6	9	1	2	4	3	7	8
4	7	8	9	3	5	1	6	2
8	9	7	2	6	3	5	4	1
1	4	6	5	8	9	2	3	7
2	3	5	4	1	7	9	8	6

Puzzle 238

4	6	8	2	3	7	1	5	9
1	2	5	8	9	4	7	6	3
3	7	9	5	1	6	8	2	4
2	8	4	3	7	9	5	1	6
9	3	1	6	8	5	4	7	2
6	5	7	1	4	2	3	9	8
7	4	2	9	5	8	6	3	1
5	1	6	4	2	3	9	8	7
8	9	3	7	6	1	2	4	5

Puzzle 239

2	3	4	9	7	5	6	1	8
7	1	6	3	2	8	9	5	4
5	9	8	1	4	6	3	2	7
4	5	7	6	9	3	1	8	2
9	8	2	5	1	4	7	6	3
1	6	3	2	8	7	4	9	5
3	4	9	8	6	2	5	7	1
8	7	1	4	5	9	2	3	6
6	2	5	7	3	1	8	4	9

Puzzle 240

3	5	7	4	9	2	1	6	8
1	6	2	7	8	3	5	4	9
9	8	4	5	6	1	2	3	7
4	7	1	2	3	8	6	9	5
5	9	6	1	7	4	8	2	3
2	3	8	9	5	6	4	7	1
8	1	9	6	4	7	3	5	2
6	2	5	3	1	9	7	8	4
7	4	3	8	2	5	9	1	6

Puzzle 241

2	4	1	5	8	7	3	6	9
3	6	7	4	1	9	2	5	8
8	9	5	3	2	6	7	4	1
6	1	9	8	5	2	4	3	7
4	5	2	9	7	3	8	1	6
7	3	8	1	6	4	5	9	2
9	8	3	7	4	1	6	2	5
5	2	4	6	9	8	1	7	3
1	7	6	2	3	5	9	8	4

Puzzle 242

7	4	9	1	6	3	5	2	8
1	2	6	7	8	5	3	4	9
5	3	8	2	9	4	1	7	6
9	8	4	5	7	6	2	1	3
6	7	2	3	4	1	9	8	5
3	1	5	8	2	9	7	6	4
2	6	3	9	1	8	4	5	7
4	5	7	6	3	2	8	9	1
8	9	1	4	5	7	6	3	2

Puzzle 243

6	3	9	8	7	1	5	2	4
2	8	4	6	9	5	3	1	7
1	5	7	2	3	4	8	9	6
3	1	5	4	6	9	7	8	2
8	4	6	7	1	2	9	3	5
9	7	2	3	5	8	4	6	1
5	9	8	1	4	6	2	7	3
7	2	1	5	8	3	6	4	9
4	6	3	9	2	7	1	5	8

Puzzle 244

7	9	4	6	8	1	2	5	3
5	1	8	9	2	3	7	4	6
3	2	6	4	7	5	1	8	9
4	7	1	3	9	2	5	6	8
8	5	2	1	6	4	3	9	7
6	3	9	8	5	7	4	1	2
1	8	7	5	3	6	9	2	4
9	4	3	2	1	8	6	7	5
2	6	5	7	4	9	8	3	1

Puzzle 245

6	7	3	9	4	2	5	8	1
8	1	5	6	3	7	4	9	2
9	2	4	1	5	8	3	6	7
3	6	7	4	9	1	2	5	8
1	4	9	8	2	5	6	7	3
2	5	8	7	6	3	1	4	9
4	3	6	2	7	9	8	1	5
5	9	1	3	8	4	7	2	6
7	8	2	5	1	6	9	3	4

Puzzle 246

9	5	1	2	8	3	4	6	7
2	7	8	6	5	4	9	1	3
3	6	4	7	9	1	5	2	8
7	1	6	9	2	8	3	5	4
8	2	9	3	4	5	6	7	1
4	3	5	1	7	6	8	9	2
1	9	3	8	6	2	7	4	5
6	4	2	5	3	7	1	8	9
5	8	7	4	1	9	2	3	6

Puzzle 247

5	3	9	6	4	8	1	2	7
8	6	1	7	5	2	3	9	4
4	7	2	3	9	1	6	5	8
3	5	4	2	8	7	9	6	1
2	9	8	1	6	5	7	4	3
7	1	6	9	3	4	5	8	2
6	8	3	4	1	9	2	7	5
9	2	5	8	7	3	4	1	6
1	4	7	5	2	6	8	3	9

Puzzle 248

5	7	2	6	3	8	9	4	1
1	8	9	4	5	7	3	6	2
3	6	4	9	2	1	5	8	7
6	3	7	1	9	2	8	5	4
4	1	8	3	6	5	7	2	9
2	9	5	7	8	4	1	3	6
9	5	3	2	7	6	4	1	8
7	4	6	8	1	3	2	9	5
8	2	1	5	4	9	6	7	3

Puzzle 249

1	8	7	2	5	4	9	6	3
2	5	3	7	6	9	8	1	4
4	9	6	1	8	3	5	2	7
6	2	1	9	7	8	3	4	5
8	4	5	6	3	1	7	9	2
7	3	9	5	4	2	6	8	1
3	1	4	8	9	5	2	7	6
5	6	8	4	2	7	1	3	9
9	7	2	3	1	6	4	5	8

Puzzle 250

8	4	1	7	2	5	6	3	9
9	7	6	4	3	1	2	8	5
3	5	2	8	6	9	4	7	1
4	8	3	1	7	2	5	9	6
2	1	7	5	9	6	3	4	8
5	6	9	3	8	4	1	2	7
7	3	5	6	4	8	9	1	2
1	9	8	2	5	3	7	6	4
6	2	4	9	1	7	8	5	3

Puzzle 251

6	1	2	9	7	5	3	4	8
8	4	7	1	2	3	6	9	5
5	9	3	6	4	8	7	2	1
4	6	9	8	3	7	1	5	2
2	5	8	4	6	1	9	7	3
7	3	1	5	9	2	4	8	6
9	2	5	3	1	4	8	6	7
1	8	6	7	5	9	2	3	4
3	7	4	2	8	6	5	1	9

Puzzle 252

1	4	9	2	8	6	5	3	7
8	3	6	7	9	5	2	1	4
5	2	7	1	3	4	6	9	8
7	1	4	5	2	9	3	8	6
9	6	2	3	1	8	7	4	5
3	8	5	4	6	7	9	2	1
6	7	8	9	4	2	1	5	3
4	9	1	6	5	3	8	7	2
2	5	3	8	7	1	4	6	9

Puzzle 253

7	5	4	8	3	2	6	1	9
1	2	6	9	5	4	8	7	3
8	3	9	7	1	6	5	4	2
2	8	3	5	6	7	4	9	1
4	9	1	2	8	3	7	6	5
5	6	7	1	4	9	2	3	8
3	1	2	6	7	8	9	5	4
6	4	8	3	9	5	1	2	7
9	7	5	4	2	1	3	8	6

Puzzle 254

2	3	4	7	9	1	8	5	6
7	5	1	3	6	8	9	2	4
6	9	8	5	4	2	7	1	3
3	8	6	1	2	5	4	9	7
4	7	5	6	8	9	1	3	2
9	1	2	4	3	7	5	6	8
8	4	3	9	5	6	2	7	1
5	6	7	2	1	4	3	8	9
1	2	9	8	7	3	6	4	5

Puzzle 255

5	9	4	1	8	3	7	6	2
1	8	6	9	2	7	4	5	3
7	3	2	4	6	5	1	9	8
6	2	1	8	7	9	5	3	4
8	5	9	2	3	4	6	7	1
3	4	7	6	5	1	8	2	9
2	6	3	5	1	8	9	4	7
9	7	8	3	4	6	2	1	5
4	1	5	7	9	2	3	8	6

Puzzle 256

3	4	5	1	9	6	2	8	7
1	8	6	2	5	7	3	9	4
9	7	2	4	8	3	5	1	6
8	3	1	6	7	5	4	2	9
5	6	4	8	2	9	1	7	3
2	9	7	3	1	4	6	5	8
4	2	8	9	3	1	7	6	5
6	5	9	7	4	2	8	3	1
7	1	3	5	6	8	9	4	2

Puzzle 257

7	2	3	9	8	4	1	5	6
5	4	6	1	7	3	2	8	9
9	1	8	5	6	2	3	4	7
4	9	7	6	5	1	8	3	2
6	8	1	2	3	9	4	7	5
3	5	2	8	4	7	6	9	1
2	6	5	4	9	8	7	1	3
1	7	4	3	2	5	9	6	8
8	3	9	7	1	6	5	2	4

Puzzle 258

7	5	2	3	1	6	9	8	4
3	6	1	4	9	8	5	7	2
4	8	9	2	7	5	6	3	1
1	2	6	5	4	7	8	9	3
8	9	4	1	6	3	7	2	5
5	7	3	8	2	9	4	1	6
9	1	8	6	5	2	3	4	7
6	4	7	9	3	1	2	5	8
2	3	5	7	8	4	1	6	9

Puzzle 259

7	3	5	8	6	1	9	4	2
4	9	6	3	7	2	5	1	8
2	1	8	4	5	9	6	7	3
9	6	4	1	2	5	8	3	7
1	8	7	9	4	3	2	6	5
5	2	3	6	8	7	4	9	1
6	4	2	7	1	8	3	5	9
8	7	9	5	3	4	1	2	6
3	5	1	2	9	6	7	8	4

Puzzle 260

5	3	9	7	6	4	1	2	8
2	4	1	3	5	8	9	6	7
8	7	6	2	1	9	3	4	5
9	6	2	5	7	3	4	8	1
7	5	8	4	9	1	6	3	2
3	1	4	6	8	2	5	7	9
4	8	3	1	2	5	7	9	6
1	2	7	9	4	6	8	5	3
6	9	5	8	3	7	2	1	4

Puzzle 261

5	7	9	1	4	2	3	6	8
3	2	6	5	7	8	4	9	1
8	4	1	9	3	6	7	5	2
9	8	2	4	5	1	6	3	7
7	6	5	3	8	9	1	2	4
4	1	3	2	6	7	5	8	9
2	3	4	8	1	5	9	7	6
6	5	8	7	9	4	2	1	3
1	9	7	6	2	3	8	4	5

Puzzle 262

9	7	4	5	6	2	8	1	3
8	3	1	7	9	4	5	6	2
5	2	6	8	3	1	4	9	7
4	8	3	1	7	5	6	2	9
2	1	7	6	8	9	3	4	5
6	5	9	4	2	3	7	8	1
1	6	8	2	5	7	9	3	4
7	9	2	3	4	6	1	5	8
3	4	5	9	1	8	2	7	6

Puzzle 263

1	9	6	8	5	2	7	4	3
2	8	4	6	3	7	5	9	1
7	5	3	4	1	9	6	2	8
9	6	2	7	8	3	1	5	4
3	4	7	1	6	5	2	8	9
5	1	8	2	9	4	3	6	7
8	2	5	9	7	1	4	3	6
6	3	1	5	4	8	9	7	2
4	7	9	3	2	6	8	1	5

Puzzle 264

1	5	7	2	8	3	4	9	6
9	2	4	7	6	5	1	3	8
6	3	8	4	1	9	7	2	5
4	8	3	6	9	1	5	7	2
5	9	2	3	7	8	6	4	1
7	1	6	5	4	2	3	8	9
3	7	9	8	5	6	2	1	4
8	4	5	1	2	7	9	6	3
2	6	1	9	3	4	8	5	7

Puzzle 265

8	9	6	4	1	2	7	5	3
4	2	5	7	9	3	1	8	6
3	7	1	5	8	6	2	4	9
5	8	9	3	4	1	6	2	7
6	4	3	9	2	7	5	1	8
7	1	2	8	6	5	3	9	4
2	3	8	1	7	4	9	6	5
9	6	7	2	5	8	4	3	1
1	5	4	6	3	9	8	7	2

Puzzle 266

2	3	7	5	1	6	4	8	9
6	5	4	3	8	9	2	1	7
8	1	9	7	2	4	6	5	3
5	8	3	4	7	1	9	2	6
9	4	1	8	6	2	7	3	5
7	6	2	9	5	3	1	4	8
1	2	5	6	3	7	8	9	4
4	7	8	2	9	5	3	6	1
3	9	6	1	4	8	5	7	2

Puzzle 267

3	6	4	9	7	2	1	8	5
7	9	1	3	5	8	2	6	4
8	2	5	6	4	1	3	9	7
9	5	6	8	2	3	7	4	1
4	1	3	7	6	9	8	5	2
2	7	8	4	1	5	6	3	9
6	4	2	5	3	7	9	1	8
1	3	9	2	8	4	5	7	6
5	8	7	1	9	6	4	2	3

Puzzle 268

6	7	8	1	9	2	5	4	3
5	9	3	8	6	4	1	2	7
1	4	2	3	5	7	8	6	9
2	8	1	6	4	9	7	3	5
9	6	5	7	2	3	4	8	1
4	3	7	5	1	8	6	9	2
8	1	4	2	3	5	9	7	6
7	2	6	9	8	1	3	5	4
3	5	9	4	7	6	2	1	8

Puzzle 269

3	5	8	6	7	9	4	2	1
7	6	2	5	1	4	8	3	9
1	4	9	8	3	2	6	5	7
4	2	3	1	8	5	7	9	6
6	8	1	3	9	7	2	4	5
5	9	7	2	4	6	3	1	8
9	3	5	4	6	8	1	7	2
8	7	4	9	2	1	5	6	3
2	1	6	7	5	3	9	8	4

Puzzle 270

9	4	3	7	2	8	5	6	1
1	7	8	9	6	5	3	2	4
2	5	6	4	3	1	8	7	9
8	3	7	6	1	9	2	4	5
4	2	9	8	5	3	7	1	6
5	6	1	2	7	4	9	8	3
6	1	2	3	9	7	4	5	8
3	8	5	1	4	2	6	9	7
7	9	4	5	8	6	1	3	2

Puzzle 271

2	4	1	7	8	6	3	5	9
5	6	3	9	4	1	8	7	2
7	9	8	3	2	5	1	4	6
9	7	5	6	1	2	4	8	3
6	3	2	4	7	8	9	1	5
8	1	4	5	9	3	2	6	7
1	5	7	2	3	4	6	9	8
4	2	9	8	6	7	5	3	1
3	8	6	1	5	9	7	2	4

Puzzle 272

4	1	5	8	9	3	2	7	6
8	2	3	1	7	6	9	4	5
9	6	7	5	4	2	8	1	3
6	4	1	9	2	7	5	3	8
5	9	8	3	6	1	7	2	4
7	3	2	4	8	5	1	6	9
1	8	4	2	3	9	6	5	7
2	7	9	6	5	4	3	8	1
3	5	6	7	1	8	4	9	2

Puzzle 273

6	9	7	5	4	8	3	2	1
4	1	8	2	6	3	5	9	7
3	2	5	1	9	7	4	8	6
9	4	6	3	1	5	8	7	2
1	7	3	8	2	6	9	5	4
8	5	2	4	7	9	6	1	3
5	8	1	7	3	4	2	6	9
2	6	4	9	5	1	7	3	8
7	3	9	6	8	2	1	4	5

Puzzle 274

8	6	4	1	7	2	5	3	9
7	3	9	6	8	5	4	1	2
5	2	1	4	9	3	7	8	6
4	7	6	8	5	1	9	2	3
9	5	2	3	6	4	8	7	1
3	1	8	7	2	9	6	4	5
1	9	5	2	4	8	3	6	7
2	8	7	5	3	6	1	9	4
6	4	3	9	1	7	2	5	8

Puzzle 275

4	9	7	3	5	2	8	1	6
6	2	1	4	8	9	7	3	5
3	5	8	6	7	1	4	9	2
2	3	9	5	4	6	1	7	8
8	1	5	7	9	3	2	6	4
7	6	4	2	1	8	3	5	9
9	8	3	1	2	5	6	4	7
5	7	6	8	3	4	9	2	1
1	4	2	9	6	7	5	8	3

Puzzle 276

5	9	2	3	4	1	8	7	6
6	7	3	8	2	9	1	4	5
4	1	8	5	6	7	2	3	9
7	4	5	6	1	3	9	2	8
9	3	1	7	8	2	5	6	4
2	8	6	4	9	5	3	1	7
3	6	9	2	7	8	4	5	1
8	2	7	1	5	4	6	9	3
1	5	4	9	3	6	7	8	2

Puzzle 277

1	7	3	8	4	5	9	2	6
5	6	8	9	2	1	3	4	7
2	4	9	3	7	6	1	8	5
3	5	2	4	6	8	7	1	9
9	1	4	7	5	3	2	6	8
7	8	6	1	9	2	4	5	3
8	9	5	2	1	7	6	3	4
6	2	7	5	3	4	8	9	1
4	3	1	6	8	9	5	7	2

Puzzle 278

5	1	9	7	4	8	6	3	2
2	7	8	9	6	3	1	5	4
3	4	6	2	5	1	9	7	8
8	2	3	5	1	9	7	4	6
9	6	7	3	8	4	5	2	1
4	5	1	6	7	2	3	8	9
7	9	4	1	2	5	8	6	3
1	8	5	4	3	6	2	9	7
6	3	2	8	9	7	4	1	5

Puzzle 279

4	3	8	5	2	6	7	9	1
9	5	7	4	3	1	2	6	8
1	2	6	7	8	9	4	5	3
8	4	9	6	1	2	5	3	7
5	7	2	3	9	4	8	1	6
3	6	1	8	5	7	9	2	4
6	1	4	2	7	5	3	8	9
7	8	5	9	6	3	1	4	2
2	9	3	1	4	8	6	7	5

Puzzle 280

1	2	7	5	3	4	8	6	9
6	4	5	2	8	9	1	3	7
9	3	8	1	7	6	4	5	2
5	8	4	3	2	7	9	1	6
2	9	3	6	5	1	7	8	4
7	6	1	9	4	8	3	2	5
8	1	2	4	9	5	6	7	3
4	5	6	7	1	3	2	9	8
3	7	9	8	6	2	5	4	1

Puzzle 281

2	9	6	1	4	8	7	3	5
1	4	8	3	5	7	9	2	6
7	5	3	9	6	2	4	1	8
3	2	7	8	9	4	5	6	1
9	6	5	2	1	3	8	4	7
8	1	4	6	7	5	3	9	2
5	7	2	4	3	1	6	8	9
6	3	1	7	8	9	2	5	4
4	8	9	5	2	6	1	7	3

Puzzle 282

2	6	5	9	3	4	7	1	8
8	4	3	6	1	7	5	9	2
1	9	7	2	5	8	4	3	6
7	5	9	4	8	1	2	6	3
3	1	2	7	6	9	8	4	5
4	8	6	3	2	5	1	7	9
9	3	8	1	7	2	6	5	4
6	2	1	5	4	3	9	8	7
5	7	4	8	9	6	3	2	1

Puzzle 283

5	8	7	4	1	9	3	2	6
4	9	3	2	7	6	1	8	5
6	1	2	3	5	8	4	9	7
8	7	6	5	3	4	9	1	2
3	5	4	9	2	1	7	6	8
9	2	1	6	8	7	5	4	3
2	6	5	1	9	3	8	7	4
1	3	8	7	4	2	6	5	9
7	4	9	8	6	5	2	3	1

Puzzle 284

7	4	8	2	1	9	3	5	6
1	9	6	8	3	5	4	2	7
2	5	3	4	7	6	1	9	8
8	2	9	6	5	4	7	3	1
6	1	7	9	2	3	8	4	5
4	3	5	7	8	1	9	6	2
5	8	4	1	9	2	6	7	3
3	6	1	5	4	7	2	8	9
9	7	2	3	6	8	5	1	4

Puzzle 285

3	9	6	7	2	1	5	8	4
7	1	8	4	6	5	9	3	2
4	2	5	3	9	8	6	1	7
8	4	3	6	5	9	2	7	1
1	5	2	8	4	7	3	9	6
6	7	9	1	3	2	4	5	8
9	6	7	2	1	3	8	4	5
2	3	1	5	8	4	7	6	9
5	8	4	9	7	6	1	2	3

Puzzle 286

1	8	9	2	7	6	4	3	5
7	5	3	4	1	9	2	8	6
2	4	6	3	5	8	7	9	1
4	1	5	8	9	7	3	6	2
3	9	7	6	2	1	8	5	4
8	6	2	5	4	3	9	1	7
6	2	1	9	3	4	5	7	8
9	7	4	1	8	5	6	2	3
5	3	8	7	6	2	1	4	9

Puzzle 287

8	9	1	2	5	3	7	4	6
4	7	2	8	1	6	3	9	5
6	3	5	9	7	4	2	8	1
1	6	4	7	8	2	9	5	3
9	8	3	6	4	5	1	7	2
5	2	7	1	3	9	4	6	8
7	4	8	3	6	1	5	2	9
3	5	9	4	2	8	6	1	7
2	1	6	5	9	7	8	3	4

Puzzle 288

4	2	5	3	1	8	9	7	6
8	3	7	2	9	6	5	4	1
9	6	1	5	4	7	2	8	3
5	8	6	7	2	9	3	1	4
3	4	9	6	5	1	8	2	7
7	1	2	4	8	3	6	5	9
1	9	3	8	7	5	4	6	2
6	5	4	1	3	2	7	9	8
2	7	8	9	6	4	1	3	5

Puzzle 289

3	6	1	7	2	9	4	8	5
5	7	8	3	6	4	9	1	2
2	4	9	5	8	1	6	7	3
7	9	3	4	5	6	8	2	1
6	1	5	2	7	8	3	4	9
8	2	4	1	9	3	5	6	7
4	8	7	9	3	2	1	5	6
9	5	6	8	1	7	2	3	4
1	3	2	6	4	5	7	9	8

Puzzle 290

6	8	1	7	4	2	5	9	3
5	2	4	3	9	6	8	1	7
3	7	9	5	1	8	4	6	2
1	3	2	8	5	9	6	7	4
8	6	5	1	7	4	3	2	9
4	9	7	6	2	3	1	8	5
2	1	8	4	3	7	9	5	6
7	5	3	9	6	1	2	4	8
9	4	6	2	8	5	7	3	1

Puzzle 291

8	3	4	7	2	9	6	5	1
6	2	7	5	4	1	3	9	8
9	5	1	3	6	8	7	2	4
7	9	3	1	8	4	5	6	2
2	6	8	9	5	3	4	1	7
4	1	5	2	7	6	8	3	9
5	4	9	8	3	2	1	7	6
3	8	2	6	1	7	9	4	5
1	7	6	4	9	5	2	8	3

Puzzle 292

3	9	8	5	4	7	2	6	1
6	7	1	9	2	8	3	5	4
4	5	2	3	6	1	8	7	9
8	3	4	2	9	5	6	1	7
9	2	5	1	7	6	4	3	8
1	6	7	4	8	3	9	2	5
7	4	6	8	5	2	1	9	3
5	8	3	6	1	9	7	4	2
2	1	9	7	3	4	5	8	6

Puzzle 293

6	4	9	3	1	5	7	2	8
2	1	3	9	7	8	6	5	4
5	8	7	2	4	6	3	9	1
3	2	5	4	8	7	1	6	9
9	7	1	6	3	2	8	4	5
4	6	8	5	9	1	2	3	7
1	9	2	8	5	3	4	7	6
8	5	6	7	2	4	9	1	3
7	3	4	1	6	9	5	8	2

Puzzle 294

7	8	4	1	9	5	2	6	3
3	6	9	7	2	8	4	1	5
2	5	1	3	6	4	8	7	9
8	3	2	5	4	6	7	9	1
4	1	7	2	3	9	5	8	6
6	9	5	8	7	1	3	2	4
9	7	8	6	5	3	1	4	2
5	2	6	4	1	7	9	3	8
1	4	3	9	8	2	6	5	7

Puzzle 295

9	6	8	3	1	4	5	7	2
3	5	2	6	7	8	1	9	4
4	7	1	5	9	2	3	8	6
7	1	3	8	4	9	6	2	5
6	8	9	1	2	5	7	4	3
2	4	5	7	3	6	9	1	8
1	2	4	9	6	3	8	5	7
8	3	7	2	5	1	4	6	9
5	9	6	4	8	7	2	3	1

Puzzle 296

4	6	9	1	3	2	5	7	8
5	2	1	6	7	8	9	4	3
7	3	8	9	4	5	1	6	2
1	5	6	7	2	9	8	3	4
8	4	2	3	6	1	7	9	5
3	9	7	5	8	4	2	1	6
2	1	4	8	9	6	3	5	7
6	7	5	2	1	3	4	8	9
9	8	3	4	5	7	6	2	1

Puzzle 297

7	2	8	5	1	4	6	3	9
1	9	3	2	6	8	5	7	4
6	5	4	3	9	7	1	8	2
3	8	7	6	5	9	4	2	1
5	4	6	7	2	1	3	9	8
9	1	2	4	8	3	7	6	5
4	6	1	9	7	2	8	5	3
2	3	5	8	4	6	9	1	7
8	7	9	1	3	5	2	4	6

Puzzle 298

6	1	3	4	9	8	7	2	5
5	4	8	3	7	2	9	1	6
7	9	2	1	6	5	8	4	3
8	7	5	9	2	1	6	3	4
9	6	1	8	3	4	5	7	2
2	3	4	6	5	7	1	8	9
3	5	7	2	8	6	4	9	1
4	2	6	7	1	9	3	5	8
1	8	9	5	4	3	2	6	7

Puzzle 299

8	9	7	2	3	4	5	6	1
6	2	5	1	9	8	7	4	3
3	4	1	6	7	5	2	8	9
4	1	9	5	2	6	8	3	7
7	6	8	3	1	9	4	2	5
5	3	2	8	4	7	1	9	6
1	5	6	9	8	2	3	7	4
2	7	3	4	6	1	9	5	8
9	8	4	7	5	3	6	1	2

Puzzle 300

5	7	3	6	4	2	8	1	9
2	1	4	9	7	8	6	5	3
9	6	8	5	3	1	7	2	4
7	3	6	8	2	5	9	4	1
1	4	9	3	6	7	2	8	5
8	5	2	1	9	4	3	7	6
6	2	7	4	5	3	1	9	8
3	8	5	2	1	9	4	6	7
4	9	1	7	8	6	5	3	2

Puzzle 301

7	2	5	3	4	1	6	8	9
4	1	9	2	8	6	5	7	3
3	6	8	7	9	5	2	1	4
2	7	1	8	3	9	4	6	5
9	5	6	4	1	7	8	3	2
8	4	3	6	5	2	7	9	1
1	3	2	5	6	8	9	4	7
6	9	7	1	2	4	3	5	8
5	8	4	9	7	3	1	2	6

Puzzle 302

2	6	1	9	7	4	8	5	3
7	9	8	1	3	5	2	6	4
5	4	3	6	2	8	7	9	1
1	2	9	3	5	6	4	8	7
4	8	6	2	9	7	1	3	5
3	5	7	4	8	1	6	2	9
8	3	4	7	6	9	5	1	2
6	1	2	5	4	3	9	7	8
9	7	5	8	1	2	3	4	6

Puzzle 303

8	9	6	2	3	5	7	1	4
3	7	4	6	1	8	5	2	9
2	1	5	4	9	7	6	3	8
1	4	7	5	8	9	2	6	3
9	5	8	3	6	2	4	7	1
6	2	3	1	7	4	9	8	5
5	8	2	7	4	3	1	9	6
4	3	1	9	2	6	8	5	7
7	6	9	8	5	1	3	4	2

Puzzle 304

1	9	7	6	3	5	8	2	4
6	5	2	9	4	8	3	7	1
8	4	3	2	7	1	9	5	6
2	7	4	8	1	3	5	6	9
3	1	6	5	9	4	2	8	7
9	8	5	7	2	6	1	4	3
4	3	8	1	5	7	6	9	2
5	2	1	4	6	9	7	3	8
7	6	9	3	8	2	4	1	5

Puzzle 305

2	1	5	6	4	9	7	8	3
7	4	8	1	2	3	9	6	5
9	3	6	5	8	7	4	1	2
4	7	3	8	1	5	2	9	6
5	6	1	3	9	2	8	7	4
8	9	2	4	7	6	5	3	1
3	5	4	9	6	8	1	2	7
1	2	9	7	3	4	6	5	8
6	8	7	2	5	1	3	4	9

Puzzle 306

1	4	2	9	3	7	6	8	5
6	5	7	8	4	1	2	9	3
9	8	3	5	6	2	1	4	7
3	2	6	1	8	5	9	7	4
7	1	4	6	9	3	5	2	8
5	9	8	2	7	4	3	6	1
4	7	5	3	2	6	8	1	9
8	6	1	7	5	9	4	3	2
2	3	9	4	1	8	7	5	6

Puzzle 307

1	8	6	7	3	5	2	4	9
7	2	4	6	8	9	5	3	1
9	3	5	4	1	2	8	7	6
8	6	3	5	7	1	4	9	2
2	9	7	8	6	4	1	5	3
5	4	1	2	9	3	7	6	8
3	5	8	1	4	6	9	2	7
4	1	9	3	2	7	6	8	5
6	7	2	9	5	8	3	1	4

Puzzle 308

7	1	6	9	8	2	4	5	3
2	9	3	5	1	4	7	6	8
8	5	4	3	6	7	9	2	1
5	3	2	6	4	1	8	9	7
1	6	9	7	3	8	5	4	2
4	8	7	2	9	5	3	1	6
3	7	5	1	2	9	6	8	4
9	4	1	8	7	6	2	3	5
6	2	8	4	5	3	1	7	9

Puzzle 309

8	3	9	4	2	5	6	1	7
7	2	5	1	6	8	9	3	4
1	6	4	9	3	7	8	5	2
6	5	7	8	9	3	2	4	1
9	4	3	2	1	6	7	8	5
2	8	1	7	5	4	3	9	6
5	1	2	6	8	9	4	7	3
4	9	6	3	7	1	5	2	8
3	7	8	5	4	2	1	6	9

Puzzle 310

3	4	2	1	8	7	5	9	6
5	7	6	2	3	9	8	1	4
1	8	9	5	4	6	3	7	2
8	2	4	9	6	1	7	3	5
6	9	5	7	2	3	4	8	1
7	3	1	4	5	8	6	2	9
4	5	7	8	1	2	9	6	3
9	1	3	6	7	5	2	4	8
2	6	8	3	9	4	1	5	7

Puzzle 311

6	4	9	2	5	8	7	3	1
3	2	7	1	4	6	9	8	5
5	1	8	7	9	3	2	4	6
1	6	2	9	8	4	3	5	7
4	7	3	5	2	1	6	9	8
9	8	5	3	6	7	4	1	2
2	3	6	8	1	9	5	7	4
8	9	4	6	7	5	1	2	3
7	5	1	4	3	2	8	6	9

Puzzle 312

2	1	6	8	5	9	3	7	4
5	4	3	1	7	6	9	2	8
8	7	9	2	3	4	1	5	6
4	5	8	3	9	1	2	6	7
6	3	7	5	8	2	4	1	9
1	9	2	4	6	7	5	8	3
7	8	5	9	1	3	6	4	2
9	6	4	7	2	5	8	3	1
3	2	1	6	4	8	7	9	5

Puzzle 313

7	3	5	4	6	1	8	9	2
4	8	2	9	7	5	3	1	6
1	9	6	3	2	8	4	5	7
3	1	8	5	9	7	2	6	4
2	6	9	8	3	4	5	7	1
5	7	4	6	1	2	9	8	3
6	2	3	7	8	9	1	4	5
8	5	7	1	4	3	6	2	9
9	4	1	2	5	6	7	3	8

Puzzle 314

8	2	7	1	4	6	5	3	9
6	1	4	5	3	9	7	2	8
9	5	3	2	7	8	1	6	4
2	8	6	7	5	4	3	9	1
7	3	5	8	9	1	2	4	6
1	4	9	6	2	3	8	7	5
4	7	8	9	1	2	6	5	3
5	9	1	3	6	7	4	8	2
3	6	2	4	8	5	9	1	7

Puzzle 315

2	6	8	5	9	3	1	7	4
3	5	7	4	2	1	6	9	8
1	9	4	6	8	7	5	3	2
6	8	2	9	3	5	7	4	1
7	4	1	8	6	2	9	5	3
5	3	9	7	1	4	8	2	6
8	2	5	1	4	9	3	6	7
9	1	3	2	7	6	4	8	5
4	7	6	3	5	8	2	1	9

Puzzle 316

9	5	3	8	1	4	6	2	7
6	1	4	3	2	7	5	9	8
2	8	7	5	6	9	3	4	1
4	2	1	9	7	5	8	6	3
8	6	9	4	3	1	2	7	5
3	7	5	6	8	2	9	1	4
7	3	8	1	9	6	4	5	2
1	4	6	2	5	3	7	8	9
5	9	2	7	4	8	1	3	6

Puzzle 317

7	3	4	1	8	6	5	9	2
8	6	5	4	9	2	7	1	3
9	2	1	5	3	7	8	6	4
3	7	2	6	1	8	9	4	5
1	4	8	3	5	9	2	7	6
6	5	9	2	7	4	3	8	1
2	8	6	9	4	3	1	5	7
4	1	7	8	2	5	6	3	9
5	9	3	7	6	1	4	2	8

Puzzle 318

2	7	1	4	3	5	6	9	8
9	3	8	7	6	1	4	2	5
4	6	5	8	9	2	3	1	7
1	5	4	2	8	6	9	7	3
6	8	7	3	4	9	2	5	1
3	9	2	5	1	7	8	4	6
5	2	9	6	7	3	1	8	4
7	4	3	1	2	8	5	6	9
8	1	6	9	5	4	7	3	2

Puzzle 319

7	9	5	2	1	3	8	6	4
8	3	6	5	7	4	1	9	2
2	1	4	8	6	9	5	7	3
6	4	1	9	8	5	2	3	7
5	7	9	3	2	1	6	4	8
3	2	8	7	4	6	9	5	1
9	6	7	1	3	2	4	8	5
1	5	3	4	9	8	7	2	6
4	8	2	6	5	7	3	1	9

Puzzle 320

6	9	7	2	8	1	5	4	3
3	2	4	7	9	5	1	8	6
1	8	5	3	6	4	2	9	7
4	7	9	5	1	8	6	3	2
2	3	8	9	7	6	4	1	5
5	1	6	4	2	3	9	7	8
9	4	3	6	5	7	8	2	1
7	5	1	8	4	2	3	6	9
8	6	2	1	3	9	7	5	4

Puzzle 321

5	8	7	4	3	9	1	2	6
4	1	6	5	7	2	3	8	9
3	2	9	1	6	8	4	7	5
8	4	1	3	5	6	7	9	2
2	9	5	7	8	4	6	3	1
6	7	3	2	9	1	8	5	4
9	6	4	8	2	7	5	1	3
1	5	8	9	4	3	2	6	7
7	3	2	6	1	5	9	4	8

Puzzle 322

7	2	5	4	1	9	3	8	6
3	1	8	2	6	7	9	4	5
9	6	4	5	8	3	2	1	7
2	9	3	1	7	6	8	5	4
5	7	1	9	4	8	6	3	2
4	8	6	3	5	2	1	7	9
6	3	7	8	9	4	5	2	1
1	4	2	6	3	5	7	9	8
8	5	9	7	2	1	4	6	3

Puzzle 323

3	4	9	1	8	2	7	6	5
8	6	5	9	3	7	4	1	2
2	1	7	5	6	4	3	9	8
1	7	2	3	9	5	6	8	4
6	3	4	8	2	1	9	5	7
9	5	8	4	7	6	2	3	1
5	9	1	7	4	3	8	2	6
4	8	6	2	1	9	5	7	3
7	2	3	6	5	8	1	4	9

Puzzle 324

5	6	3	9	7	2	4	1	8
1	8	4	3	5	6	2	7	9
2	9	7	8	1	4	5	3	6
4	2	9	5	3	8	1	6	7
6	5	8	1	2	7	9	4	3
7	3	1	4	6	9	8	5	2
3	4	6	2	8	5	7	9	1
8	7	5	6	9	1	3	2	4
9	1	2	7	4	3	6	8	5

Puzzle 325

3	1	7	6	5	9	2	4	8
2	4	9	7	8	1	5	6	3
5	6	8	4	2	3	7	9	1
8	7	6	3	9	5	4	1	2
9	5	3	2	1	4	8	7	6
4	2	1	8	6	7	9	3	5
6	9	4	5	3	8	1	2	7
7	8	2	1	4	6	3	5	9
1	3	5	9	7	2	6	8	4

Puzzle 326

9	4	6	2	1	3	5	7	8
8	5	2	9	4	7	3	1	6
7	3	1	5	6	8	2	4	9
5	2	4	3	7	6	8	9	1
3	7	8	4	9	1	6	5	2
1	6	9	8	2	5	4	3	7
4	9	7	6	3	2	1	8	5
6	8	3	1	5	9	7	2	4
2	1	5	7	8	4	9	6	3

Puzzle 327

8	7	2	4	5	6	9	3	1
9	6	5	7	3	1	4	2	8
1	4	3	2	9	8	5	6	7
6	9	4	5	1	7	3	8	2
3	5	8	6	2	4	7	1	9
2	1	7	9	8	3	6	5	4
5	8	9	3	7	2	1	4	6
7	2	6	1	4	5	8	9	3
4	3	1	8	6	9	2	7	5

Puzzle 328

1	5	2	8	4	6	3	7	9
9	4	6	5	3	7	2	1	8
7	8	3	9	1	2	5	6	4
2	3	1	6	7	9	8	4	5
4	9	5	3	8	1	7	2	6
6	7	8	2	5	4	9	3	1
5	2	7	4	6	8	1	9	3
8	6	9	1	2	3	4	5	7
3	1	4	7	9	5	6	8	2

Puzzle 329

1	8	5	2	9	6	7	3	4
9	3	2	8	7	4	6	5	1
4	6	7	3	5	1	8	9	2
8	2	1	6	4	9	3	7	5
5	9	3	1	2	7	4	6	8
7	4	6	5	3	8	2	1	9
2	1	9	4	6	3	5	8	7
3	7	4	9	8	5	1	2	6
6	5	8	7	1	2	9	4	3

Puzzle 330

3	8	1	2	9	4	5	7	6
5	7	2	8	6	3	4	1	9
6	4	9	1	5	7	3	2	8
8	6	5	3	7	2	9	4	1
1	2	7	5	4	9	8	6	3
4	9	3	6	1	8	7	5	2
7	1	8	4	3	6	2	9	5
9	3	6	7	2	5	1	8	4
2	5	4	9	8	1	6	3	7

Puzzle 331

2	8	4	7	9	5	1	6	3
6	1	3	8	4	2	9	7	5
9	5	7	6	3	1	8	2	4
4	2	9	3	6	8	7	5	1
3	7	5	9	1	4	6	8	2
8	6	1	2	5	7	3	4	9
7	4	8	1	2	3	5	9	6
1	9	2	5	8	6	4	3	7
5	3	6	4	7	9	2	1	8

Puzzle 332

9	6	8	5	3	4	7	2	1
1	3	4	8	7	2	9	6	5
5	2	7	1	9	6	4	8	3
8	1	5	4	2	3	6	7	9
2	9	6	7	1	5	8	3	4
4	7	3	9	6	8	5	1	2
3	4	1	6	8	9	2	5	7
6	5	2	3	4	7	1	9	8
7	8	9	2	5	1	3	4	6

Puzzle 333

3	5	2	7	8	4	9	6	1
6	9	1	2	5	3	4	7	8
7	8	4	6	9	1	3	5	2
4	2	5	3	7	8	1	9	6
8	3	7	9	1	6	2	4	5
9	1	6	5	4	2	8	3	7
2	4	9	8	6	7	5	1	3
1	6	8	4	3	5	7	2	9
5	7	3	1	2	9	6	8	4

Puzzle 334

8	2	5	1	9	6	7	4	3
9	1	7	2	4	3	5	6	8
4	3	6	5	8	7	9	2	1
5	8	3	9	1	4	2	7	6
1	9	2	6	7	8	4	3	5
7	6	4	3	5	2	1	8	9
3	4	9	8	2	1	6	5	7
6	7	1	4	3	5	8	9	2
2	5	8	7	6	9	3	1	4

Puzzle 335

6	8	5	9	4	3	2	1	7
4	9	1	8	7	2	3	6	5
2	7	3	6	1	5	4	9	8
1	5	9	2	6	7	8	3	4
7	4	8	3	5	1	9	2	6
3	2	6	4	8	9	5	7	1
5	3	7	1	2	8	6	4	9
9	1	4	5	3	6	7	8	2
8	6	2	7	9	4	1	5	3

Puzzle 336

4	7	2	5	6	1	8	3	9
3	5	1	9	8	2	6	7	4
8	9	6	3	7	4	1	5	2
9	4	7	8	2	5	3	1	6
6	8	5	4	1	3	2	9	7
2	1	3	6	9	7	5	4	8
5	3	8	2	4	9	7	6	1
7	6	4	1	3	8	9	2	5
1	2	9	7	5	6	4	8	3

Puzzle 337

8	4	9	5	7	6	3	1	2
6	7	5	2	3	1	8	9	4
2	3	1	8	4	9	7	6	5
9	6	3	4	1	5	2	8	7
1	8	7	6	2	3	4	5	9
5	2	4	9	8	7	6	3	1
4	9	6	7	5	8	1	2	3
3	5	2	1	6	4	9	7	8
7	1	8	3	9	2	5	4	6

Puzzle 338

8	7	3	5	2	4	1	9	6
6	2	1	9	7	8	3	4	5
5	9	4	3	6	1	8	7	2
2	8	6	7	5	9	4	3	1
9	1	5	8	4	3	2	6	7
4	3	7	6	1	2	5	8	9
3	6	2	4	9	5	7	1	8
1	4	9	2	8	7	6	5	3
7	5	8	1	3	6	9	2	4

Puzzle 339

9	4	5	7	2	3	6	1	8
2	8	6	9	4	1	5	7	3
1	3	7	6	8	5	4	2	9
3	5	8	4	9	2	1	6	7
7	1	2	8	3	6	9	5	4
4	6	9	1	5	7	3	8	2
6	9	4	2	1	8	7	3	5
5	2	1	3	7	9	8	4	6
8	7	3	5	6	4	2	9	1

Puzzle 340

4	8	9	1	7	5	3	2	6
1	5	3	9	6	2	8	4	7
2	7	6	4	3	8	9	5	1
9	1	5	6	4	3	2	7	8
7	3	2	5	8	1	4	6	9
8	6	4	2	9	7	1	3	5
3	4	8	7	1	6	5	9	2
6	2	1	3	5	9	7	8	4
5	9	7	8	2	4	6	1	3

Puzzle 341

1	4	2	5	3	8	6	7	9
7	3	9	2	6	4	5	1	8
6	8	5	7	1	9	2	3	4
2	5	8	3	7	6	9	4	1
3	9	6	4	2	1	7	8	5
4	1	7	9	8	5	3	2	6
9	6	3	1	4	7	8	5	2
5	2	4	8	9	3	1	6	7
8	7	1	6	5	2	4	9	3

Puzzle 342

2	5	9	1	3	6	7	4	8
8	7	6	2	4	5	9	3	1
4	3	1	7	9	8	5	6	2
7	1	4	8	6	2	3	5	9
9	2	5	4	7	3	1	8	6
3	6	8	9	5	1	2	7	4
6	4	3	5	2	9	8	1	7
1	9	7	3	8	4	6	2	5
5	8	2	6	1	7	4	9	3

Puzzle 343

3	6	1	7	8	5	2	4	9
2	4	9	6	3	1	7	5	8
8	7	5	4	2	9	6	3	1
5	8	2	1	7	4	3	9	6
7	3	6	8	9	2	5	1	4
1	9	4	5	6	3	8	2	7
9	2	8	3	1	7	4	6	5
6	5	3	9	4	8	1	7	2
4	1	7	2	5	6	9	8	3

Puzzle 344

5	7	4	2	1	6	3	8	9
8	1	6	4	3	9	2	5	7
2	3	9	5	8	7	1	6	4
9	8	7	6	5	3	4	1	2
3	6	1	7	2	4	5	9	8
4	5	2	1	9	8	7	3	6
6	9	5	3	4	2	8	7	1
1	2	8	9	7	5	6	4	3
7	4	3	8	6	1	9	2	5

Puzzle 345

2	5	7	4	9	8	1	3	6
8	4	3	5	1	6	7	2	9
6	1	9	3	2	7	5	4	8
7	6	5	8	3	4	9	1	2
1	3	4	9	7	2	6	8	5
9	8	2	1	6	5	4	7	3
4	7	8	6	5	3	2	9	1
3	9	6	2	4	1	8	5	7
5	2	1	7	8	9	3	6	4

Puzzle 346

1	7	5	9	4	6	8	2	3
4	8	3	2	7	5	9	1	6
2	9	6	3	1	8	4	7	5
9	5	8	1	2	7	3	6	4
7	1	4	5	6	3	2	8	9
6	3	2	8	9	4	7	5	1
5	4	7	6	3	2	1	9	8
3	6	1	7	8	9	5	4	2
8	2	9	4	5	1	6	3	7

Puzzle 347

1	5	4	3	6	2	7	8	9
8	6	3	4	9	7	1	5	2
9	7	2	1	8	5	6	4	3
2	9	7	6	1	8	5	3	4
6	4	8	5	3	9	2	7	1
3	1	5	7	2	4	9	6	8
5	3	6	9	4	1	8	2	7
7	8	1	2	5	3	4	9	6
4	2	9	8	7	6	3	1	5

Puzzle 348

8	1	6	9	3	4	2	5	7
9	7	2	1	5	6	4	8	3
4	5	3	2	7	8	6	1	9
5	9	7	3	6	2	1	4	8
3	2	1	4	8	9	5	7	6
6	8	4	7	1	5	3	9	2
7	4	9	5	2	3	8	6	1
2	6	5	8	9	1	7	3	4
1	3	8	6	4	7	9	2	5

Puzzle 349

2	8	4	1	7	3	9	5	6
1	5	3	6	8	9	4	7	2
7	9	6	5	2	4	8	3	1
4	2	5	3	1	7	6	8	9
8	7	9	2	6	5	1	4	3
6	3	1	4	9	8	7	2	5
5	1	7	9	4	2	3	6	8
3	6	8	7	5	1	2	9	4
9	4	2	8	3	6	5	1	7

Puzzle 350

2	4	9	6	1	5	8	7	3
7	6	5	4	3	8	9	1	2
1	8	3	2	7	9	5	4	6
3	2	1	9	6	7	4	5	8
8	9	7	3	5	4	2	6	1
6	5	4	8	2	1	7	3	9
5	7	2	1	9	3	6	8	4
9	1	8	5	4	6	3	2	7
4	3	6	7	8	2	1	9	5

Puzzle 351

2	9	6	1	5	3	4	7	8
8	1	7	2	9	4	6	5	3
3	4	5	8	6	7	2	1	9
5	2	1	4	7	8	3	9	6
6	7	8	5	3	9	1	4	2
4	3	9	6	2	1	7	8	5
7	6	3	9	4	5	8	2	1
9	8	4	3	1	2	5	6	7
1	5	2	7	8	6	9	3	4

Puzzle 352

7	9	3	8	1	5	4	6	2
2	8	1	6	9	4	5	7	3
4	6	5	3	2	7	8	1	9
8	4	6	9	3	2	7	5	1
3	7	9	4	5	1	2	8	6
5	1	2	7	6	8	9	3	4
9	2	8	1	7	6	3	4	5
6	3	4	5	8	9	1	2	7
1	5	7	2	4	3	6	9	8

Puzzle 353

3	7	4	5	2	8	9	6	1
8	1	5	4	9	6	7	3	2
2	9	6	1	7	3	8	4	5
1	2	3	9	5	4	6	8	7
7	6	9	2	8	1	4	5	3
5	4	8	3	6	7	2	1	9
6	3	2	7	4	5	1	9	8
4	5	7	8	1	9	3	2	6
9	8	1	6	3	2	5	7	4

Puzzle 354

7	6	9	2	1	8	3	5	4
1	8	2	3	5	4	7	9	6
5	3	4	6	9	7	2	1	8
8	5	6	4	2	9	1	3	7
4	7	3	1	8	5	9	6	2
2	9	1	7	6	3	8	4	5
6	2	7	9	4	1	5	8	3
9	4	5	8	3	2	6	7	1
3	1	8	5	7	6	4	2	9

Puzzle 355

5	1	6	7	3	2	9	8	4
9	3	8	5	6	4	1	7	2
7	2	4	8	9	1	5	3	6
3	9	1	2	7	5	6	4	8
6	7	5	4	8	3	2	9	1
8	4	2	6	1	9	3	5	7
1	5	7	3	4	6	8	2	9
4	6	3	9	2	8	7	1	5
2	8	9	1	5	7	4	6	3

Puzzle 356

7	3	5	4	2	1	9	8	6
2	6	1	7	8	9	5	3	4
4	9	8	5	6	3	2	1	7
6	2	3	8	9	5	4	7	1
8	1	4	3	7	2	6	5	9
9	5	7	1	4	6	8	2	3
3	7	9	6	5	8	1	4	2
5	4	2	9	1	7	3	6	8
1	8	6	2	3	4	7	9	5

Puzzle 357

2	8	6	9	3	5	7	4	1
4	7	3	6	8	1	9	5	2
5	9	1	7	4	2	6	3	8
7	6	2	1	9	3	4	8	5
1	4	9	5	2	8	3	6	7
8	3	5	4	6	7	2	1	9
9	1	7	3	5	4	8	2	6
3	5	8	2	7	6	1	9	4
6	2	4	8	1	9	5	7	3

Puzzle 358

5	2	6	8	9	4	3	1	7
1	3	9	6	5	7	2	8	4
8	7	4	3	2	1	6	9	5
4	1	8	2	7	9	5	3	6
2	5	7	1	3	6	8	4	9
9	6	3	5	4	8	1	7	2
7	8	5	9	1	2	4	6	3
6	4	2	7	8	3	9	5	1
3	9	1	4	6	5	7	2	8

Puzzle 359

7	3	2	6	1	9	4	8	5
8	4	6	7	3	5	1	2	9
1	9	5	4	8	2	6	3	7
3	6	7	2	9	1	5	4	8
4	8	1	3	5	7	2	9	6
2	5	9	8	6	4	7	1	3
9	7	3	1	4	6	8	5	2
6	1	8	5	2	3	9	7	4
5	2	4	9	7	8	3	6	1

Puzzle 360

5	3	2	4	7	1	9	8	6
6	1	9	5	2	8	7	3	4
8	7	4	3	9	6	1	5	2
4	5	3	8	1	2	6	7	9
9	8	7	6	3	4	2	1	5
2	6	1	7	5	9	3	4	8
3	2	8	9	4	7	5	6	1
1	4	5	2	6	3	8	9	7
7	9	6	1	8	5	4	2	3

Puzzle 361

7	4	9	2	1	8	5	3	6
8	2	5	9	6	3	4	1	7
3	6	1	5	7	4	8	9	2
1	3	6	7	8	9	2	4	5
9	7	8	4	2	5	1	6	3
2	5	4	6	3	1	9	7	8
4	8	7	3	9	2	6	5	1
6	9	2	1	5	7	3	8	4
5	1	3	8	4	6	7	2	9

Puzzle 362

6	2	7	4	3	9	5	8	1
4	3	5	7	8	1	6	2	9
8	1	9	2	5	6	3	7	4
2	6	4	3	9	5	7	1	8
3	7	8	1	2	4	9	5	6
5	9	1	6	7	8	2	4	3
9	5	3	8	4	2	1	6	7
1	8	2	9	6	7	4	3	5
7	4	6	5	1	3	8	9	2

Puzzle 363

4	6	8	2	9	3	1	5	7
9	7	1	5	4	6	2	8	3
5	2	3	8	7	1	9	4	6
7	8	2	6	5	4	3	9	1
1	3	9	7	8	2	4	6	5
6	5	4	1	3	9	7	2	8
2	9	7	3	6	5	8	1	4
8	4	5	9	1	7	6	3	2
3	1	6	4	2	8	5	7	9

Puzzle 364

7	4	1	3	6	8	2	5	9
3	2	8	4	9	5	6	7	1
9	6	5	1	2	7	8	3	4
1	7	9	8	3	4	5	6	2
6	5	2	9	7	1	4	8	3
8	3	4	2	5	6	1	9	7
4	1	6	7	8	3	9	2	5
5	9	3	6	1	2	7	4	8
2	8	7	5	4	9	3	1	6

Puzzle 365

6	7	1	4	9	8	5	3	2
2	8	4	3	1	5	9	6	7
5	9	3	2	7	6	4	8	1
4	1	9	6	2	7	3	5	8
8	3	5	9	4	1	7	2	6
7	2	6	5	8	3	1	9	4
9	5	2	1	6	4	8	7	3
3	4	7	8	5	2	6	1	9
1	6	8	7	3	9	2	4	5

Puzzle 366

2	7	8	9	1	5	4	6	3
9	4	5	2	3	6	1	8	7
1	3	6	4	7	8	5	9	2
6	9	7	8	2	4	3	5	1
8	1	4	7	5	3	9	2	6
3	5	2	1	6	9	7	4	8
5	8	1	3	4	2	6	7	9
7	6	9	5	8	1	2	3	4
4	2	3	6	9	7	8	1	5

Puzzle 367

7	4	9	6	3	5	1	8	2
8	3	2	1	9	7	4	6	5
1	5	6	4	8	2	3	7	9
5	8	4	7	1	6	2	9	3
9	6	3	8	2	4	7	5	1
2	1	7	3	5	9	8	4	6
3	9	8	5	4	1	6	2	7
6	2	1	9	7	8	5	3	4
4	7	5	2	6	3	9	1	8

Puzzle 368

1	6	3	4	8	5	2	7	9
4	5	2	1	7	9	6	3	8
7	8	9	2	3	6	4	1	5
8	1	7	3	5	4	9	6	2
9	3	6	8	2	1	7	5	4
5	2	4	6	9	7	3	8	1
6	9	5	7	1	2	8	4	3
2	7	8	5	4	3	1	9	6
3	4	1	9	6	8	5	2	7

Puzzle 369

4	8	5	9	1	7	3	2	6
7	1	3	8	6	2	5	4	9
2	9	6	5	3	4	1	7	8
8	4	2	7	9	3	6	1	5
5	7	9	1	8	6	4	3	2
6	3	1	4	2	5	9	8	7
9	6	4	3	7	8	2	5	1
3	2	7	6	5	1	8	9	4
1	5	8	2	4	9	7	6	3

Puzzle 370

1	3	7	4	9	6	5	8	2
2	4	8	1	7	5	9	6	3
9	5	6	3	8	2	1	7	4
4	9	2	5	1	7	6	3	8
7	1	5	8	6	3	2	4	9
8	6	3	2	4	9	7	1	5
5	8	1	7	2	4	3	9	6
6	2	4	9	3	1	8	5	7
3	7	9	6	5	8	4	2	1

Puzzle 371

3	6	8	4	2	1	7	9	5
4	9	2	6	7	5	3	1	8
1	7	5	9	8	3	4	2	6
9	4	7	2	6	8	5	3	1
2	1	6	3	5	7	9	8	4
8	5	3	1	9	4	2	6	7
5	3	4	8	1	9	6	7	2
7	2	1	5	3	6	8	4	9
6	8	9	7	4	2	1	5	3

Puzzle 372

1	7	5	6	3	9	8	2	4
6	9	3	2	4	8	1	7	5
4	2	8	5	1	7	3	9	6
9	1	6	7	2	3	5	4	8
5	3	4	8	9	6	7	1	2
7	8	2	4	5	1	9	6	3
2	5	7	9	8	4	6	3	1
3	4	9	1	6	5	2	8	7
8	6	1	3	7	2	4	5	9

Puzzle 373

3	9	1	7	2	6	8	4	5
6	7	8	3	4	5	2	1	9
5	2	4	1	9	8	3	7	6
1	4	3	8	7	9	5	6	2
2	8	5	4	6	1	9	3	7
9	6	7	2	5	3	1	8	4
7	1	6	5	3	2	4	9	8
8	5	9	6	1	4	7	2	3
4	3	2	9	8	7	6	5	1

Puzzle 374

5	3	9	2	6	4	7	8	1
2	6	8	3	7	1	9	4	5
4	7	1	9	5	8	6	2	3
7	4	6	5	3	9	2	1	8
1	8	3	7	4	2	5	9	6
9	2	5	8	1	6	4	3	7
6	1	2	4	8	5	3	7	9
3	5	4	1	9	7	8	6	2
8	9	7	6	2	3	1	5	4

Puzzle 375

2	9	5	8	6	1	7	3	4
4	7	8	5	3	2	9	6	1
6	3	1	9	4	7	8	2	5
3	1	9	7	5	6	4	8	2
7	6	2	1	8	4	3	5	9
5	8	4	3	2	9	6	1	7
9	2	3	6	7	5	1	4	8
8	4	7	2	1	3	5	9	6
1	5	6	4	9	8	2	7	3

Puzzle 376

9	1	7	5	3	6	8	2	4
4	5	6	8	7	2	9	1	3
3	8	2	1	9	4	7	6	5
7	9	4	2	8	3	1	5	6
6	2	1	9	4	5	3	7	8
8	3	5	6	1	7	2	4	9
2	7	3	4	6	9	5	8	1
5	4	8	3	2	1	6	9	7
1	6	9	7	5	8	4	3	2

Puzzle 377

8	7	4	3	1	9	5	6	2
9	5	3	4	2	6	1	8	7
2	6	1	7	5	8	3	9	4
7	4	6	2	9	5	8	3	1
1	3	2	8	6	7	4	5	9
5	8	9	1	4	3	7	2	6
3	1	8	6	7	2	9	4	5
6	9	7	5	3	4	2	1	8
4	2	5	9	8	1	6	7	3

Puzzle 378

8	4	9	7	2	1	3	6	5
3	5	6	8	9	4	2	1	7
7	2	1	5	3	6	4	8	9
1	8	4	9	6	3	5	7	2
6	7	2	4	8	5	1	9	3
9	3	5	1	7	2	8	4	6
2	6	8	3	1	9	7	5	4
5	9	7	2	4	8	6	3	1
4	1	3	6	5	7	9	2	8

Puzzle 379

6	4	5	7	1	9	2	3	8
7	1	9	3	8	2	4	5	6
8	2	3	5	6	4	7	9	1
9	6	7	8	5	3	1	2	4
4	3	8	1	2	7	5	6	9
2	5	1	9	4	6	3	8	7
3	8	2	6	7	1	9	4	5
1	9	6	4	3	5	8	7	2
5	7	4	2	9	8	6	1	3

Puzzle 380

9	3	7	8	5	4	2	1	6
5	6	4	9	1	2	3	8	7
1	2	8	7	3	6	4	5	9
6	9	2	5	7	8	1	3	4
7	5	1	4	6	3	9	2	8
4	8	3	2	9	1	7	6	5
2	7	6	3	8	9	5	4	1
8	4	9	1	2	5	6	7	3
3	1	5	6	4	7	8	9	2

Puzzle 381

3	2	4	7	8	5	1	6	9
9	8	1	3	4	6	5	2	7
7	6	5	2	9	1	4	3	8
5	3	7	1	6	4	9	8	2
8	1	6	9	5	2	3	7	4
2	4	9	8	3	7	6	1	5
6	9	2	4	1	8	7	5	3
4	5	8	6	7	3	2	9	1
1	7	3	5	2	9	8	4	6

Puzzle 382

4	8	1	7	6	9	3	2	5
7	9	2	4	3	5	6	1	8
6	3	5	2	8	1	7	4	9
8	1	4	3	5	6	2	9	7
3	6	7	1	9	2	8	5	4
5	2	9	8	7	4	1	3	6
1	5	3	6	4	8	9	7	2
9	7	8	5	2	3	4	6	1
2	4	6	9	1	7	5	8	3

Puzzle 383

7	6	8	1	9	5	4	3	2
1	2	5	3	6	4	8	9	7
3	4	9	2	8	7	6	1	5
2	7	6	5	3	9	1	4	8
9	1	4	6	7	8	5	2	3
5	8	3	4	1	2	9	7	6
4	3	7	9	5	6	2	8	1
8	5	2	7	4	1	3	6	9
6	9	1	8	2	3	7	5	4

Puzzle 384

1	6	7	2	4	8	5	9	3
3	4	9	7	5	6	1	2	8
5	8	2	9	1	3	4	7	6
6	7	5	8	3	2	9	4	1
8	2	3	1	9	4	7	6	5
4	9	1	6	7	5	8	3	2
2	5	6	4	8	7	3	1	9
9	3	4	5	2	1	6	8	7
7	1	8	3	6	9	2	5	4

Puzzle 385

9	1	6	5	3	2	7	4	8
5	2	8	7	1	4	6	9	3
3	7	4	8	9	6	1	5	2
1	9	3	4	8	5	2	7	6
6	8	2	3	7	9	5	1	4
4	5	7	6	2	1	3	8	9
8	4	1	2	5	3	9	6	7
7	3	5	9	6	8	4	2	1
2	6	9	1	4	7	8	3	5

Puzzle 386

8	7	2	9	6	4	1	3	5
3	9	5	2	7	1	8	4	6
1	4	6	8	5	3	2	7	9
6	8	1	3	4	5	7	9	2
5	3	4	7	2	9	6	1	8
9	2	7	6	1	8	4	5	3
4	6	3	5	8	7	9	2	1
2	1	9	4	3	6	5	8	7
7	5	8	1	9	2	3	6	4

Puzzle 387

7	2	3	5	4	8	9	6	1
5	9	4	7	6	1	3	8	2
1	6	8	2	9	3	5	4	7
9	7	5	1	2	6	8	3	4
2	4	6	8	3	5	1	7	9
8	3	1	9	7	4	2	5	6
3	5	2	4	1	7	6	9	8
4	8	9	6	5	2	7	1	3
6	1	7	3	8	9	4	2	5

Puzzle 388

6	8	7	1	5	2	3	4	9
2	1	3	6	4	9	8	5	7
5	9	4	8	7	3	6	1	2
7	3	1	5	8	6	2	9	4
9	2	8	3	1	4	5	7	6
4	5	6	9	2	7	1	3	8
8	7	5	2	9	1	4	6	3
3	4	2	7	6	5	9	8	1
1	6	9	4	3	8	7	2	5

Puzzle 389

9	4	3	5	1	6	2	7	8
8	5	1	4	2	7	6	3	9
7	2	6	3	8	9	5	4	1
4	9	8	6	7	5	3	1	2
6	1	5	2	9	3	4	8	7
3	7	2	8	4	1	9	6	5
1	8	4	9	3	2	7	5	6
5	3	9	7	6	8	1	2	4
2	6	7	1	5	4	8	9	3

Puzzle 390

7	1	2	4	3	5	8	6	9
4	6	3	7	9	8	5	2	1
9	8	5	1	2	6	3	7	4
8	2	4	9	1	3	7	5	6
1	9	6	5	8	7	4	3	2
5	3	7	2	6	4	9	1	8
6	5	9	3	4	1	2	8	7
3	4	1	8	7	2	6	9	5
2	7	8	6	5	9	1	4	3

Puzzle 391

2	1	6	9	7	4	3	5	8
3	7	5	1	8	2	9	4	6
9	4	8	6	3	5	1	2	7
7	9	4	2	5	3	6	8	1
5	8	3	4	6	1	7	9	2
1	6	2	8	9	7	4	3	5
6	3	7	5	4	8	2	1	9
4	5	1	7	2	9	8	6	3
8	2	9	3	1	6	5	7	4

Puzzle 392

8	5	9	2	1	3	7	4	6
7	4	3	8	5	6	9	2	1
2	6	1	7	9	4	3	5	8
9	7	5	3	6	1	4	8	2
6	1	4	9	8	2	5	3	7
3	2	8	5	4	7	6	1	9
4	9	2	1	7	5	8	6	3
5	3	7	6	2	8	1	9	4
1	8	6	4	3	9	2	7	5

Puzzle 393

9	6	7	8	5	2	4	3	1
1	5	3	4	9	7	8	6	2
2	8	4	1	6	3	7	9	5
3	1	8	9	2	6	5	7	4
7	9	5	3	4	8	1	2	6
4	2	6	7	1	5	3	8	9
6	3	1	5	8	9	2	4	7
8	4	9	2	7	1	6	5	3
5	7	2	6	3	4	9	1	8

Puzzle 394

1	8	3	4	7	2	6	9	5
4	6	2	8	9	5	1	7	3
5	9	7	6	1	3	8	2	4
2	3	5	9	4	1	7	8	6
9	7	6	5	2	8	3	4	1
8	4	1	7	3	6	2	5	9
7	5	8	3	6	4	9	1	2
3	2	9	1	5	7	4	6	8
6	1	4	2	8	9	5	3	7

Puzzle 395

8	6	2	7	3	4	5	9	1
1	9	3	8	2	5	4	6	7
4	5	7	1	6	9	2	3	8
3	8	1	9	7	2	6	4	5
2	7	5	4	8	6	9	1	3
9	4	6	3	5	1	8	7	2
5	2	9	6	1	3	7	8	4
6	1	8	2	4	7	3	5	9
7	3	4	5	9	8	1	2	6

Puzzle 396

1	5	6	9	3	8	2	4	7
3	8	4	7	2	5	6	9	1
9	2	7	1	4	6	5	8	3
5	1	3	4	9	2	7	6	8
2	6	8	5	7	1	4	3	9
7	4	9	6	8	3	1	2	5
6	3	1	8	5	4	9	7	2
8	7	5	2	6	9	3	1	4
4	9	2	3	1	7	8	5	6

Puzzle 397

1	6	9	3	4	2	7	8	5
2	5	8	6	9	7	1	4	3
7	4	3	5	1	8	6	2	9
6	7	1	8	5	3	2	9	4
3	8	4	7	2	9	5	1	6
9	2	5	1	6	4	3	7	8
4	9	7	2	3	5	8	6	1
8	3	6	4	7	1	9	5	2
5	1	2	9	8	6	4	3	7

Puzzle 398

7	4	5	6	9	2	3	8	1
9	6	8	7	1	3	4	5	2
1	2	3	4	5	8	7	6	9
5	8	9	2	6	7	1	4	3
6	7	4	5	3	1	2	9	8
2	3	1	9	8	4	5	7	6
4	5	6	3	2	9	8	1	7
3	1	7	8	4	6	9	2	5
8	9	2	1	7	5	6	3	4

Puzzle 399

2	8	1	4	5	6	9	7	3
6	5	4	3	7	9	8	1	2
7	9	3	8	1	2	6	4	5
4	6	9	1	8	3	5	2	7
8	3	7	2	9	5	4	6	1
1	2	5	6	4	7	3	9	8
3	1	2	9	6	8	7	5	4
5	4	6	7	3	1	2	8	9
9	7	8	5	2	4	1	3	6

Puzzle 400

7	6	4	3	1	2	8	9	5
9	2	8	7	6	5	4	3	1
5	3	1	8	9	4	2	6	7
1	4	9	6	7	3	5	2	8
8	5	2	1	4	9	6	7	3
6	7	3	5	2	8	1	4	9
4	8	5	2	3	7	9	1	6
3	9	6	4	5	1	7	8	2
2	1	7	9	8	6	3	5	4

Puzzle 401

5	4	9	7	8	2	1	6	3
7	2	6	1	5	3	4	9	8
3	1	8	9	4	6	2	5	7
1	7	5	3	9	8	6	2	4
4	9	3	6	2	7	8	1	5
6	8	2	5	1	4	3	7	9
9	6	4	2	3	5	7	8	1
2	3	1	8	7	9	5	4	6
8	5	7	4	6	1	9	3	2

Puzzle 402

2	9	7	8	1	4	6	5	3
8	1	4	3	5	6	2	9	7
5	6	3	7	9	2	1	4	8
6	3	9	5	4	1	7	8	2
4	5	1	2	8	7	3	6	9
7	2	8	9	6	3	4	1	5
9	7	6	1	2	8	5	3	4
3	4	5	6	7	9	8	2	1
1	8	2	4	3	5	9	7	6

Puzzle 403

4	9	7	5	2	1	6	3	8
6	1	5	9	3	8	4	2	7
8	3	2	6	4	7	1	5	9
9	5	6	1	7	4	3	8	2
2	7	3	8	9	6	5	4	1
1	4	8	3	5	2	9	7	6
3	6	4	2	8	9	7	1	5
7	2	9	4	1	5	8	6	3
5	8	1	7	6	3	2	9	4

Puzzle 404

9	2	6	8	3	5	4	1	7
5	3	1	9	7	4	8	2	6
4	8	7	1	2	6	5	9	3
1	5	3	6	9	2	7	4	8
8	9	2	5	4	7	3	6	1
6	7	4	3	8	1	9	5	2
2	6	8	4	5	3	1	7	9
7	4	9	2	1	8	6	3	5
3	1	5	7	6	9	2	8	4

Puzzle 405

9	5	2	4	6	7	3	1	8
6	7	8	1	3	9	5	2	4
3	1	4	8	2	5	6	7	9
2	4	9	5	1	6	8	3	7
8	6	7	3	4	2	9	5	1
5	3	1	9	7	8	4	6	2
7	2	3	6	8	4	1	9	5
4	9	6	7	5	1	2	8	3
1	8	5	2	9	3	7	4	6

Puzzle 406

4	8	7	5	6	3	9	2	1
6	5	2	7	9	1	4	8	3
1	3	9	4	2	8	6	5	7
5	7	1	2	4	6	8	3	9
8	4	3	1	7	9	5	6	2
2	9	6	3	8	5	7	1	4
7	1	4	6	5	2	3	9	8
3	6	8	9	1	4	2	7	5
9	2	5	8	3	7	1	4	6

Puzzle 407

3	6	9	8	2	5	1	7	4
7	8	4	6	9	1	2	3	5
5	2	1	7	4	3	9	6	8
6	4	8	5	3	2	7	1	9
1	9	7	4	8	6	5	2	3
2	3	5	9	1	7	4	8	6
9	7	3	2	6	4	8	5	1
8	5	6	1	7	9	3	4	2
4	1	2	3	5	8	6	9	7

Puzzle 408

2	8	4	1	3	6	9	5	7
5	6	9	2	4	7	1	8	3
1	7	3	9	5	8	6	4	2
7	4	8	5	9	3	2	1	6
3	1	6	7	2	4	8	9	5
9	5	2	6	8	1	3	7	4
4	9	7	8	6	2	5	3	1
8	2	1	3	7	5	4	6	9
6	3	5	4	1	9	7	2	8

Puzzle 409

8	1	5	7	2	9	6	4	3
2	4	3	6	5	1	7	9	8
9	7	6	8	3	4	5	1	2
7	3	1	4	9	6	8	2	5
6	5	2	1	8	3	9	7	4
4	9	8	2	7	5	3	6	1
1	6	9	3	4	8	2	5	7
3	2	4	5	6	7	1	8	9
5	8	7	9	1	2	4	3	6

Puzzle 410

5	1	3	2	6	7	4	8	9
4	8	6	3	9	5	1	2	7
2	9	7	4	8	1	5	3	6
9	4	5	1	7	2	8	6	3
3	6	2	9	4	8	7	1	5
1	7	8	5	3	6	2	9	4
8	3	9	7	2	4	6	5	1
6	5	4	8	1	9	3	7	2
7	2	1	6	5	3	9	4	8

Puzzle 411

4	9	2	6	8	3	5	1	7
5	8	1	4	9	7	3	2	6
6	7	3	1	5	2	8	4	9
9	4	6	5	2	8	1	7	3
2	5	7	9	3	1	4	6	8
3	1	8	7	4	6	2	9	5
1	3	5	2	6	9	7	8	4
7	6	4	8	1	5	9	3	2
8	2	9	3	7	4	6	5	1

Puzzle 412

8	7	2	4	9	5	3	1	6
3	1	4	6	8	7	2	9	5
6	5	9	2	3	1	8	7	4
9	4	3	1	5	2	7	6	8
7	6	1	3	4	8	9	5	2
5	2	8	7	6	9	4	3	1
1	9	5	8	2	3	6	4	7
4	8	7	9	1	6	5	2	3
2	3	6	5	7	4	1	8	9

Puzzle 413

6	2	1	4	5	3	9	7	8
7	9	3	6	8	2	4	5	1
4	8	5	1	7	9	3	6	2
5	4	6	8	9	7	1	2	3
3	1	8	2	4	6	7	9	5
2	7	9	3	1	5	8	4	6
9	3	7	5	2	8	6	1	4
1	6	2	7	3	4	5	8	9
8	5	4	9	6	1	2	3	7

Puzzle 414

8	5	4	6	9	7	2	1	3
2	1	9	8	3	5	6	7	4
6	7	3	4	2	1	9	5	8
9	8	5	7	1	2	4	3	6
4	3	1	5	8	6	7	2	9
7	6	2	9	4	3	1	8	5
1	4	8	2	5	9	3	6	7
5	2	7	3	6	4	8	9	1
3	9	6	1	7	8	5	4	2

Puzzle 415

9	5	6	1	3	2	8	4	7
1	8	7	6	9	4	2	3	5
2	3	4	5	7	8	6	1	9
8	6	2	4	5	1	7	9	3
7	4	9	2	6	3	5	8	1
5	1	3	9	8	7	4	6	2
4	9	1	7	2	6	3	5	8
3	2	5	8	4	9	1	7	6
6	7	8	3	1	5	9	2	4

Puzzle 416

9	2	6	1	4	8	3	7	5
3	7	4	5	6	2	1	9	8
1	8	5	9	7	3	4	2	6
4	3	2	7	1	6	8	5	9
7	5	8	4	2	9	6	1	3
6	1	9	8	3	5	2	4	7
5	4	1	3	8	7	9	6	2
2	9	3	6	5	1	7	8	4
8	6	7	2	9	4	5	3	1

Puzzle 417

3	5	1	2	4	8	6	7	9
8	6	7	1	9	5	4	3	2
9	4	2	3	6	7	5	1	8
4	7	5	8	1	6	2	9	3
2	8	3	4	7	9	1	6	5
1	9	6	5	3	2	7	8	4
7	3	8	6	2	4	9	5	1
5	2	9	7	8	1	3	4	6
6	1	4	9	5	3	8	2	7

Puzzle 418

3	2	4	1	9	5	6	8	7
1	8	9	4	7	6	3	2	5
5	7	6	8	3	2	1	9	4
9	4	2	6	8	7	5	1	3
6	3	8	9	5	1	4	7	2
7	1	5	2	4	3	8	6	9
8	9	1	3	2	4	7	5	6
2	5	3	7	6	8	9	4	1
4	6	7	5	1	9	2	3	8

Puzzle 419

8	1	9	5	4	3	7	2	6
4	3	6	7	9	2	1	5	8
2	5	7	8	1	6	9	4	3
6	4	5	3	8	1	2	7	9
9	8	3	4	2	7	5	6	1
1	7	2	6	5	9	8	3	4
3	6	1	2	7	8	4	9	5
5	2	8	9	3	4	6	1	7
7	9	4	1	6	5	3	8	2

Puzzle 420

3	4	6	1	8	2	5	7	9
9	5	1	6	3	7	4	8	2
2	7	8	4	9	5	1	3	6
7	8	3	9	1	6	2	4	5
4	9	2	8	5	3	7	6	1
6	1	5	2	7	4	8	9	3
1	6	7	3	2	8	9	5	4
5	2	4	7	6	9	3	1	8
8	3	9	5	4	1	6	2	7

Puzzle 421

9	6	4	2	8	7	5	1	3
5	7	1	6	3	4	2	8	9
8	3	2	9	1	5	4	7	6
3	2	5	4	7	1	6	9	8
1	4	6	3	9	8	7	5	2
7	8	9	5	2	6	1	3	4
6	1	8	7	4	3	9	2	5
4	9	7	8	5	2	3	6	1
2	5	3	1	6	9	8	4	7

Puzzle 422

9	6	7	1	8	2	4	5	3
5	4	1	6	9	3	2	8	7
2	3	8	5	7	4	1	6	9
6	1	5	2	4	9	7	3	8
8	2	3	7	1	6	5	9	4
4	7	9	8	3	5	6	2	1
7	8	6	9	2	1	3	4	5
3	9	2	4	5	7	8	1	6
1	5	4	3	6	8	9	7	2

Puzzle 423

7	3	9	6	8	2	1	5	4
6	8	4	5	9	1	7	2	3
5	1	2	7	3	4	8	6	9
4	7	1	9	6	8	5	3	2
2	9	6	1	5	3	4	7	8
8	5	3	2	4	7	9	1	6
1	4	5	8	2	6	3	9	7
3	2	7	4	1	9	6	8	5
9	6	8	3	7	5	2	4	1

Puzzle 424

9	5	8	7	6	3	4	1	2
1	6	3	5	4	2	9	8	7
2	4	7	8	9	1	5	6	3
7	3	2	6	8	5	1	9	4
4	1	6	2	7	9	3	5	8
8	9	5	1	3	4	7	2	6
5	2	4	3	1	6	8	7	9
3	7	1	9	2	8	6	4	5
6	8	9	4	5	7	2	3	1

Puzzle 425

5	1	3	6	9	4	8	2	7
4	8	9	7	3	2	5	1	6
2	6	7	5	8	1	4	3	9
7	4	6	8	5	3	2	9	1
9	5	1	2	4	6	3	7	8
8	3	2	1	7	9	6	4	5
6	9	5	3	2	7	1	8	4
1	2	4	9	6	8	7	5	3
3	7	8	4	1	5	9	6	2

Puzzle 426

2	3	7	4	1	9	6	8	5
1	5	9	8	6	3	2	7	4
8	6	4	5	2	7	1	9	3
3	7	2	1	8	4	5	6	9
5	8	6	3	9	2	4	1	7
9	4	1	6	7	5	8	3	2
4	2	8	7	3	6	9	5	1
6	9	3	2	5	1	7	4	8
7	1	5	9	4	8	3	2	6

Puzzle 427

1	3	5	2	9	4	8	6	7
4	6	9	1	7	8	3	5	2
7	2	8	6	5	3	1	4	9
2	5	7	8	1	9	4	3	6
3	4	6	7	2	5	9	8	1
9	8	1	3	4	6	7	2	5
6	9	3	5	8	1	2	7	4
8	1	2	4	6	7	5	9	3
5	7	4	9	3	2	6	1	8

Puzzle 428

7	4	6	9	8	2	1	5	3
3	1	2	4	6	5	9	7	8
5	9	8	3	7	1	2	4	6
9	7	3	5	1	8	4	6	2
1	8	5	2	4	6	3	9	7
6	2	4	7	9	3	8	1	5
2	5	9	6	3	4	7	8	1
8	6	7	1	2	9	5	3	4
4	3	1	8	5	7	6	2	9

Puzzle 429

1	3	8	7	9	4	5	2	6
9	7	4	6	5	2	8	1	3
2	6	5	1	3	8	4	9	7
8	2	6	5	4	9	3	7	1
5	9	7	3	2	1	6	4	8
4	1	3	8	6	7	2	5	9
3	4	2	9	7	6	1	8	5
7	5	1	2	8	3	9	6	4
6	8	9	4	1	5	7	3	2

Puzzle 430

4	1	7	5	8	6	9	2	3
3	8	6	7	9	2	1	4	5
5	2	9	3	4	1	6	7	8
9	6	3	4	2	8	7	5	1
2	7	4	1	5	9	8	3	6
8	5	1	6	7	3	4	9	2
6	3	5	9	1	7	2	8	4
1	9	8	2	3	4	5	6	7
7	4	2	8	6	5	3	1	9

Puzzle 431

5	2	4	1	6	7	8	9	3
8	7	6	2	3	9	4	5	1
9	3	1	8	5	4	2	7	6
1	8	5	6	7	3	9	2	4
7	6	3	4	9	2	5	1	8
4	9	2	5	8	1	6	3	7
3	4	8	7	2	5	1	6	9
2	1	7	9	4	6	3	8	5
6	5	9	3	1	8	7	4	2

Puzzle 432

7	4	6	1	8	2	3	9	5
8	9	2	3	6	5	1	7	4
1	3	5	7	4	9	2	6	8
6	5	1	2	9	3	8	4	7
9	2	8	5	7	4	6	1	3
3	7	4	6	1	8	9	5	2
4	1	9	8	3	7	5	2	6
2	6	3	4	5	1	7	8	9
5	8	7	9	2	6	4	3	1

Puzzle 433

2	8	9	3	4	1	5	6	7
3	6	5	8	2	7	1	9	4
7	4	1	5	6	9	8	3	2
8	5	7	6	9	3	2	4	1
1	9	3	2	5	4	6	7	8
6	2	4	1	7	8	3	5	9
4	1	8	7	3	6	9	2	5
5	7	6	9	1	2	4	8	3
9	3	2	4	8	5	7	1	6

Puzzle 434

5	1	9	7	8	6	4	2	3
2	3	7	5	9	4	1	6	8
8	6	4	2	3	1	9	7	5
7	4	5	8	2	9	6	3	1
1	8	6	4	5	3	2	9	7
3	9	2	6	1	7	5	8	4
9	7	1	3	6	5	8	4	2
4	5	8	9	7	2	3	1	6
6	2	3	1	4	8	7	5	9

Puzzle 435

5	1	9	7	8	3	6	4	2
7	2	8	1	4	6	5	9	3
4	6	3	5	9	2	1	8	7
8	5	6	2	3	4	7	1	9
2	4	1	9	5	7	3	6	8
9	3	7	6	1	8	2	5	4
1	9	2	4	7	5	8	3	6
6	8	4	3	2	1	9	7	5
3	7	5	8	6	9	4	2	1

Puzzle 436

1	9	6	7	2	8	3	4	5
5	3	8	4	6	9	2	7	1
7	2	4	1	3	5	6	8	9
6	7	5	8	1	2	9	3	4
3	1	2	5	9	4	7	6	8
4	8	9	3	7	6	5	1	2
8	4	3	9	5	7	1	2	6
2	5	7	6	4	1	8	9	3
9	6	1	2	8	3	4	5	7

Puzzle 437

1	9	8	6	7	4	2	3	5
4	2	5	8	3	1	9	7	6
3	7	6	2	9	5	1	4	8
6	8	2	5	4	3	7	9	1
5	1	7	9	8	2	3	6	4
9	4	3	1	6	7	5	8	2
7	6	1	3	5	8	4	2	9
2	3	9	4	1	6	8	5	7
8	5	4	7	2	9	6	1	3

Puzzle 438

3	2	8	5	6	7	1	4	9
5	9	1	2	8	4	3	6	7
6	7	4	3	1	9	8	2	5
1	4	9	6	2	5	7	8	3
7	3	6	1	9	8	2	5	4
2	8	5	7	4	3	6	9	1
8	6	3	9	5	1	4	7	2
9	1	2	4	7	6	5	3	8
4	5	7	8	3	2	9	1	6

Puzzle 439

5	8	7	4	9	1	2	3	6
1	2	6	7	8	3	5	4	9
9	4	3	2	6	5	8	1	7
2	7	1	5	3	6	4	9	8
3	5	4	8	2	9	7	6	1
8	6	9	1	4	7	3	2	5
7	9	5	3	1	4	6	8	2
4	1	8	6	7	2	9	5	3
6	3	2	9	5	8	1	7	4

Puzzle 440

4	3	2	6	7	5	9	1	8
9	8	1	2	4	3	6	5	7
5	6	7	9	1	8	3	4	2
2	9	5	8	6	7	1	3	4
6	7	8	1	3	4	2	9	5
1	4	3	5	9	2	7	8	6
7	5	6	3	8	1	4	2	9
3	2	9	4	5	6	8	7	1
8	1	4	7	2	9	5	6	3

Puzzle 441

5	7	9	1	6	8	2	4	3
1	6	8	2	3	4	9	7	5
4	3	2	7	9	5	1	6	8
6	4	5	8	1	3	7	9	2
8	9	7	5	2	6	4	3	1
3	2	1	9	4	7	8	5	6
9	5	4	6	8	2	3	1	7
2	1	6	3	7	9	5	8	4
7	8	3	4	5	1	6	2	9

Puzzle 442

7	5	3	2	8	4	6	9	1
2	6	9	7	3	1	5	8	4
1	8	4	6	9	5	7	2	3
3	4	1	8	2	7	9	5	6
9	2	8	5	4	6	1	3	7
6	7	5	9	1	3	2	4	8
5	1	2	3	6	8	4	7	9
8	9	6	4	7	2	3	1	5
4	3	7	1	5	9	8	6	2

Puzzle 443

5	4	9	6	1	8	3	7	2
2	6	3	5	9	7	4	8	1
1	8	7	4	2	3	9	6	5
9	3	8	1	5	4	6	2	7
7	2	6	8	3	9	1	5	4
4	5	1	7	6	2	8	3	9
8	9	5	2	4	6	7	1	3
6	1	4	3	7	5	2	9	8
3	7	2	9	8	1	5	4	6

Puzzle 444

1	9	2	4	7	5	6	8	3
5	6	7	9	8	3	1	2	4
3	4	8	1	2	6	9	7	5
9	7	3	5	6	2	8	4	1
2	1	4	8	9	7	5	3	6
8	5	6	3	1	4	2	9	7
7	2	1	6	3	9	4	5	8
6	3	5	2	4	8	7	1	9
4	8	9	7	5	1	3	6	2

Puzzle 445

1	8	3	2	5	9	7	4	6
2	6	9	1	4	7	3	8	5
7	4	5	6	3	8	9	1	2
5	7	1	3	2	4	6	9	8
4	9	2	8	1	6	5	3	7
6	3	8	9	7	5	4	2	1
8	1	6	7	9	3	2	5	4
9	2	4	5	6	1	8	7	3
3	5	7	4	8	2	1	6	9

Puzzle 446

3	9	7	2	4	1	8	5	6
8	1	2	9	5	6	4	3	7
4	6	5	8	3	7	9	1	2
9	3	6	7	1	2	5	4	8
5	4	8	6	9	3	7	2	1
7	2	1	5	8	4	6	9	3
2	5	4	1	6	8	3	7	9
1	8	3	4	7	9	2	6	5
6	7	9	3	2	5	1	8	4

Puzzle 447

2	8	4	7	6	1	5	9	3
1	5	9	3	8	4	7	6	2
6	7	3	5	2	9	1	4	8
7	1	8	4	3	2	9	5	6
3	6	5	9	1	8	4	2	7
9	4	2	6	7	5	8	3	1
5	3	7	8	9	6	2	1	4
8	9	1	2	4	3	6	7	5
4	2	6	1	5	7	3	8	9

Puzzle 448

7	8	9	5	4	3	6	2	1
4	2	5	8	6	1	7	9	3
1	6	3	7	9	2	4	8	5
3	7	8	6	1	4	9	5	2
2	9	1	3	7	5	8	6	4
6	5	4	2	8	9	3	1	7
5	1	6	9	3	7	2	4	8
9	3	2	4	5	8	1	7	6
8	4	7	1	2	6	5	3	9

Puzzle 449

3	2	7	6	5	4	8	1	9
1	4	6	3	8	9	5	7	2
9	8	5	7	2	1	3	4	6
7	1	2	9	4	5	6	8	3
6	9	3	2	7	8	4	5	1
4	5	8	1	3	6	9	2	7
8	3	9	4	1	7	2	6	5
5	6	1	8	9	2	7	3	4
2	7	4	5	6	3	1	9	8

Puzzle 450

2	4	7	5	9	1	6	3	8
1	9	3	6	8	4	7	2	5
8	5	6	7	3	2	4	1	9
7	6	9	1	2	8	5	4	3
4	8	1	9	5	3	2	6	7
3	2	5	4	6	7	8	9	1
9	7	2	8	1	6	3	5	4
6	1	4	3	7	5	9	8	2
5	3	8	2	4	9	1	7	6

Puzzle 451

1	5	8	7	9	3	2	4	6
6	2	9	4	1	8	7	5	3
4	3	7	6	2	5	1	9	8
2	7	1	8	6	4	5	3	9
3	8	6	2	5	9	4	1	7
9	4	5	1	3	7	6	8	2
7	9	2	3	4	1	8	6	5
5	6	4	9	8	2	3	7	1
8	1	3	5	7	6	9	2	4

Puzzle 452

7	6	8	1	5	9	3	4	2
1	5	4	2	3	8	9	7	6
2	3	9	7	6	4	8	1	5
3	8	5	6	9	7	1	2	4
6	2	7	4	1	3	5	8	9
4	9	1	5	8	2	6	3	7
5	4	3	8	7	6	2	9	1
8	7	6	9	2	1	4	5	3
9	1	2	3	4	5	7	6	8

Puzzle 453

4	1	7	6	5	9	3	2	8
6	9	2	3	7	8	5	1	4
8	5	3	2	4	1	6	9	7
5	2	9	1	8	4	7	6	3
3	6	1	7	2	5	8	4	9
7	8	4	9	6	3	2	5	1
2	3	6	4	1	7	9	8	5
9	4	8	5	3	2	1	7	6
1	7	5	8	9	6	4	3	2

Puzzle 454

2	6	5	9	7	4	8	1	3
3	7	8	6	1	5	2	4	9
1	9	4	8	3	2	6	5	7
7	3	1	5	6	9	4	8	2
6	4	9	2	8	7	5	3	1
8	5	2	1	4	3	9	7	6
5	2	7	3	9	8	1	6	4
4	8	6	7	2	1	3	9	5
9	1	3	4	5	6	7	2	8

Puzzle 455

2	8	3	6	9	5	4	1	7
1	5	9	7	2	4	8	3	6
7	4	6	8	3	1	2	9	5
4	1	8	2	6	7	9	5	3
6	2	7	9	5	3	1	8	4
3	9	5	4	1	8	6	7	2
9	6	1	5	7	2	3	4	8
8	7	2	3	4	9	5	6	1
5	3	4	1	8	6	7	2	9

Puzzle 456

4	5	3	8	6	9	1	2	7
6	8	7	5	1	2	9	4	3
1	9	2	4	7	3	8	6	5
3	6	9	1	4	7	5	8	2
8	2	4	3	9	5	7	1	6
7	1	5	2	8	6	3	9	4
2	4	1	7	3	8	6	5	9
5	7	6	9	2	1	4	3	8
9	3	8	6	5	4	2	7	1

Puzzle 457

7	3	8	4	5	2	6	9	1
5	1	6	3	7	9	2	4	8
9	4	2	8	1	6	3	5	7
2	5	7	9	3	8	4	1	6
1	6	3	5	2	4	8	7	9
8	9	4	7	6	1	5	3	2
4	2	1	6	9	5	7	8	3
3	8	9	2	4	7	1	6	5
6	7	5	1	8	3	9	2	4

Puzzle 458

3	5	4	8	6	2	9	7	1
8	7	2	1	4	9	6	5	3
6	9	1	3	7	5	4	2	8
4	1	3	7	5	8	2	6	9
5	2	9	4	1	6	8	3	7
7	6	8	9	2	3	5	1	4
2	4	7	5	9	1	3	8	6
1	8	6	2	3	4	7	9	5
9	3	5	6	8	7	1	4	2

Puzzle 459

8	1	6	2	3	9	4	7	5
2	5	9	6	4	7	1	3	8
7	3	4	5	1	8	9	6	2
5	4	1	3	9	2	6	8	7
3	8	2	1	7	6	5	9	4
6	9	7	4	8	5	2	1	3
1	6	5	8	2	3	7	4	9
9	2	8	7	6	4	3	5	1
4	7	3	9	5	1	8	2	6

Puzzle 460

6	7	2	8	3	9	5	1	4
8	4	9	5	1	6	2	7	3
3	1	5	2	4	7	8	6	9
4	5	6	3	7	8	1	9	2
1	3	7	6	9	2	4	8	5
2	9	8	1	5	4	7	3	6
7	6	1	9	2	5	3	4	8
5	8	4	7	6	3	9	2	1
9	2	3	4	8	1	6	5	7

Puzzle 461

6	4	5	7	3	2	9	8	1
7	2	8	6	9	1	4	5	3
3	1	9	5	4	8	2	6	7
5	8	6	4	2	3	7	1	9
2	9	7	8	1	5	3	4	6
4	3	1	9	6	7	8	2	5
1	5	4	2	7	9	6	3	8
9	6	3	1	8	4	5	7	2
8	7	2	3	5	6	1	9	4

Puzzle 462

4	5	1	8	7	6	2	9	3
8	6	3	9	5	2	4	7	1
9	7	2	4	3	1	5	8	6
1	3	7	2	8	9	6	5	4
2	9	6	5	4	7	3	1	8
5	4	8	6	1	3	7	2	9
6	8	5	1	2	4	9	3	7
7	1	9	3	6	5	8	4	2
3	2	4	7	9	8	1	6	5

Puzzle 463

2	4	3	7	1	6	9	8	5
8	1	6	4	9	5	2	3	7
9	5	7	3	8	2	6	4	1
4	2	1	8	6	7	3	5	9
3	9	5	1	2	4	8	7	6
7	6	8	9	5	3	1	2	4
6	7	2	5	3	9	4	1	8
5	8	9	2	4	1	7	6	3
1	3	4	6	7	8	5	9	2

Puzzle 464

3	9	4	5	1	7	8	2	6
6	5	1	4	8	2	7	3	9
8	7	2	6	9	3	1	5	4
1	6	3	9	4	8	2	7	5
9	8	7	3	2	5	4	6	1
2	4	5	1	7	6	3	9	8
7	1	9	2	5	4	6	8	3
5	2	6	8	3	1	9	4	7
4	3	8	7	6	9	5	1	2

Puzzle 465

5	2	4	3	6	7	1	8	9
7	3	6	9	8	1	2	5	4
1	8	9	5	2	4	3	7	6
2	1	8	6	3	9	7	4	5
3	9	5	4	7	8	6	2	1
6	4	7	1	5	2	8	9	3
4	6	2	8	9	3	5	1	7
9	7	3	2	1	5	4	6	8
8	5	1	7	4	6	9	3	2

Puzzle 466

7	1	4	9	6	2	5	8	3
8	2	9	3	4	5	1	6	7
6	3	5	8	1	7	2	4	9
1	8	7	4	9	3	6	5	2
9	6	3	5	2	1	8	7	4
4	5	2	7	8	6	9	3	1
5	4	8	2	3	9	7	1	6
3	9	6	1	7	8	4	2	5
2	7	1	6	5	4	3	9	8

Puzzle 467

5	4	3	7	2	1	8	9	6
6	7	9	3	8	5	4	2	1
2	8	1	4	6	9	3	5	7
3	2	6	9	5	7	1	8	4
1	9	4	6	3	8	5	7	2
8	5	7	1	4	2	9	6	3
4	6	2	8	9	3	7	1	5
9	1	5	2	7	4	6	3	8
7	3	8	5	1	6	2	4	9

Puzzle 468

9	8	1	7	2	6	5	4	3
3	7	6	9	5	4	2	8	1
4	2	5	3	1	8	9	7	6
5	3	2	1	9	7	4	6	8
8	1	4	2	6	3	7	5	9
6	9	7	4	8	5	3	1	2
7	6	3	8	4	2	1	9	5
1	4	8	5	3	9	6	2	7
2	5	9	6	7	1	8	3	4

Puzzle 469

1	4	2	3	8	5	9	6	7
5	9	6	1	7	2	8	3	4
7	3	8	4	9	6	5	1	2
2	7	9	5	3	1	4	8	6
8	6	1	7	2	4	3	9	5
4	5	3	8	6	9	2	7	1
6	8	7	2	5	3	1	4	9
3	2	4	9	1	7	6	5	8
9	1	5	6	4	8	7	2	3

Puzzle 470

7	8	5	6	1	3	2	4	9
4	9	3	2	7	5	6	8	1
6	1	2	9	4	8	3	5	7
1	7	4	5	8	2	9	3	6
2	5	9	3	6	4	1	7	8
3	6	8	1	9	7	4	2	5
8	2	6	7	3	1	5	9	4
5	4	1	8	2	9	7	6	3
9	3	7	4	5	6	8	1	2

Puzzle 471

6	9	2	5	1	7	3	8	4
1	5	4	9	8	3	2	6	7
7	8	3	4	6	2	5	9	1
8	3	7	1	5	4	9	2	6
5	1	6	2	9	8	4	7	3
4	2	9	7	3	6	8	1	5
9	7	1	8	4	5	6	3	2
3	4	8	6	2	1	7	5	9
2	6	5	3	7	9	1	4	8

Puzzle 472

1	9	6	4	8	5	7	3	2
2	7	8	9	6	3	5	4	1
4	5	3	7	2	1	8	6	9
8	4	1	5	9	2	6	7	3
7	6	9	3	1	8	4	2	5
5	3	2	6	7	4	9	1	8
6	2	5	8	3	7	1	9	4
3	8	7	1	4	9	2	5	6
9	1	4	2	5	6	3	8	7

Puzzle 473

9	1	6	3	4	5	8	2	7
4	7	3	1	8	2	5	9	6
8	5	2	6	9	7	3	4	1
3	4	8	2	1	9	7	6	5
5	2	9	8	7	6	4	1	3
1	6	7	5	3	4	2	8	9
2	9	4	7	5	1	6	3	8
6	3	5	9	2	8	1	7	4
7	8	1	4	6	3	9	5	2

Puzzle 474

6	1	2	3	5	7	9	8	4
5	3	7	9	4	8	2	1	6
9	8	4	2	6	1	7	3	5
4	7	3	5	1	2	6	9	8
2	6	1	7	8	9	5	4	3
8	9	5	6	3	4	1	7	2
3	2	9	8	7	5	4	6	1
7	4	8	1	2	6	3	5	9
1	5	6	4	9	3	8	2	7

Puzzle 475

5	8	6	4	3	7	2	9	1
9	2	1	5	6	8	7	4	3
3	4	7	2	1	9	5	6	8
6	1	3	7	2	5	9	8	4
8	9	5	6	4	3	1	2	7
4	7	2	9	8	1	6	3	5
7	6	4	3	5	2	8	1	9
2	5	8	1	9	4	3	7	6
1	3	9	8	7	6	4	5	2

Puzzle 476

4	7	2	8	3	6	5	1	9
8	3	9	1	2	5	6	7	4
5	1	6	4	9	7	3	8	2
3	4	7	9	6	2	8	5	1
9	5	8	3	1	4	7	2	6
2	6	1	7	5	8	9	4	3
7	8	3	2	4	9	1	6	5
1	2	5	6	8	3	4	9	7
6	9	4	5	7	1	2	3	8

Puzzle 477

5	9	2	8	3	4	7	6	1
4	6	3	1	5	7	8	2	9
8	7	1	9	6	2	5	3	4
2	3	6	4	7	5	1	9	8
1	8	5	3	2	9	6	4	7
7	4	9	6	8	1	2	5	3
6	2	4	7	9	8	3	1	5
3	1	7	5	4	6	9	8	2
9	5	8	2	1	3	4	7	6

Puzzle 478

6	5	1	7	3	9	8	4	2
3	9	4	2	5	8	1	6	7
7	8	2	1	4	6	9	5	3
9	3	8	4	7	5	6	2	1
4	7	5	6	1	2	3	9	8
1	2	6	8	9	3	4	7	5
2	6	7	3	8	4	5	1	9
5	1	3	9	6	7	2	8	4
8	4	9	5	2	1	7	3	6

Puzzle 479

4	1	7	8	9	3	2	5	6
8	6	5	1	7	2	9	3	4
2	9	3	4	6	5	7	1	8
7	4	6	5	1	9	8	2	3
3	5	9	7	2	8	6	4	1
1	2	8	3	4	6	5	9	7
5	3	1	9	8	7	4	6	2
9	7	2	6	3	4	1	8	5
6	8	4	2	5	1	3	7	9

Puzzle 480

4	1	2	9	8	6	7	3	5
9	6	7	1	3	5	8	4	2
8	3	5	4	2	7	1	6	9
7	8	9	3	1	4	2	5	6
5	2	3	7	6	8	4	9	1
1	4	6	5	9	2	3	8	7
2	9	8	6	7	3	5	1	4
3	5	1	2	4	9	6	7	8
6	7	4	8	5	1	9	2	3

Puzzle 481

8	2	3	5	7	9	1	4	6
7	4	6	1	3	2	8	5	9
5	1	9	6	4	8	3	2	7
6	7	1	9	8	4	2	3	5
9	8	5	3	2	1	6	7	4
2	3	4	7	5	6	9	1	8
3	5	2	8	9	7	4	6	1
1	9	7	4	6	3	5	8	2
4	6	8	2	1	5	7	9	3

Puzzle 482

8	9	4	7	2	5	3	6	1
1	3	7	8	6	9	2	5	4
5	2	6	3	1	4	8	9	7
7	6	2	9	3	8	1	4	5
4	1	9	2	5	7	6	8	3
3	8	5	1	4	6	9	7	2
2	5	8	4	9	1	7	3	6
9	4	1	6	7	3	5	2	8
6	7	3	5	8	2	4	1	9

Puzzle 483

4	1	5	3	2	8	6	9	7
6	2	7	1	4	9	5	3	8
9	8	3	7	6	5	1	4	2
5	9	1	8	3	7	4	2	6
3	4	6	9	5	2	7	8	1
8	7	2	4	1	6	3	5	9
1	5	8	2	7	4	9	6	3
7	6	9	5	8	3	2	1	4
2	3	4	6	9	1	8	7	5

Puzzle 484

7	1	2	4	5	9	6	8	3
6	4	8	7	3	2	5	1	9
5	9	3	1	8	6	7	4	2
3	7	1	8	2	5	4	9	6
4	6	5	9	7	3	8	2	1
8	2	9	6	1	4	3	7	5
2	5	4	3	9	7	1	6	8
1	3	6	2	4	8	9	5	7
9	8	7	5	6	1	2	3	4

Puzzle 485

6	9	5	2	4	8	7	3	1
2	1	3	7	6	9	5	4	8
8	4	7	3	5	1	9	6	2
3	6	9	5	1	2	8	7	4
1	8	4	9	3	7	2	5	6
7	5	2	4	8	6	1	9	3
9	3	1	6	2	5	4	8	7
4	7	8	1	9	3	6	2	5
5	2	6	8	7	4	3	1	9

Puzzle 486

7	1	9	8	2	6	5	3	4
8	4	6	7	5	3	9	1	2
2	5	3	4	1	9	7	6	8
1	2	4	5	6	7	8	9	3
6	7	5	3	9	8	4	2	1
3	9	8	2	4	1	6	5	7
9	8	1	6	3	4	2	7	5
5	3	7	9	8	2	1	4	6
4	6	2	1	7	5	3	8	9

Puzzle 487

4	2	3	7	6	8	5	1	9
1	9	5	4	3	2	6	7	8
8	6	7	9	5	1	3	2	4
3	4	9	8	2	7	1	5	6
6	1	8	3	4	5	7	9	2
5	7	2	1	9	6	8	4	3
2	8	4	5	7	3	9	6	1
7	3	6	2	1	9	4	8	5
9	5	1	6	8	4	2	3	7

Puzzle 488

6	4	5	7	1	8	9	3	2
3	7	2	4	6	9	1	8	5
1	9	8	5	3	2	6	4	7
7	8	6	2	5	4	3	9	1
2	1	4	6	9	3	7	5	8
5	3	9	1	8	7	2	6	4
8	2	1	9	4	6	5	7	3
4	6	7	3	2	5	8	1	9
9	5	3	8	7	1	4	2	6

Puzzle 489

3	7	5	4	8	6	9	1	2
1	2	4	5	9	3	6	8	7
6	9	8	7	2	1	4	5	3
7	4	6	2	3	5	1	9	8
2	1	9	6	7	8	3	4	5
5	8	3	1	4	9	7	2	6
8	5	7	9	6	4	2	3	1
9	6	1	3	5	2	8	7	4
4	3	2	8	1	7	5	6	9

Puzzle 490

9	2	8	6	4	7	1	3	5
5	6	3	9	1	2	8	7	4
1	7	4	5	3	8	2	9	6
4	3	6	7	8	5	9	1	2
7	1	2	4	6	9	5	8	3
8	9	5	3	2	1	4	6	7
3	4	9	8	5	6	7	2	1
6	8	1	2	7	4	3	5	9
2	5	7	1	9	3	6	4	8

Puzzle 491

3	1	5	9	6	2	7	8	4
2	8	4	3	7	5	9	6	1
7	6	9	8	4	1	5	2	3
1	4	2	6	5	3	8	7	9
6	3	7	2	8	9	4	1	5
5	9	8	4	1	7	6	3	2
8	7	3	5	2	4	1	9	6
9	5	1	7	3	6	2	4	8
4	2	6	1	9	8	3	5	7

Puzzle 492

5	3	9	1	6	4	2	7	8
4	8	1	5	2	7	6	9	3
6	7	2	8	3	9	5	4	1
9	6	7	2	8	3	1	5	4
2	4	3	9	5	1	8	6	7
8	1	5	7	4	6	9	3	2
1	5	6	4	7	8	3	2	9
3	9	4	6	1	2	7	8	5
7	2	8	3	9	5	4	1	6

Puzzle 493

2	8	5	4	6	9	3	1	7
9	3	6	2	7	1	4	5	8
1	4	7	3	5	8	2	9	6
5	9	1	8	2	4	6	7	3
7	2	4	6	1	3	5	8	9
8	6	3	7	9	5	1	2	4
6	1	2	9	3	7	8	4	5
4	5	9	1	8	6	7	3	2
3	7	8	5	4	2	9	6	1

Puzzle 494

6	5	7	8	4	1	9	3	2
2	4	8	9	6	3	7	1	5
3	1	9	7	2	5	6	8	4
1	9	2	3	8	4	5	6	7
5	8	3	2	7	6	4	9	1
4	7	6	1	5	9	8	2	3
9	2	4	6	3	7	1	5	8
8	6	5	4	1	2	3	7	9
7	3	1	5	9	8	2	4	6

Puzzle 495

2	8	7	9	3	4	1	5	6
6	4	9	2	1	5	8	7	3
1	5	3	6	8	7	2	4	9
3	1	8	5	4	6	9	2	7
5	7	2	8	9	1	6	3	4
9	6	4	7	2	3	5	8	1
8	2	6	3	7	9	4	1	5
4	3	5	1	6	2	7	9	8
7	9	1	4	5	8	3	6	2

Puzzle 496

8	6	1	7	3	5	9	2	4
3	7	9	2	4	8	5	1	6
4	5	2	9	6	1	8	3	7
2	1	3	6	5	7	4	9	8
9	8	6	3	1	4	7	5	2
5	4	7	8	2	9	3	6	1
1	3	5	4	8	6	2	7	9
7	2	4	1	9	3	6	8	5
6	9	8	5	7	2	1	4	3

Puzzle 497

7	3	1	4	5	9	8	6	2
8	2	6	3	1	7	5	9	4
5	9	4	6	8	2	7	1	3
2	8	9	1	3	4	6	7	5
3	1	5	8	7	6	2	4	9
6	4	7	9	2	5	3	8	1
1	5	3	7	4	8	9	2	6
4	6	8	2	9	3	1	5	7
9	7	2	5	6	1	4	3	8

Puzzle 498

7	9	4	1	2	6	5	3	8
3	8	2	5	7	9	4	1	6
1	6	5	8	4	3	7	9	2
8	1	9	3	5	2	6	4	7
6	2	7	9	1	4	3	8	5
4	5	3	6	8	7	9	2	1
2	3	8	4	6	5	1	7	9
5	4	1	7	9	8	2	6	3
9	7	6	2	3	1	8	5	4

Puzzle 499

8	1	3	4	5	9	2	6	7
9	6	5	1	7	2	4	8	3
4	7	2	6	8	3	5	9	1
3	2	4	8	1	6	9	7	5
5	8	6	3	9	7	1	2	4
7	9	1	5	2	4	6	3	8
6	5	9	7	3	1	8	4	2
1	4	7	2	6	8	3	5	9
2	3	8	9	4	5	7	1	6

Puzzle 500

6	9	7	8	4	1	5	3	2
4	3	2	5	7	9	1	8	6
1	8	5	3	2	6	7	9	4
9	7	3	6	5	2	8	4	1
8	2	1	4	9	3	6	7	5
5	4	6	1	8	7	3	2	9
3	1	9	2	6	8	4	5	7
7	6	4	9	3	5	2	1	8
2	5	8	7	1	4	9	6	3

Puzzle 501

3	6	9	4	8	2	7	5	1
5	1	4	3	6	7	2	9	8
7	2	8	9	5	1	3	4	6
2	7	3	5	4	6	1	8	9
8	5	1	2	7	9	6	3	4
4	9	6	1	3	8	5	7	2
9	3	5	6	2	4	8	1	7
6	4	7	8	1	5	9	2	3
1	8	2	7	9	3	4	6	5

Puzzle 502

3	6	8	1	2	7	9	5	4
7	2	4	8	9	5	1	3	6
9	1	5	3	6	4	7	2	8
1	5	7	2	4	3	8	6	9
4	9	2	7	8	6	3	1	5
6	8	3	9	5	1	2	4	7
8	7	6	4	1	2	5	9	3
2	4	9	5	3	8	6	7	1
5	3	1	6	7	9	4	8	2

Puzzle 503

8	1	4	7	5	6	2	3	9
6	5	7	9	2	3	8	1	4
2	3	9	8	4	1	7	6	5
5	7	1	3	6	2	4	9	8
9	8	6	4	7	5	1	2	3
3	4	2	1	8	9	6	5	7
1	2	8	5	3	7	9	4	6
7	9	3	6	1	4	5	8	2
4	6	5	2	9	8	3	7	1

Puzzle 504

9	5	2	1	7	3	6	8	4
1	6	4	9	5	8	3	7	2
7	8	3	2	6	4	9	1	5
5	3	1	7	4	9	2	6	8
8	9	7	6	1	2	5	4	3
4	2	6	3	8	5	1	9	7
6	7	8	5	3	1	4	2	9
3	4	9	8	2	6	7	5	1
2	1	5	4	9	7	8	3	6

Puzzle 505

3	1	2	8	5	6	4	9	7
9	8	6	4	1	7	5	2	3
5	7	4	2	9	3	6	1	8
7	9	1	5	8	2	3	6	4
6	5	8	1	3	4	2	7	9
4	2	3	6	7	9	1	8	5
2	4	5	7	6	8	9	3	1
8	6	9	3	4	1	7	5	2
1	3	7	9	2	5	8	4	6

Puzzle 506

4	9	2	5	6	7	3	1	8
7	1	8	3	9	2	5	6	4
6	5	3	8	1	4	7	9	2
8	3	5	9	7	1	4	2	6
1	2	7	6	4	5	8	3	9
9	4	6	2	3	8	1	7	5
2	7	9	4	8	3	6	5	1
3	6	4	1	5	9	2	8	7
5	8	1	7	2	6	9	4	3

Puzzle 507

1	9	7	2	3	8	5	6	4
6	4	5	1	7	9	3	2	8
2	3	8	6	4	5	7	1	9
3	5	4	7	6	1	9	8	2
7	6	2	9	8	3	4	5	1
8	1	9	5	2	4	6	7	3
5	7	3	8	9	2	1	4	6
9	2	1	4	5	6	8	3	7
4	8	6	3	1	7	2	9	5

Puzzle 508

6	7	8	1	4	3	5	9	2
3	4	2	5	9	6	1	8	7
9	1	5	8	7	2	6	4	3
2	6	1	3	8	5	4	7	9
4	9	3	6	2	7	8	1	5
5	8	7	4	1	9	3	2	6
7	5	4	2	3	1	9	6	8
8	3	9	7	6	4	2	5	1
1	2	6	9	5	8	7	3	4

Puzzle 509

4	1	7	5	6	9	8	3	2
2	3	8	7	1	4	5	9	6
6	5	9	8	3	2	7	4	1
3	7	1	6	5	8	9	2	4
9	6	2	4	7	3	1	5	8
8	4	5	9	2	1	6	7	3
1	2	6	3	9	7	4	8	5
5	9	4	2	8	6	3	1	7
7	8	3	1	4	5	2	6	9

Puzzle 510

1	4	7	6	2	9	3	5	8
9	3	8	7	4	5	2	1	6
2	5	6	8	1	3	4	9	7
5	8	4	2	9	6	1	7	3
6	1	2	4	3	7	5	8	9
3	7	9	1	5	8	6	4	2
7	6	1	5	8	2	9	3	4
4	2	3	9	7	1	8	6	5
8	9	5	3	6	4	7	2	1

Puzzle 511

8	3	1	9	5	7	2	6	4
2	5	4	3	8	6	1	9	7
6	7	9	1	4	2	8	5	3
9	1	6	8	2	4	7	3	5
3	8	7	5	1	9	6	4	2
4	2	5	7	6	3	9	8	1
1	9	8	4	7	5	3	2	6
7	4	2	6	3	8	5	1	9
5	6	3	2	9	1	4	7	8

Puzzle 512

5	7	6	8	1	9	4	3	2
1	9	3	5	2	4	8	7	6
4	8	2	7	3	6	1	9	5
2	4	8	1	7	5	3	6	9
6	3	9	4	8	2	7	5	1
7	1	5	9	6	3	2	8	4
8	5	1	2	9	7	6	4	3
3	2	4	6	5	8	9	1	7
9	6	7	3	4	1	5	2	8

Puzzle 513

7	2	9	1	6	8	5	3	4
3	8	6	9	4	5	7	1	2
5	4	1	3	2	7	8	9	6
8	5	7	6	3	4	9	2	1
1	6	4	8	9	2	3	7	5
2	9	3	5	7	1	6	4	8
6	1	2	7	8	9	4	5	3
9	3	5	4	1	6	2	8	7
4	7	8	2	5	3	1	6	9

Puzzle 514

4	8	9	1	5	3	2	6	7
3	6	1	8	2	7	4	5	9
2	5	7	9	6	4	3	8	1
9	1	5	2	3	8	6	7	4
7	2	8	6	4	1	9	3	5
6	4	3	7	9	5	8	1	2
5	9	2	3	1	6	7	4	8
1	7	6	4	8	2	5	9	3
8	3	4	5	7	9	1	2	6

Puzzle 515

3	6	5	1	9	7	4	2	8
9	1	4	2	6	8	3	7	5
2	8	7	5	3	4	1	6	9
1	2	9	8	7	5	6	3	4
5	7	6	9	4	3	2	8	1
8	4	3	6	1	2	9	5	7
4	3	1	7	8	6	5	9	2
6	5	8	4	2	9	7	1	3
7	9	2	3	5	1	8	4	6

Puzzle 516

3	5	8	2	7	9	4	6	1
9	7	4	8	1	6	3	5	2
1	2	6	3	4	5	7	9	8
6	4	5	7	9	8	1	2	3
7	9	1	4	3	2	6	8	5
8	3	2	5	6	1	9	4	7
5	1	7	9	2	4	8	3	6
2	6	9	1	8	3	5	7	4
4	8	3	6	5	7	2	1	9

Puzzle 517

4	7	2	5	8	3	9	6	1
9	6	3	1	4	7	2	5	8
5	1	8	9	2	6	7	3	4
7	9	5	6	1	4	3	8	2
2	4	6	7	3	8	1	9	5
3	8	1	2	5	9	4	7	6
1	2	7	3	6	5	8	4	9
6	3	4	8	9	2	5	1	7
8	5	9	4	7	1	6	2	3

Puzzle 518

8	1	6	7	9	4	2	3	5
3	5	7	8	2	1	6	9	4
9	2	4	5	6	3	7	1	8
4	8	2	6	7	9	3	5	1
1	3	9	2	4	5	8	7	6
6	7	5	1	3	8	4	2	9
7	9	3	4	1	6	5	8	2
2	6	8	9	5	7	1	4	3
5	4	1	3	8	2	9	6	7

Puzzle 519

4	2	9	5	7	3	1	8	6
5	8	7	9	1	6	3	4	2
6	3	1	8	4	2	7	5	9
2	1	8	4	6	5	9	7	3
7	5	6	3	2	9	8	1	4
3	9	4	7	8	1	2	6	5
8	4	3	2	5	7	6	9	1
9	6	5	1	3	8	4	2	7
1	7	2	6	9	4	5	3	8

Puzzle 520

9	7	5	2	4	8	6	3	1
2	1	3	6	5	7	4	9	8
6	8	4	1	3	9	2	5	7
5	3	2	4	7	6	1	8	9
8	4	6	5	9	1	3	7	2
7	9	1	3	8	2	5	4	6
1	5	7	9	6	4	8	2	3
3	6	9	8	2	5	7	1	4
4	2	8	7	1	3	9	6	5

Puzzle 521

3	8	9	1	2	7	5	4	6
4	7	1	8	6	5	3	2	9
2	5	6	9	3	4	1	7	8
1	9	7	6	4	8	2	5	3
6	2	3	5	7	9	4	8	1
5	4	8	3	1	2	9	6	7
9	6	4	7	5	3	8	1	2
7	3	5	2	8	1	6	9	4
8	1	2	4	9	6	7	3	5

Puzzle 522

4	1	8	6	7	9	5	2	3
3	9	5	4	8	2	7	6	1
2	7	6	3	5	1	9	4	8
7	8	4	9	3	5	6	1	2
5	2	9	8	1	6	3	7	4
6	3	1	7	2	4	8	9	5
9	5	2	1	6	8	4	3	7
1	6	3	5	4	7	2	8	9
8	4	7	2	9	3	1	5	6

Puzzle 523

2	8	7	3	5	9	4	1	6
1	5	4	7	8	6	3	2	9
6	3	9	4	2	1	5	8	7
4	1	8	6	3	5	9	7	2
5	6	2	8	9	7	1	3	4
9	7	3	2	1	4	8	6	5
3	2	6	9	4	8	7	5	1
8	4	1	5	7	2	6	9	3
7	9	5	1	6	3	2	4	8

Puzzle 524

2	5	7	4	3	8	6	1	9
9	8	4	1	5	6	3	7	2
1	3	6	9	2	7	5	8	4
8	4	5	6	1	9	2	3	7
3	1	2	7	8	5	4	9	6
7	6	9	3	4	2	1	5	8
4	7	3	8	6	1	9	2	5
5	9	1	2	7	4	8	6	3
6	2	8	5	9	3	7	4	1

Puzzle 525

8	3	9	4	2	1	5	6	7
2	1	7	5	3	6	8	9	4
6	4	5	8	7	9	3	2	1
9	7	4	3	1	8	2	5	6
3	6	1	7	5	2	4	8	9
5	8	2	6	9	4	1	7	3
1	2	6	9	8	3	7	4	5
4	5	3	2	6	7	9	1	8
7	9	8	1	4	5	6	3	2

Puzzle 526

1	8	6	2	4	5	9	3	7
7	9	5	1	6	3	8	2	4
4	2	3	9	8	7	1	5	6
5	4	1	3	2	9	7	6	8
2	7	9	6	1	8	5	4	3
3	6	8	7	5	4	2	1	9
6	3	7	5	9	2	4	8	1
8	1	2	4	7	6	3	9	5
9	5	4	8	3	1	6	7	2

Puzzle 527

9	1	8	2	4	3	5	6	7
3	6	7	1	5	8	9	4	2
5	4	2	7	9	6	3	1	8
2	3	9	8	6	4	1	7	5
6	5	4	3	7	1	2	8	9
8	7	1	5	2	9	4	3	6
4	2	5	6	1	7	8	9	3
1	8	6	9	3	5	7	2	4
7	9	3	4	8	2	6	5	1

Puzzle 528

2	1	8	9	7	4	5	3	6
4	5	7	3	2	6	9	8	1
9	3	6	5	1	8	2	4	7
8	7	2	4	6	1	3	9	5
5	4	1	7	9	3	8	6	2
3	6	9	2	8	5	1	7	4
1	2	4	8	3	7	6	5	9
7	9	3	6	5	2	4	1	8
6	8	5	1	4	9	7	2	3

Puzzle 529

9	3	1	7	4	5	2	6	8
4	8	5	1	6	2	7	3	9
2	7	6	3	8	9	4	5	1
5	4	7	9	3	6	8	1	2
8	2	3	4	5	1	6	9	7
6	1	9	2	7	8	5	4	3
7	5	2	6	1	3	9	8	4
3	6	4	8	9	7	1	2	5
1	9	8	5	2	4	3	7	6

Puzzle 530

6	3	8	5	9	7	1	2	4
4	7	5	2	1	3	9	6	8
1	9	2	4	8	6	3	5	7
9	6	1	8	2	4	7	3	5
8	5	3	7	6	9	4	1	2
2	4	7	1	3	5	8	9	6
7	8	6	3	5	1	2	4	9
3	2	9	6	4	8	5	7	1
5	1	4	9	7	2	6	8	3

Puzzle 531

1	9	5	4	3	8	7	2	6
3	4	7	1	6	2	9	8	5
8	2	6	5	9	7	1	3	4
7	1	2	8	5	4	3	6	9
9	5	4	6	7	3	2	1	8
6	8	3	9	2	1	4	5	7
5	6	1	2	4	9	8	7	3
2	7	9	3	8	6	5	4	1
4	3	8	7	1	5	6	9	2

Puzzle 532

4	2	3	9	1	5	8	7	6
1	5	6	3	7	8	9	2	4
8	7	9	6	4	2	5	1	3
3	1	8	5	9	4	2	6	7
9	4	5	7	2	6	3	8	1
2	6	7	8	3	1	4	9	5
7	3	1	4	8	9	6	5	2
5	8	4	2	6	7	1	3	9
6	9	2	1	5	3	7	4	8

Puzzle 533

6	9	2	8	7	3	5	1	4
1	5	3	6	4	9	7	8	2
8	4	7	2	5	1	6	3	9
3	6	9	7	1	4	2	5	8
5	8	1	9	6	2	3	4	7
7	2	4	3	8	5	1	9	6
2	7	5	4	3	8	9	6	1
9	3	8	1	2	6	4	7	5
4	1	6	5	9	7	8	2	3

Puzzle 534

8	3	1	2	5	6	9	4	7
4	9	2	7	1	8	5	6	3
6	7	5	3	9	4	8	1	2
5	4	8	1	3	9	7	2	6
2	1	7	4	6	5	3	8	9
9	6	3	8	2	7	1	5	4
3	5	6	9	4	1	2	7	8
7	2	4	5	8	3	6	9	1
1	8	9	6	7	2	4	3	5

Puzzle 535

2	4	1	7	9	3	8	6	5
3	5	7	2	6	8	9	4	1
8	6	9	1	5	4	3	2	7
4	7	6	9	2	5	1	3	8
1	9	3	8	4	7	6	5	2
5	2	8	3	1	6	4	7	9
7	1	2	4	3	9	5	8	6
6	8	4	5	7	1	2	9	3
9	3	5	6	8	2	7	1	4

Puzzle 536

6	5	9	3	2	8	1	7	4
7	3	1	9	4	5	8	6	2
2	8	4	1	6	7	9	5	3
1	4	3	7	5	6	2	8	9
9	7	6	2	8	1	4	3	5
5	2	8	4	3	9	6	1	7
4	6	2	5	1	3	7	9	8
3	1	7	8	9	2	5	4	6
8	9	5	6	7	4	3	2	1

Puzzle 537

9	1	3	8	5	6	7	2	4
6	2	7	3	4	1	8	5	9
4	5	8	7	2	9	1	3	6
5	7	4	9	6	3	2	8	1
2	6	9	1	8	5	4	7	3
3	8	1	4	7	2	6	9	5
8	4	5	6	3	7	9	1	2
7	9	2	5	1	4	3	6	8
1	3	6	2	9	8	5	4	7

Puzzle 538

9	8	1	7	5	4	2	3	6
7	6	4	1	2	3	8	9	5
2	5	3	9	8	6	4	1	7
3	7	8	4	6	1	9	5	2
4	9	2	8	7	5	3	6	1
5	1	6	3	9	2	7	8	4
8	3	5	6	4	7	1	2	9
1	2	7	5	3	9	6	4	8
6	4	9	2	1	8	5	7	3

Puzzle 539

8	7	9	2	6	1	4	5	3
2	6	3	7	5	4	1	8	9
5	1	4	3	9	8	6	7	2
1	8	5	4	2	7	9	3	6
7	3	6	1	8	9	5	2	4
9	4	2	6	3	5	7	1	8
4	5	8	9	1	3	2	6	7
6	9	1	8	7	2	3	4	5
3	2	7	5	4	6	8	9	1

Puzzle 540

3	9	6	4	8	7	2	1	5
4	5	8	9	2	1	6	7	3
1	2	7	6	3	5	8	4	9
7	1	5	3	9	6	4	8	2
9	8	3	1	4	2	7	5	6
2	6	4	5	7	8	9	3	1
6	3	2	7	1	4	5	9	8
5	7	1	8	6	9	3	2	4
8	4	9	2	5	3	1	6	7

Puzzle 541

8	9	6	4	2	1	7	3	5
3	4	7	8	6	5	1	9	2
2	1	5	9	3	7	8	4	6
7	8	1	5	9	6	4	2	3
6	2	9	1	4	3	5	8	7
5	3	4	2	7	8	9	6	1
9	7	8	6	1	2	3	5	4
4	6	3	7	5	9	2	1	8
1	5	2	3	8	4	6	7	9

Puzzle 542

2	4	7	6	9	8	5	1	3
6	5	1	7	2	3	8	4	9
8	9	3	5	1	4	2	7	6
1	3	4	8	6	7	9	5	2
5	7	6	9	4	2	1	3	8
9	8	2	3	5	1	4	6	7
7	2	5	4	8	6	3	9	1
3	1	9	2	7	5	6	8	4
4	6	8	1	3	9	7	2	5

Puzzle 543

5	2	7	4	3	9	8	6	1
1	9	3	8	6	5	4	7	2
6	4	8	7	1	2	3	5	9
3	1	6	9	2	4	7	8	5
8	5	2	6	7	1	9	4	3
4	7	9	5	8	3	1	2	6
7	8	5	1	9	6	2	3	4
9	3	4	2	5	7	6	1	8
2	6	1	3	4	8	5	9	7

Puzzle 544

2	3	9	5	6	7	8	1	4
5	4	6	9	1	8	7	2	3
7	1	8	3	4	2	9	5	6
8	9	4	1	5	3	2	6	7
1	2	3	7	8	6	5	4	9
6	5	7	4	2	9	3	8	1
4	7	2	8	9	1	6	3	5
3	6	5	2	7	4	1	9	8
9	8	1	6	3	5	4	7	2

Puzzle 545

2	9	6	8	4	1	7	3	5
8	4	5	6	3	7	9	2	1
1	3	7	5	2	9	8	6	4
5	7	8	9	6	4	2	1	3
4	6	9	2	1	3	5	8	7
3	2	1	7	8	5	4	9	6
6	5	2	3	7	8	1	4	9
7	8	4	1	9	6	3	5	2
9	1	3	4	5	2	6	7	8

Puzzle 546

9	4	1	2	8	7	3	5	6
5	6	7	1	4	3	9	8	2
3	2	8	5	9	6	7	4	1
7	3	9	6	1	5	8	2	4
4	8	6	9	3	2	1	7	5
1	5	2	8	7	4	6	3	9
2	9	4	7	6	8	5	1	3
8	1	5	3	2	9	4	6	7
6	7	3	4	5	1	2	9	8

Puzzle 547

2	9	1	6	5	4	7	3	8
7	6	8	2	3	1	5	4	9
3	5	4	9	7	8	6	2	1
1	3	7	8	6	5	2	9	4
5	8	9	7	4	2	1	6	3
4	2	6	1	9	3	8	5	7
9	7	3	5	8	6	4	1	2
6	4	2	3	1	7	9	8	5
8	1	5	4	2	9	3	7	6

Puzzle 548

1	2	6	7	3	4	9	8	5
8	3	5	9	1	2	6	4	7
4	7	9	6	5	8	1	2	3
7	9	8	3	6	1	4	5	2
6	1	4	8	2	5	3	7	9
2	5	3	4	7	9	8	1	6
3	6	2	1	8	7	5	9	4
9	8	7	5	4	3	2	6	1
5	4	1	2	9	6	7	3	8

Puzzle 549

6	2	4	9	3	5	7	8	1
5	3	7	6	8	1	9	4	2
9	1	8	4	7	2	3	5	6
7	8	2	3	1	4	6	9	5
1	5	3	2	6	9	4	7	8
4	9	6	8	5	7	1	2	3
8	6	9	7	2	3	5	1	4
2	4	5	1	9	6	8	3	7
3	7	1	5	4	8	2	6	9

Puzzle 550

4	3	6	7	5	1	9	2	8
2	8	1	6	4	9	5	7	3
5	9	7	8	2	3	1	6	4
3	2	4	9	1	5	6	8	7
1	7	9	3	8	6	4	5	2
8	6	5	4	7	2	3	1	9
9	5	8	2	6	4	7	3	1
7	1	3	5	9	8	2	4	6
6	4	2	1	3	7	8	9	5

Puzzle 551

3	8	5	6	9	1	2	7	4
1	2	9	7	4	3	8	6	5
7	6	4	5	2	8	3	9	1
8	9	1	2	7	4	6	5	3
4	7	6	9	3	5	1	2	8
5	3	2	1	8	6	7	4	9
2	1	3	4	6	9	5	8	7
6	4	8	3	5	7	9	1	2
9	5	7	8	1	2	4	3	6

Puzzle 552

7	1	9	5	8	3	6	2	4
4	5	3	1	2	6	9	7	8
6	2	8	4	7	9	1	3	5
9	8	1	7	3	4	5	6	2
3	7	4	2	6	5	8	9	1
5	6	2	9	1	8	7	4	3
8	4	5	3	9	7	2	1	6
2	9	6	8	4	1	3	5	7
1	3	7	6	5	2	4	8	9

Puzzle 553

3	4	9	6	7	1	5	2	8
1	2	8	3	4	5	7	6	9
7	6	5	8	9	2	4	3	1
2	8	1	5	6	7	9	4	3
4	7	3	1	8	9	2	5	6
5	9	6	2	3	4	8	1	7
8	1	2	7	5	6	3	9	4
9	5	7	4	1	3	6	8	2
6	3	4	9	2	8	1	7	5

Puzzle 554

7	6	4	2	9	3	5	1	8
1	5	8	6	4	7	9	2	3
2	9	3	5	1	8	7	4	6
3	2	1	4	6	5	8	9	7
9	4	7	8	3	1	6	5	2
5	8	6	7	2	9	4	3	1
6	1	9	3	8	4	2	7	5
4	7	2	1	5	6	3	8	9
8	3	5	9	7	2	1	6	4

Puzzle 555

4	2	9	6	5	1	7	8	3
5	8	3	9	7	2	4	6	1
7	1	6	4	3	8	5	9	2
2	7	5	3	1	6	8	4	9
9	3	4	2	8	7	1	5	6
8	6	1	5	4	9	3	2	7
1	5	2	8	6	3	9	7	4
6	4	7	1	9	5	2	3	8
3	9	8	7	2	4	6	1	5

Puzzle 556

3	2	4	7	8	1	5	6	9
9	5	1	3	6	2	4	7	8
6	7	8	4	5	9	1	3	2
8	9	3	6	7	5	2	1	4
5	6	2	8	1	4	3	9	7
4	1	7	2	9	3	6	8	5
1	8	9	5	2	6	7	4	3
7	3	5	1	4	8	9	2	6
2	4	6	9	3	7	8	5	1

Puzzle 557

6	8	7	2	3	5	1	4	9
9	1	4	6	7	8	3	2	5
5	2	3	4	9	1	6	7	8
8	9	6	3	5	7	4	1	2
1	4	5	9	6	2	8	3	7
3	7	2	8	1	4	9	5	6
4	5	8	1	2	9	7	6	3
2	3	9	7	4	6	5	8	1
7	6	1	5	8	3	2	9	4

Puzzle 558

3	9	1	2	4	7	5	6	8
4	6	5	9	3	8	7	2	1
8	2	7	6	1	5	9	3	4
7	4	6	1	5	2	8	9	3
5	8	9	7	6	3	1	4	2
2	1	3	8	9	4	6	5	7
9	7	2	4	8	6	3	1	5
1	3	8	5	2	9	4	7	6
6	5	4	3	7	1	2	8	9

Puzzle 559

5	4	6	9	8	2	3	7	1
2	3	7	4	1	5	8	6	9
1	8	9	7	3	6	4	2	5
3	1	2	6	9	8	7	5	4
4	9	8	2	5	7	1	3	6
6	7	5	3	4	1	2	9	8
9	2	3	1	6	4	5	8	7
8	6	4	5	7	3	9	1	2
7	5	1	8	2	9	6	4	3

Puzzle 560

7	1	3	5	8	4	6	2	9
4	6	9	1	2	7	8	3	5
2	5	8	6	9	3	1	4	7
3	4	1	7	6	5	2	9	8
9	2	5	3	4	8	7	6	1
8	7	6	9	1	2	3	5	4
5	3	2	4	7	1	9	8	6
1	9	4	8	3	6	5	7	2
6	8	7	2	5	9	4	1	3

Puzzle 561

1	9	7	6	5	3	8	2	4
4	8	3	1	2	7	5	9	6
2	5	6	4	8	9	7	3	1
7	6	5	2	9	4	1	8	3
3	2	4	8	6	1	9	7	5
8	1	9	7	3	5	6	4	2
6	4	1	3	7	8	2	5	9
5	3	8	9	1	2	4	6	7
9	7	2	5	4	6	3	1	8

Puzzle 562

5	9	8	4	3	7	2	1	6
1	6	4	9	8	2	7	3	5
3	2	7	5	6	1	9	8	4
8	1	9	3	2	5	4	6	7
6	4	5	1	7	9	3	2	8
7	3	2	6	4	8	5	9	1
4	8	6	2	5	3	1	7	9
2	5	1	7	9	6	8	4	3
9	7	3	8	1	4	6	5	2

Puzzle 563

4	2	7	9	1	3	8	5	6
6	3	8	7	2	5	9	1	4
1	9	5	6	8	4	7	3	2
9	8	6	1	4	7	3	2	5
3	1	2	8	5	9	6	4	7
5	7	4	3	6	2	1	9	8
2	6	3	4	9	8	5	7	1
7	4	1	5	3	6	2	8	9
8	5	9	2	7	1	4	6	3

Puzzle 564

7	4	8	2	5	9	3	1	6
1	3	2	4	7	6	9	5	8
9	6	5	3	1	8	4	7	2
3	5	1	8	2	4	7	6	9
8	9	4	5	6	7	2	3	1
6	2	7	9	3	1	8	4	5
5	1	3	7	8	2	6	9	4
2	7	9	6	4	5	1	8	3
4	8	6	1	9	3	5	2	7

Puzzle 565

9	8	3	7	2	6	4	5	1
6	4	5	8	3	1	7	2	9
7	1	2	4	9	5	3	8	6
2	3	4	1	8	9	6	7	5
5	7	8	6	4	3	1	9	2
1	9	6	2	5	7	8	4	3
3	5	7	9	1	4	2	6	8
8	6	9	3	7	2	5	1	4
4	2	1	5	6	8	9	3	7

Puzzle 566

6	5	1	3	9	7	2	4	8
9	2	7	6	8	4	1	3	5
8	3	4	2	1	5	9	7	6
2	6	3	9	4	1	5	8	7
5	1	9	8	7	3	4	6	2
4	7	8	5	2	6	3	9	1
7	9	6	4	5	2	8	1	3
1	4	2	7	3	8	6	5	9
3	8	5	1	6	9	7	2	4

Puzzle 567

9	5	4	8	7	1	2	3	6
3	1	2	6	4	9	5	8	7
6	7	8	5	3	2	1	9	4
1	8	7	3	5	6	9	4	2
4	2	3	1	9	7	8	6	5
5	6	9	2	8	4	7	1	3
7	3	5	9	6	8	4	2	1
8	4	1	7	2	3	6	5	9
2	9	6	4	1	5	3	7	8

Puzzle 568

3	5	4	1	7	6	9	2	8
7	8	2	3	9	4	1	6	5
9	6	1	2	5	8	7	4	3
2	1	7	4	8	5	6	3	9
8	3	9	6	1	7	4	5	2
6	4	5	9	2	3	8	7	1
5	7	6	8	3	1	2	9	4
1	2	3	7	4	9	5	8	6
4	9	8	5	6	2	3	1	7

Puzzle 569

8	6	1	7	4	5	2	9	3
7	2	4	9	1	3	6	8	5
3	9	5	2	6	8	1	4	7
9	3	2	8	5	1	7	6	4
1	7	6	4	9	2	3	5	8
5	4	8	6	3	7	9	1	2
6	5	9	3	7	4	8	2	1
2	1	3	5	8	6	4	7	9
4	8	7	1	2	9	5	3	6

Puzzle 570

8	1	5	9	7	4	6	2	3
2	9	3	1	5	6	7	8	4
7	4	6	2	3	8	9	5	1
9	6	1	3	4	2	5	7	8
5	8	2	6	1	7	3	4	9
3	7	4	5	8	9	1	6	2
1	5	8	4	6	3	2	9	7
6	2	7	8	9	1	4	3	5
4	3	9	7	2	5	8	1	6

Puzzle 571

1	7	6	5	2	8	4	9	3
4	3	2	1	9	6	8	7	5
9	8	5	4	3	7	1	6	2
7	2	3	9	8	5	6	4	1
8	5	4	6	1	3	7	2	9
6	9	1	7	4	2	5	3	8
5	6	9	3	7	1	2	8	4
3	1	8	2	6	4	9	5	7
2	4	7	8	5	9	3	1	6

Puzzle 572

5	3	2	7	8	4	9	1	6
8	9	7	6	5	1	3	2	4
1	4	6	3	2	9	7	8	5
6	8	1	9	3	2	4	5	7
2	5	4	1	6	7	8	3	9
9	7	3	8	4	5	2	6	1
7	1	8	5	9	3	6	4	2
4	6	5	2	7	8	1	9	3
3	2	9	4	1	6	5	7	8

Puzzle 573

1	8	5	2	4	6	9	7	3
9	7	3	5	1	8	6	2	4
4	2	6	3	7	9	5	1	8
2	5	9	8	6	7	4	3	1
3	6	7	4	5	1	2	8	9
8	1	4	9	2	3	7	5	6
7	9	1	6	8	2	3	4	5
5	3	2	1	9	4	8	6	7
6	4	8	7	3	5	1	9	2

Puzzle 574

8	6	1	7	2	3	5	9	4
4	3	7	9	8	5	2	6	1
2	5	9	6	4	1	7	8	3
7	4	6	3	9	8	1	5	2
9	2	5	4	1	6	8	3	7
1	8	3	5	7	2	6	4	9
5	9	4	1	6	7	3	2	8
3	7	8	2	5	9	4	1	6
6	1	2	8	3	4	9	7	5

Puzzle 575

3	8	1	6	4	5	7	9	2
5	4	2	7	8	9	1	3	6
9	6	7	3	2	1	4	5	8
8	7	5	9	3	2	6	4	1
4	9	6	1	5	7	2	8	3
2	1	3	8	6	4	9	7	5
1	3	4	5	7	6	8	2	9
6	2	8	4	9	3	5	1	7
7	5	9	2	1	8	3	6	4

Puzzle 576

6	4	1	8	5	9	3	2	7
2	9	8	3	6	7	1	5	4
5	3	7	4	1	2	9	6	8
4	1	6	9	7	3	5	8	2
7	2	5	6	8	1	4	9	3
3	8	9	2	4	5	7	1	6
8	5	4	1	3	6	2	7	9
9	7	3	5	2	8	6	4	1
1	6	2	7	9	4	8	3	5

Puzzle 577

3	8	2	1	9	4	7	5	6
5	4	6	8	2	7	1	3	9
7	1	9	5	3	6	8	2	4
6	9	8	3	7	1	5	4	2
1	7	4	6	5	2	9	8	3
2	3	5	9	4	8	6	7	1
8	5	1	2	6	3	4	9	7
4	6	3	7	8	9	2	1	5
9	2	7	4	1	5	3	6	8

Puzzle 578

9	3	5	2	7	4	8	1	6
6	7	2	1	5	8	4	3	9
8	1	4	9	6	3	2	7	5
1	2	8	4	3	5	6	9	7
5	6	3	8	9	7	1	4	2
4	9	7	6	2	1	5	8	3
3	4	9	5	1	6	7	2	8
7	5	1	3	8	2	9	6	4
2	8	6	7	4	9	3	5	1

Puzzle 579

4	7	2	1	9	5	6	8	3
3	6	9	2	8	7	4	5	1
1	8	5	4	3	6	7	2	9
8	3	7	9	2	4	5	1	6
2	5	1	8	6	3	9	7	4
9	4	6	5	7	1	8	3	2
5	1	8	3	4	9	2	6	7
6	9	3	7	5	2	1	4	8
7	2	4	6	1	8	3	9	5

Puzzle 580

7	4	8	6	1	2	9	5	3
6	3	2	8	9	5	4	7	1
5	9	1	4	7	3	6	8	2
3	8	7	5	6	9	2	1	4
2	6	4	7	3	1	8	9	5
1	5	9	2	8	4	3	6	7
8	7	5	3	4	6	1	2	9
9	2	3	1	5	8	7	4	6
4	1	6	9	2	7	5	3	8

Puzzle 581

8	3	2	4	5	1	6	7	9
1	5	9	7	6	2	3	8	4
4	7	6	8	9	3	5	2	1
9	1	4	2	3	5	8	6	7
5	8	3	6	7	4	9	1	2
6	2	7	1	8	9	4	3	5
2	9	5	3	1	6	7	4	8
7	6	1	9	4	8	2	5	3
3	4	8	5	2	7	1	9	6

Puzzle 582

8	7	6	9	5	4	1	2	3
4	5	1	3	8	2	9	6	7
2	9	3	6	7	1	8	4	5
3	8	9	2	1	6	7	5	4
6	2	4	7	9	5	3	8	1
5	1	7	4	3	8	6	9	2
7	3	5	8	2	9	4	1	6
1	6	8	5	4	7	2	3	9
9	4	2	1	6	3	5	7	8

Puzzle 583

1	6	7	5	3	8	4	9	2
4	5	3	1	2	9	6	8	7
8	9	2	7	6	4	1	5	3
7	2	1	9	4	5	3	6	8
9	4	6	3	8	7	5	2	1
5	3	8	2	1	6	7	4	9
3	7	9	6	5	2	8	1	4
6	1	4	8	9	3	2	7	5
2	8	5	4	7	1	9	3	6

Puzzle 584

8	1	6	9	4	3	7	5	2
4	9	2	6	7	5	1	8	3
7	5	3	1	2	8	4	9	6
3	7	4	5	1	6	8	2	9
1	2	8	4	3	9	5	6	7
9	6	5	2	8	7	3	1	4
6	4	1	7	5	2	9	3	8
2	3	7	8	9	1	6	4	5
5	8	9	3	6	4	2	7	1

Puzzle 585

2	3	8	9	6	7	4	1	5
7	5	1	8	3	4	9	6	2
6	4	9	2	5	1	7	3	8
9	8	5	4	1	3	2	7	6
1	7	6	5	9	2	8	4	3
4	2	3	6	7	8	1	5	9
8	6	2	1	4	5	3	9	7
3	9	4	7	2	6	5	8	1
5	1	7	3	8	9	6	2	4

Puzzle 586

9	3	1	5	8	6	2	4	7
6	7	5	2	3	4	9	8	1
8	2	4	9	7	1	5	3	6
2	5	6	3	4	8	7	1	9
7	9	3	6	1	2	4	5	8
4	1	8	7	5	9	6	2	3
1	6	7	8	2	5	3	9	4
5	4	9	1	6	3	8	7	2
3	8	2	4	9	7	1	6	5

Puzzle 587

2	6	8	3	9	4	1	7	5
9	1	3	6	5	7	2	8	4
7	4	5	1	8	2	6	3	9
1	7	4	9	3	8	5	2	6
5	3	9	7	2	6	8	4	1
6	8	2	4	1	5	3	9	7
4	2	1	8	6	9	7	5	3
3	5	7	2	4	1	9	6	8
8	9	6	5	7	3	4	1	2

Puzzle 588

3	7	8	4	9	2	1	5	6
9	1	4	5	7	6	3	8	2
5	6	2	8	1	3	4	9	7
6	9	1	2	4	8	7	3	5
2	4	5	1	3	7	8	6	9
8	3	7	9	6	5	2	4	1
1	5	9	7	8	4	6	2	3
7	8	3	6	2	9	5	1	4
4	2	6	3	5	1	9	7	8

Puzzle 589

4	9	7	3	6	8	2	5	1
5	1	8	2	9	4	7	6	3
2	6	3	7	1	5	8	9	4
8	7	1	9	4	2	5	3	6
6	3	5	8	7	1	4	2	9
9	2	4	5	3	6	1	8	7
3	5	6	1	8	7	9	4	2
1	4	2	6	5	9	3	7	8
7	8	9	4	2	3	6	1	5

Puzzle 590

3	7	6	9	4	1	8	5	2
1	2	8	7	5	3	9	6	4
4	5	9	8	2	6	7	1	3
6	1	3	4	7	8	2	9	5
7	9	5	3	6	2	4	8	1
8	4	2	5	1	9	6	3	7
9	6	7	1	3	4	5	2	8
2	3	4	6	8	5	1	7	9
5	8	1	2	9	7	3	4	6

Puzzle 591

1	5	4	3	2	9	7	6	8
3	8	6	1	7	5	4	2	9
2	7	9	8	4	6	5	1	3
6	4	3	2	5	8	9	7	1
5	9	7	6	3	1	8	4	2
8	2	1	4	9	7	3	5	6
7	1	8	9	6	4	2	3	5
4	6	2	5	8	3	1	9	7
9	3	5	7	1	2	6	8	4

Puzzle 592

4	9	1	6	8	5	3	2	7
8	7	3	2	9	4	5	1	6
2	6	5	3	1	7	4	9	8
9	2	7	8	3	1	6	5	4
6	1	8	5	4	9	7	3	2
5	3	4	7	2	6	1	8	9
7	5	2	1	6	8	9	4	3
3	4	6	9	5	2	8	7	1
1	8	9	4	7	3	2	6	5

Puzzle 593

8	6	2	1	9	4	7	3	5
3	1	5	6	7	2	9	8	4
7	4	9	3	8	5	2	6	1
6	9	1	7	3	8	5	4	2
4	3	7	2	5	1	6	9	8
5	2	8	9	4	6	3	1	7
1	5	3	4	2	9	8	7	6
9	8	4	5	6	7	1	2	3
2	7	6	8	1	3	4	5	9

Puzzle 594

1	6	3	4	7	8	2	9	5
8	7	4	5	2	9	6	1	3
9	2	5	1	6	3	7	8	4
3	5	8	6	4	7	1	2	9
2	9	6	8	5	1	3	4	7
4	1	7	9	3	2	5	6	8
7	4	1	2	8	5	9	3	6
6	3	2	7	9	4	8	5	1
5	8	9	3	1	6	4	7	2

Puzzle 595

5	9	3	1	7	8	2	6	4
7	8	1	6	4	2	9	5	3
2	4	6	9	3	5	8	7	1
1	2	5	7	6	3	4	9	8
3	7	9	4	8	1	5	2	6
8	6	4	5	2	9	3	1	7
6	3	7	2	9	4	1	8	5
4	1	2	8	5	7	6	3	9
9	5	8	3	1	6	7	4	2

Puzzle 596

4	7	8	2	6	9	1	3	5
6	5	3	8	7	1	2	9	4
9	1	2	3	5	4	6	7	8
2	8	9	6	3	7	4	5	1
7	6	4	5	1	2	3	8	9
5	3	1	9	4	8	7	2	6
1	4	5	7	8	3	9	6	2
8	9	7	4	2	6	5	1	3
3	2	6	1	9	5	8	4	7

Puzzle 597

2	9	7	8	5	4	6	1	3
5	1	8	3	6	7	2	4	9
3	4	6	9	1	2	7	5	8
9	6	5	1	2	8	3	7	4
8	2	4	7	3	6	1	9	5
1	7	3	5	4	9	8	6	2
6	3	2	4	9	1	5	8	7
4	8	1	2	7	5	9	3	6
7	5	9	6	8	3	4	2	1

Puzzle 598

2	6	7	5	3	9	8	4	1
3	5	1	6	4	8	7	2	9
4	9	8	1	7	2	3	6	5
8	3	6	4	2	1	9	5	7
1	7	5	9	8	6	2	3	4
9	2	4	3	5	7	6	1	8
6	8	9	2	1	5	4	7	3
5	4	2	7	9	3	1	8	6
7	1	3	8	6	4	5	9	2

Puzzle 599

3	9	2	7	6	1	4	8	5
1	8	5	3	9	4	6	7	2
4	7	6	5	8	2	3	1	9
2	5	9	1	7	6	8	3	4
7	6	4	2	3	8	9	5	1
8	1	3	4	5	9	2	6	7
9	4	8	6	1	7	5	2	3
6	3	1	9	2	5	7	4	8
5	2	7	8	4	3	1	9	6

Puzzle 600

9	2	6	1	7	3	5	4	8
7	5	8	2	9	4	1	6	3
1	3	4	5	8	6	7	2	9
4	9	7	3	2	8	6	5	1
3	6	5	4	1	7	8	9	2
8	1	2	6	5	9	3	7	4
2	8	9	7	6	1	4	3	5
5	7	3	8	4	2	9	1	6
6	4	1	9	3	5	2	8	7

Puzzle 601

3	5	7	6	2	4	9	1	8
8	2	4	7	1	9	6	5	3
6	9	1	3	5	8	2	4	7
5	3	9	8	7	1	4	2	6
7	8	2	9	4	6	5	3	1
4	1	6	2	3	5	7	8	9
2	6	8	5	9	3	1	7	4
1	7	3	4	6	2	8	9	5
9	4	5	1	8	7	3	6	2

Puzzle 602

5	1	6	7	2	9	3	4	8
2	4	3	8	1	5	7	6	9
7	9	8	6	3	4	1	5	2
1	7	2	3	4	8	6	9	5
4	6	9	2	5	7	8	3	1
8	3	5	9	6	1	4	2	7
9	2	4	1	7	6	5	8	3
3	5	7	4	8	2	9	1	6
6	8	1	5	9	3	2	7	4

Puzzle 603

4	3	7	5	9	6	8	2	1
2	8	5	3	4	1	7	6	9
6	1	9	2	7	8	3	5	4
7	6	2	8	1	9	4	3	5
8	4	3	6	5	7	1	9	2
5	9	1	4	3	2	6	7	8
9	2	8	7	6	4	5	1	3
3	7	4	1	2	5	9	8	6
1	5	6	9	8	3	2	4	7

Puzzle 604

5	7	3	9	2	4	1	8	6
9	2	4	8	6	1	3	7	5
6	8	1	3	5	7	4	2	9
8	5	2	1	9	3	7	6	4
3	1	9	7	4	6	2	5	8
4	6	7	5	8	2	9	3	1
1	9	5	2	7	8	6	4	3
2	4	8	6	3	9	5	1	7
7	3	6	4	1	5	8	9	2

Puzzle 605

4	6	7	8	1	3	9	2	5
9	2	8	6	4	5	1	7	3
3	5	1	7	9	2	4	8	6
6	7	9	3	2	8	5	4	1
8	1	4	9	5	7	3	6	2
5	3	2	4	6	1	7	9	8
1	4	6	5	8	9	2	3	7
2	8	3	1	7	4	6	5	9
7	9	5	2	3	6	8	1	4

Puzzle 606

9	2	1	3	8	7	5	6	4
5	3	4	6	2	1	8	7	9
7	6	8	9	4	5	3	1	2
6	7	2	4	5	3	9	8	1
4	9	3	7	1	8	2	5	6
8	1	5	2	9	6	4	3	7
3	5	9	1	7	2	6	4	8
1	4	6	8	3	9	7	2	5
2	8	7	5	6	4	1	9	3

Puzzle 607

5	9	1	8	6	2	4	7	3
7	2	3	4	5	9	6	8	1
4	6	8	1	7	3	2	5	9
6	4	7	2	1	8	3	9	5
2	8	5	9	3	6	7	1	4
3	1	9	7	4	5	8	6	2
9	3	2	6	8	1	5	4	7
8	5	4	3	9	7	1	2	6
1	7	6	5	2	4	9	3	8

Puzzle 608

5	1	4	3	2	6	9	8	7
3	9	2	5	8	7	6	1	4
6	7	8	4	1	9	3	5	2
7	3	5	2	4	8	1	6	9
9	4	6	1	7	5	2	3	8
2	8	1	6	9	3	4	7	5
4	5	7	9	3	1	8	2	6
1	6	9	8	5	2	7	4	3
8	2	3	7	6	4	5	9	1

Puzzle 609

1	3	9	5	2	7	6	4	8
6	2	8	9	4	1	5	7	3
5	4	7	8	6	3	9	1	2
8	6	4	3	5	2	1	9	7
9	1	5	7	8	4	2	3	6
3	7	2	1	9	6	4	8	5
2	8	6	4	7	9	3	5	1
7	9	3	2	1	5	8	6	4
4	5	1	6	3	8	7	2	9

Puzzle 610

5	6	8	9	7	2	4	3	1
2	4	9	1	6	3	8	7	5
7	3	1	8	4	5	9	2	6
3	2	4	7	8	6	5	1	9
9	7	5	4	2	1	6	8	3
1	8	6	5	3	9	2	4	7
6	9	2	3	1	8	7	5	4
4	5	3	2	9	7	1	6	8
8	1	7	6	5	4	3	9	2

Puzzle 611

9	3	6	1	5	2	7	8	4
1	7	2	8	3	4	5	6	9
5	8	4	7	9	6	3	2	1
4	6	5	3	1	9	8	7	2
2	1	7	6	8	5	9	4	3
8	9	3	2	4	7	1	5	6
7	5	1	4	6	3	2	9	8
3	4	9	5	2	8	6	1	7
6	2	8	9	7	1	4	3	5

Puzzle 612

9	1	3	7	2	4	6	8	5
4	5	2	6	8	1	7	3	9
8	7	6	5	3	9	4	1	2
1	8	9	4	6	3	2	5	7
3	2	5	1	9	7	8	6	4
7	6	4	8	5	2	1	9	3
6	3	7	2	1	5	9	4	8
2	9	1	3	4	8	5	7	6
5	4	8	9	7	6	3	2	1

Puzzle 613

1	4	2	8	9	5	7	3	6
6	9	3	2	7	4	1	5	8
7	8	5	3	1	6	9	4	2
9	5	6	1	3	8	2	7	4
3	7	8	9	4	2	6	1	5
4	2	1	6	5	7	8	9	3
5	6	7	4	8	1	3	2	9
2	3	4	7	6	9	5	8	1
8	1	9	5	2	3	4	6	7

Puzzle 614

3	1	8	9	5	4	6	2	7
6	4	5	8	2	7	1	9	3
9	7	2	3	1	6	4	8	5
5	8	6	7	3	1	9	4	2
4	2	3	6	8	9	5	7	1
7	9	1	5	4	2	3	6	8
1	5	4	2	9	8	7	3	6
8	6	9	1	7	3	2	5	4
2	3	7	4	6	5	8	1	9

Puzzle 615

8	5	1	6	2	3	9	7	4
6	4	7	1	9	5	8	3	2
9	2	3	4	8	7	1	6	5
3	8	2	9	4	6	7	5	1
7	9	5	8	3	1	4	2	6
4	1	6	5	7	2	3	9	8
2	7	4	3	5	8	6	1	9
5	6	8	7	1	9	2	4	3
1	3	9	2	6	4	5	8	7

Puzzle 616

2	3	5	4	9	6	7	8	1
4	7	8	1	2	3	5	9	6
9	6	1	7	5	8	4	2	3
1	9	4	3	7	2	6	5	8
7	8	3	6	4	5	2	1	9
6	5	2	9	8	1	3	7	4
5	4	6	2	1	9	8	3	7
8	1	7	5	3	4	9	6	2
3	2	9	8	6	7	1	4	5

Puzzle 617

5	9	6	4	1	7	8	3	2
4	3	8	9	2	5	6	1	7
7	1	2	3	6	8	4	5	9
1	6	3	5	4	9	2	7	8
8	5	9	2	7	6	3	4	1
2	4	7	8	3	1	9	6	5
9	7	5	6	8	3	1	2	4
3	8	4	1	5	2	7	9	6
6	2	1	7	9	4	5	8	3

Puzzle 618

5	1	6	3	8	4	2	9	7
7	4	9	6	5	2	8	1	3
2	8	3	7	9	1	5	6	4
8	3	7	2	1	5	6	4	9
4	9	1	8	6	3	7	5	2
6	2	5	9	4	7	1	3	8
3	5	8	4	7	6	9	2	1
1	7	2	5	3	9	4	8	6
9	6	4	1	2	8	3	7	5

Puzzle 619

2	6	3	8	4	9	5	7	1
9	5	4	1	3	7	6	2	8
8	7	1	2	5	6	3	9	4
4	9	6	3	2	5	1	8	7
1	8	7	6	9	4	2	5	3
5	3	2	7	1	8	9	4	6
7	1	5	9	8	3	4	6	2
3	4	8	5	6	2	7	1	9
6	2	9	4	7	1	8	3	5

Puzzle 620

8	7	2	9	1	3	4	5	6
1	3	6	4	8	5	9	7	2
9	5	4	6	2	7	8	1	3
2	6	1	7	5	9	3	8	4
5	9	3	8	4	6	7	2	1
7	4	8	2	3	1	6	9	5
3	8	5	1	7	4	2	6	9
4	2	9	5	6	8	1	3	7
6	1	7	3	9	2	5	4	8

Puzzle 621

8	3	2	7	9	4	5	1	6
1	5	4	3	2	6	9	8	7
6	9	7	8	5	1	4	3	2
5	1	3	6	8	2	7	9	4
9	2	6	1	4	7	8	5	3
7	4	8	9	3	5	6	2	1
3	6	1	5	7	9	2	4	8
4	8	5	2	6	3	1	7	9
2	7	9	4	1	8	3	6	5

Puzzle 622

1	7	3	9	8	2	5	4	6
9	6	5	3	7	4	2	8	1
2	8	4	6	5	1	3	7	9
6	2	7	4	1	5	9	3	8
3	4	8	7	9	6	1	5	2
5	1	9	8	2	3	4	6	7
8	5	2	1	4	7	6	9	3
7	3	1	5	6	9	8	2	4
4	9	6	2	3	8	7	1	5

Puzzle 623

4	1	8	2	5	9	6	3	7
3	5	9	7	4	6	2	8	1
2	6	7	8	1	3	4	9	5
6	7	4	1	9	5	3	2	8
5	2	1	6	3	8	7	4	9
9	8	3	4	2	7	1	5	6
8	9	6	3	7	2	5	1	4
7	4	2	5	8	1	9	6	3
1	3	5	9	6	4	8	7	2

Puzzle 624

2	3	9	1	7	5	8	4	6
5	8	7	2	4	6	1	9	3
4	1	6	8	3	9	5	2	7
8	5	2	4	1	3	7	6	9
9	6	3	7	5	2	4	8	1
7	4	1	9	6	8	2	3	5
6	2	4	5	9	7	3	1	8
3	7	8	6	2	1	9	5	4
1	9	5	3	8	4	6	7	2

Puzzle 625

4	5	8	2	7	3	9	6	1
3	7	1	6	8	9	2	5	4
6	9	2	4	5	1	7	8	3
9	3	4	5	2	7	8	1	6
2	8	6	1	3	4	5	7	9
7	1	5	9	6	8	3	4	2
5	6	7	3	1	2	4	9	8
8	2	9	7	4	6	1	3	5
1	4	3	8	9	5	6	2	7

Puzzle 626

7	3	4	1	9	6	8	5	2
2	9	6	5	3	8	4	1	7
1	5	8	7	4	2	6	3	9
3	8	1	6	5	9	2	7	4
4	6	7	8	2	3	5	9	1
9	2	5	4	7	1	3	8	6
8	7	2	3	1	4	9	6	5
5	4	3	9	6	7	1	2	8
6	1	9	2	8	5	7	4	3

Puzzle 627

4	9	6	1	7	2	3	5	8
3	1	7	4	8	5	9	2	6
2	5	8	9	6	3	1	7	4
5	2	3	7	9	6	8	4	1
7	4	1	3	2	8	6	9	5
8	6	9	5	1	4	2	3	7
6	3	2	8	4	7	5	1	9
9	7	5	6	3	1	4	8	2
1	8	4	2	5	9	7	6	3

Puzzle 628

1	2	8	5	4	6	3	7	9
5	7	6	1	3	9	8	2	4
3	4	9	8	7	2	6	5	1
2	8	1	6	5	4	7	9	3
7	6	5	9	1	3	2	4	8
9	3	4	7	2	8	5	1	6
6	1	2	3	9	5	4	8	7
8	5	7	4	6	1	9	3	2
4	9	3	2	8	7	1	6	5

Puzzle 629

5	8	6	2	4	7	3	1	9
7	1	4	8	9	3	5	6	2
9	2	3	5	6	1	4	8	7
6	7	1	3	2	5	8	9	4
2	3	5	9	8	4	1	7	6
4	9	8	7	1	6	2	3	5
3	6	2	4	7	8	9	5	1
1	5	9	6	3	2	7	4	8
8	4	7	1	5	9	6	2	3

Puzzle 630

7	5	4	9	8	6	3	1	2
6	9	3	2	4	1	7	5	8
2	1	8	3	7	5	6	4	9
8	3	6	4	2	9	5	7	1
1	2	5	6	3	7	8	9	4
9	4	7	5	1	8	2	6	3
3	6	2	7	9	4	1	8	5
4	7	1	8	5	2	9	3	6
5	8	9	1	6	3	4	2	7

Puzzle 631

3	9	1	6	5	8	4	2	7
7	2	5	1	3	4	8	9	6
8	6	4	2	7	9	5	1	3
2	4	3	5	1	7	6	8	9
6	8	7	3	9	2	1	5	4
1	5	9	8	4	6	3	7	2
5	7	6	9	8	3	2	4	1
9	1	2	4	6	5	7	3	8
4	3	8	7	2	1	9	6	5

Puzzle 632

1	4	7	3	9	6	2	5	8
8	5	6	2	4	1	9	7	3
2	9	3	8	5	7	6	1	4
3	2	5	4	7	8	1	6	9
6	7	1	9	2	3	4	8	5
4	8	9	6	1	5	7	3	2
7	3	2	5	6	9	8	4	1
5	6	4	1	8	2	3	9	7
9	1	8	7	3	4	5	2	6

Puzzle 633

4	1	5	9	8	6	2	3	7
7	9	6	2	4	3	5	1	8
3	2	8	7	1	5	4	9	6
6	5	9	1	3	8	7	2	4
8	4	2	6	7	9	3	5	1
1	3	7	4	5	2	6	8	9
2	7	3	8	9	4	1	6	5
9	6	1	5	2	7	8	4	3
5	8	4	3	6	1	9	7	2

Puzzle 634

9	1	4	3	6	2	7	5	8
2	7	3	9	5	8	1	4	6
5	8	6	7	1	4	2	9	3
8	2	1	4	9	6	5	3	7
4	5	7	8	2	3	6	1	9
6	3	9	1	7	5	8	2	4
3	4	2	6	8	1	9	7	5
1	9	8	5	3	7	4	6	2
7	6	5	2	4	9	3	8	1

Puzzle 635

5	3	4	8	7	2	9	1	6
7	9	2	3	6	1	5	4	8
1	8	6	5	9	4	2	3	7
4	2	9	6	3	5	8	7	1
3	1	5	4	8	7	6	2	9
6	7	8	1	2	9	4	5	3
2	4	7	9	1	8	3	6	5
9	5	3	7	4	6	1	8	2
8	6	1	2	5	3	7	9	4

Puzzle 636

6	8	2	1	5	3	7	9	4
7	5	9	6	8	4	3	2	1
3	1	4	7	2	9	5	6	8
1	6	5	4	3	8	2	7	9
8	4	3	9	7	2	1	5	6
9	2	7	5	1	6	8	4	3
4	7	8	2	9	1	6	3	5
5	3	6	8	4	7	9	1	2
2	9	1	3	6	5	4	8	7

Puzzle 637

3	8	2	6	7	9	4	5	1
7	9	4	2	5	1	3	6	8
1	6	5	8	3	4	9	2	7
4	1	8	5	6	2	7	3	9
5	7	9	3	1	8	6	4	2
2	3	6	9	4	7	1	8	5
8	2	3	7	9	6	5	1	4
9	5	1	4	8	3	2	7	6
6	4	7	1	2	5	8	9	3

Puzzle 638

9	3	4	8	5	7	2	6	1
2	7	1	3	9	6	8	4	5
8	5	6	1	4	2	3	9	7
6	2	9	4	8	1	7	5	3
1	8	7	5	6	3	9	2	4
3	4	5	7	2	9	1	8	6
4	1	8	9	3	5	6	7	2
7	9	2	6	1	4	5	3	8
5	6	3	2	7	8	4	1	9

Puzzle 639

9	7	1	6	5	2	4	3	8
2	5	6	8	3	4	9	7	1
3	4	8	1	7	9	2	6	5
1	3	5	4	8	7	6	2	9
7	6	4	2	9	5	1	8	3
8	2	9	3	1	6	5	4	7
6	8	7	9	4	1	3	5	2
4	9	3	5	2	8	7	1	6
5	1	2	7	6	3	8	9	4

Puzzle 640

7	8	2	5	6	1	9	3	4
5	3	9	8	7	4	1	2	6
6	4	1	2	3	9	7	5	8
8	9	3	7	2	5	6	4	1
2	7	6	1	4	3	8	9	5
1	5	4	6	9	8	2	7	3
3	2	7	4	8	6	5	1	9
9	1	8	3	5	7	4	6	2
4	6	5	9	1	2	3	8	7

Puzzle 641

3	6	4	2	9	8	7	1	5
5	1	7	6	3	4	2	9	8
2	8	9	1	5	7	3	4	6
8	2	1	4	7	5	6	3	9
7	5	6	9	2	3	1	8	4
9	4	3	8	6	1	5	7	2
4	7	2	5	1	9	8	6	3
6	3	8	7	4	2	9	5	1
1	9	5	3	8	6	4	2	7

Puzzle 642

5	9	7	8	6	2	1	3	4
8	4	3	7	9	1	5	6	2
1	6	2	3	5	4	9	8	7
6	3	5	9	4	7	2	1	8
2	8	4	1	3	6	7	9	5
9	7	1	5	2	8	6	4	3
4	5	6	2	1	3	8	7	9
7	1	9	4	8	5	3	2	6
3	2	8	6	7	9	4	5	1

Puzzle 643

1	2	7	5	8	3	4	9	6
5	4	6	1	9	7	8	2	3
9	3	8	2	6	4	1	7	5
8	1	5	4	7	2	3	6	9
3	9	2	8	5	6	7	4	1
7	6	4	3	1	9	2	5	8
6	5	1	7	4	8	9	3	2
4	8	3	9	2	5	6	1	7
2	7	9	6	3	1	5	8	4

Puzzle 644

6	5	1	8	2	4	3	9	7
9	2	8	6	7	3	1	4	5
7	3	4	1	5	9	2	6	8
1	8	5	2	6	7	4	3	9
2	7	3	4	9	5	8	1	6
4	9	6	3	1	8	7	5	2
5	6	2	7	3	1	9	8	4
3	4	7	9	8	6	5	2	1
8	1	9	5	4	2	6	7	3

Puzzle 645

2	8	1	3	4	6	9	7	5
9	3	5	8	2	7	6	1	4
7	6	4	5	1	9	8	2	3
8	5	6	1	9	4	7	3	2
1	9	2	6	7	3	5	4	8
4	7	3	2	8	5	1	6	9
5	4	8	7	6	2	3	9	1
3	2	7	9	5	1	4	8	6
6	1	9	4	3	8	2	5	7

Puzzle 646

3	4	6	8	5	7	9	1	2
2	9	7	3	6	1	4	5	8
1	5	8	9	2	4	3	7	6
6	1	9	2	4	5	7	8	3
8	2	3	1	7	9	5	6	4
5	7	4	6	3	8	2	9	1
4	8	1	7	9	3	6	2	5
7	3	2	5	1	6	8	4	9
9	6	5	4	8	2	1	3	7

Puzzle 647

9	3	2	8	1	5	4	6	7
4	8	1	6	7	3	9	5	2
5	7	6	4	2	9	3	8	1
7	1	5	2	8	4	6	9	3
8	9	3	7	5	6	1	2	4
2	6	4	3	9	1	8	7	5
6	5	8	1	3	7	2	4	9
3	2	7	9	4	8	5	1	6
1	4	9	5	6	2	7	3	8

Puzzle 648

8	5	2	1	6	7	4	3	9
9	7	1	4	3	8	6	2	5
3	4	6	5	9	2	8	1	7
5	2	3	8	7	4	9	6	1
6	8	9	3	1	5	7	4	2
7	1	4	6	2	9	3	5	8
2	9	5	7	4	6	1	8	3
4	3	8	9	5	1	2	7	6
1	6	7	2	8	3	5	9	4

Puzzle 649

1	8	9	6	7	2	4	3	5
2	4	6	5	8	3	7	9	1
3	5	7	9	4	1	6	8	2
7	3	5	4	9	8	2	1	6
4	9	2	1	3	6	5	7	8
8	6	1	2	5	7	9	4	3
6	7	8	3	2	9	1	5	4
9	2	4	8	1	5	3	6	7
5	1	3	7	6	4	8	2	9

Puzzle 650

3	2	7	1	6	8	5	9	4
9	1	8	5	3	4	2	6	7
6	5	4	9	2	7	3	1	8
1	4	6	3	8	9	7	2	5
2	7	3	4	5	1	6	8	9
5	8	9	6	7	2	4	3	1
8	6	2	7	1	5	9	4	3
4	3	5	8	9	6	1	7	2
7	9	1	2	4	3	8	5	6

Puzzle 651

1	9	3	6	4	8	5	7	2
4	2	8	9	5	7	1	6	3
7	5	6	2	1	3	9	4	8
9	8	5	4	3	6	7	2	1
3	7	4	1	9	2	6	8	5
6	1	2	7	8	5	3	9	4
8	4	7	3	6	1	2	5	9
2	3	9	5	7	4	8	1	6
5	6	1	8	2	9	4	3	7

Puzzle 652

3	1	4	7	8	5	2	6	9
2	6	7	3	1	9	4	8	5
5	8	9	4	2	6	3	7	1
4	9	8	2	3	1	6	5	7
1	7	3	6	5	4	8	9	2
6	5	2	9	7	8	1	3	4
7	3	6	1	9	2	5	4	8
8	4	1	5	6	7	9	2	3
9	2	5	8	4	3	7	1	6

Puzzle 653

1	3	5	8	7	2	9	4	6
4	6	2	9	5	1	7	8	3
7	9	8	3	4	6	5	1	2
9	4	6	2	3	7	1	5	8
3	2	7	1	8	5	6	9	4
8	5	1	6	9	4	3	2	7
6	7	4	5	1	8	2	3	9
2	1	9	4	6	3	8	7	5
5	8	3	7	2	9	4	6	1

Puzzle 654

8	1	5	6	4	3	7	2	9
3	4	2	8	9	7	1	5	6
9	6	7	5	1	2	3	4	8
2	3	6	4	5	9	8	7	1
5	9	8	7	6	1	2	3	4
1	7	4	3	2	8	6	9	5
6	8	9	2	7	4	5	1	3
4	2	3	1	8	5	9	6	7
7	5	1	9	3	6	4	8	2

Puzzle 655

3	4	9	7	8	6	5	1	2
6	7	5	2	3	1	9	4	8
2	8	1	9	4	5	7	3	6
9	2	6	8	7	3	1	5	4
8	3	7	1	5	4	2	6	9
1	5	4	6	2	9	8	7	3
4	9	2	3	1	7	6	8	5
5	1	8	4	6	2	3	9	7
7	6	3	5	9	8	4	2	1

Puzzle 656

8	1	7	4	9	2	5	3	6
5	3	9	1	8	6	4	2	7
2	6	4	5	3	7	1	9	8
4	7	2	6	5	1	3	8	9
9	5	3	8	2	4	6	7	1
6	8	1	3	7	9	2	5	4
1	2	8	7	4	3	9	6	5
7	9	6	2	1	5	8	4	3
3	4	5	9	6	8	7	1	2

Puzzle 657

9	8	4	7	1	2	3	6	5
7	3	5	8	9	6	1	4	2
2	6	1	4	3	5	7	8	9
1	9	2	6	4	8	5	3	7
5	7	8	3	2	9	4	1	6
6	4	3	5	7	1	9	2	8
4	2	7	9	8	3	6	5	1
8	5	9	1	6	4	2	7	3
3	1	6	2	5	7	8	9	4

Puzzle 658

9	8	1	5	2	4	3	7	6
3	4	7	8	6	9	5	2	1
6	5	2	7	1	3	9	4	8
5	1	3	6	4	8	7	9	2
8	7	4	2	9	1	6	5	3
2	9	6	3	5	7	1	8	4
4	3	9	1	8	5	2	6	7
1	2	8	9	7	6	4	3	5
7	6	5	4	3	2	8	1	9

Puzzle 659

8	1	6	9	4	5	2	3	7
4	2	7	6	3	8	5	1	9
5	3	9	7	2	1	8	6	4
9	6	1	2	8	7	3	4	5
2	5	4	1	6	3	7	9	8
7	8	3	5	9	4	6	2	1
1	7	2	4	5	6	9	8	3
6	4	8	3	7	9	1	5	2
3	9	5	8	1	2	4	7	6

Puzzle 660

1	7	8	4	6	3	9	2	5
3	5	9	2	7	1	4	6	8
2	4	6	9	8	5	7	1	3
5	1	2	6	4	8	3	9	7
7	9	4	1	3	2	8	5	6
6	8	3	5	9	7	2	4	1
4	6	7	3	1	9	5	8	2
9	3	5	8	2	6	1	7	4
8	2	1	7	5	4	6	3	9

Puzzle 661

2	1	6	8	5	4	7	9	3
9	7	3	6	1	2	5	8	4
8	4	5	7	9	3	6	2	1
1	3	8	5	7	9	2	4	6
5	9	2	4	3	6	8	1	7
7	6	4	1	2	8	3	5	9
3	2	1	9	6	5	4	7	8
4	5	9	3	8	7	1	6	2
6	8	7	2	4	1	9	3	5

Puzzle 662

5	9	1	2	4	7	6	3	8
6	7	8	9	1	3	4	2	5
4	3	2	8	6	5	1	7	9
9	1	4	7	3	6	5	8	2
3	8	5	1	2	4	7	9	6
7	2	6	5	8	9	3	4	1
8	5	7	3	9	1	2	6	4
2	6	3	4	5	8	9	1	7
1	4	9	6	7	2	8	5	3

Puzzle 663

3	4	7	2	1	5	9	6	8
8	5	2	9	4	6	7	3	1
1	6	9	3	8	7	5	4	2
7	3	8	1	2	9	4	5	6
2	1	4	6	5	3	8	7	9
6	9	5	4	7	8	2	1	3
5	2	3	8	6	4	1	9	7
9	7	1	5	3	2	6	8	4
4	8	6	7	9	1	3	2	5

Puzzle 664

2	3	1	4	9	6	8	7	5
7	4	9	5	2	8	1	6	3
8	6	5	1	7	3	4	9	2
4	9	3	6	8	2	7	5	1
1	7	2	3	4	5	6	8	9
6	5	8	7	1	9	3	2	4
5	8	6	2	3	4	9	1	7
9	1	4	8	5	7	2	3	6
3	2	7	9	6	1	5	4	8

Puzzle 665

5	1	2	3	9	4	8	7	6
3	4	7	6	8	1	5	2	9
9	6	8	5	7	2	3	4	1
6	7	4	9	1	8	2	5	3
8	9	3	2	4	5	6	1	7
2	5	1	7	3	6	9	8	4
7	2	9	1	5	3	4	6	8
4	3	5	8	6	7	1	9	2
1	8	6	4	2	9	7	3	5

Puzzle 666

4	1	2	6	8	7	9	5	3
7	8	3	9	5	2	6	1	4
9	5	6	3	1	4	8	2	7
2	4	1	7	3	8	5	9	6
5	7	9	4	2	6	1	3	8
6	3	8	5	9	1	7	4	2
3	2	5	8	6	9	4	7	1
8	9	7	1	4	3	2	6	5
1	6	4	2	7	5	3	8	9

Puzzle 667

6	4	5	7	9	1	2	8	3
2	8	1	5	3	6	7	9	4
9	3	7	8	2	4	1	5	6
4	6	3	9	7	2	5	1	8
7	2	8	3	1	5	4	6	9
1	5	9	6	4	8	3	7	2
8	9	2	4	5	7	6	3	1
5	1	6	2	8	3	9	4	7
3	7	4	1	6	9	8	2	5

Puzzle 668

4	9	7	8	2	5	6	1	3
2	6	1	3	9	4	8	7	5
5	8	3	6	7	1	9	2	4
1	5	2	7	3	9	4	8	6
3	4	6	1	8	2	5	9	7
8	7	9	5	4	6	1	3	2
9	3	8	4	5	7	2	6	1
6	2	4	9	1	3	7	5	8
7	1	5	2	6	8	3	4	9

Puzzle 669

4	3	9	8	1	6	2	5	7
5	7	6	2	9	4	3	8	1
1	8	2	7	3	5	6	9	4
8	9	3	5	6	1	4	7	2
7	4	1	3	2	8	5	6	9
6	2	5	9	4	7	1	3	8
3	5	7	1	8	2	9	4	6
2	6	8	4	5	9	7	1	3
9	1	4	6	7	3	8	2	5

Puzzle 670

9	4	2	8	5	7	6	3	1
6	5	3	4	2	1	9	7	8
1	8	7	3	6	9	2	5	4
8	3	5	6	7	2	1	4	9
2	1	4	5	9	8	7	6	3
7	6	9	1	4	3	5	8	2
3	9	8	7	1	5	4	2	6
5	2	6	9	3	4	8	1	7
4	7	1	2	8	6	3	9	5

Puzzle 671

7	1	9	8	6	3	4	5	2
6	5	4	7	1	2	8	3	9
3	2	8	5	4	9	1	6	7
8	4	7	3	2	6	5	9	1
1	3	5	9	7	8	2	4	6
2	9	6	1	5	4	7	8	3
4	7	1	6	9	5	3	2	8
9	8	2	4	3	1	6	7	5
5	6	3	2	8	7	9	1	4

Puzzle 672

8	2	7	1	3	5	9	4	6
6	3	1	9	7	4	5	8	2
5	9	4	8	2	6	3	1	7
2	6	8	4	5	7	1	3	9
7	4	9	3	8	1	2	6	5
1	5	3	2	6	9	8	7	4
9	1	2	6	4	8	7	5	3
4	8	5	7	9	3	6	2	1
3	7	6	5	1	2	4	9	8

Puzzle 673

7	1	2	4	6	9	3	5	8
4	6	9	8	3	5	7	2	1
5	3	8	1	7	2	9	4	6
2	4	3	5	1	6	8	9	7
8	9	7	3	2	4	1	6	5
1	5	6	7	9	8	2	3	4
3	7	5	2	4	1	6	8	9
6	8	1	9	5	3	4	7	2
9	2	4	6	8	7	5	1	3

Puzzle 674

5	1	9	6	2	3	7	4	8
7	6	2	4	9	8	1	5	3
8	3	4	1	5	7	2	9	6
3	9	8	2	4	5	6	1	7
6	7	5	3	8	1	9	2	4
4	2	1	7	6	9	8	3	5
2	4	3	9	7	6	5	8	1
9	5	6	8	1	4	3	7	2
1	8	7	5	3	2	4	6	9

Puzzle 675

7	6	8	1	9	3	4	2	5
2	3	5	7	4	6	8	9	1
4	1	9	2	8	5	3	7	6
3	4	2	9	7	1	6	5	8
9	8	7	6	5	4	1	3	2
6	5	1	3	2	8	7	4	9
8	7	6	5	3	2	9	1	4
1	2	3	4	6	9	5	8	7
5	9	4	8	1	7	2	6	3

Puzzle 676

7	6	9	8	5	2	3	4	1
2	1	5	7	4	3	6	9	8
8	4	3	1	9	6	5	2	7
5	2	4	9	7	1	8	3	6
9	3	1	5	6	8	4	7	2
6	8	7	2	3	4	9	1	5
4	7	6	3	2	5	1	8	9
3	9	8	6	1	7	2	5	4
1	5	2	4	8	9	7	6	3

Puzzle 677

7	2	8	5	3	9	1	4	6
3	4	9	2	6	1	8	7	5
1	5	6	8	7	4	9	3	2
6	7	5	3	4	8	2	1	9
9	3	4	6	1	2	7	5	8
8	1	2	7	9	5	3	6	4
4	6	3	9	2	7	5	8	1
5	9	1	4	8	3	6	2	7
2	8	7	1	5	6	4	9	3

Puzzle 678

4	5	6	2	1	7	9	8	3
9	1	8	4	5	3	6	2	7
2	7	3	9	8	6	1	5	4
8	6	5	3	7	1	4	9	2
7	9	1	5	2	4	3	6	8
3	2	4	8	6	9	7	1	5
5	4	7	1	9	2	8	3	6
6	8	9	7	3	5	2	4	1
1	3	2	6	4	8	5	7	9

Puzzle 679

3	4	8	1	5	6	9	2	7
1	6	5	7	9	2	3	4	8
2	7	9	4	3	8	1	6	5
6	2	7	5	4	1	8	3	9
5	1	3	2	8	9	4	7	6
9	8	4	6	7	3	5	1	2
4	5	6	8	1	7	2	9	3
8	3	2	9	6	4	7	5	1
7	9	1	3	2	5	6	8	4

Puzzle 680

1	9	3	8	6	7	5	2	4
7	2	6	1	5	4	3	9	8
8	4	5	9	2	3	6	1	7
2	3	1	7	9	8	4	5	6
5	7	8	6	4	1	9	3	2
4	6	9	2	3	5	8	7	1
9	8	7	3	1	6	2	4	5
6	5	2	4	7	9	1	8	3
3	1	4	5	8	2	7	6	9

Puzzle 681

6	1	7	5	9	4	2	8	3
8	4	3	1	2	6	7	9	5
9	5	2	7	3	8	4	1	6
2	9	4	8	6	7	3	5	1
1	6	8	4	5	3	9	2	7
3	7	5	2	1	9	8	6	4
4	3	6	9	8	5	1	7	2
7	2	9	6	4	1	5	3	8
5	8	1	3	7	2	6	4	9

Puzzle 682

6	1	2	7	8	4	5	9	3
7	5	4	6	9	3	8	1	2
9	3	8	1	5	2	4	7	6
4	2	9	3	1	8	6	5	7
3	7	6	5	4	9	2	8	1
5	8	1	2	6	7	9	3	4
8	6	5	4	7	1	3	2	9
1	4	3	9	2	5	7	6	8
2	9	7	8	3	6	1	4	5

Puzzle 683

7	3	2	1	4	9	6	5	8
5	6	4	2	3	8	7	9	1
8	9	1	7	5	6	3	2	4
6	1	5	8	7	2	9	4	3
2	4	7	6	9	3	1	8	5
3	8	9	4	1	5	2	6	7
9	2	3	5	8	1	4	7	6
4	5	6	3	2	7	8	1	9
1	7	8	9	6	4	5	3	2

Puzzle 684

6	3	1	5	2	9	7	8	4
4	9	8	3	7	6	2	5	1
2	7	5	8	4	1	6	9	3
9	4	7	2	1	3	8	6	5
8	2	3	6	9	5	4	1	7
5	1	6	7	8	4	3	2	9
3	8	2	1	5	7	9	4	6
1	6	9	4	3	8	5	7	2
7	5	4	9	6	2	1	3	8

Puzzle 685

4	3	8	5	6	7	9	2	1
9	1	6	8	3	2	7	5	4
5	2	7	9	4	1	8	3	6
8	6	9	1	5	3	2	4	7
3	4	2	6	7	8	5	1	9
1	7	5	2	9	4	6	8	3
7	5	4	3	8	6	1	9	2
2	9	3	7	1	5	4	6	8
6	8	1	4	2	9	3	7	5

Puzzle 686

7	1	4	5	9	6	8	3	2
5	3	6	7	8	2	9	1	4
8	2	9	4	3	1	7	5	6
6	7	3	8	4	5	2	9	1
2	5	1	6	7	9	3	4	8
4	9	8	1	2	3	6	7	5
3	4	5	2	6	7	1	8	9
1	6	7	9	5	8	4	2	3
9	8	2	3	1	4	5	6	7

Puzzle 687

3	8	5	1	6	9	2	7	4
4	1	7	8	3	2	6	9	5
9	6	2	5	4	7	3	8	1
5	2	9	6	8	3	1	4	7
6	7	1	4	2	5	8	3	9
8	3	4	9	7	1	5	6	2
2	5	6	7	9	8	4	1	3
1	9	8	3	5	4	7	2	6
7	4	3	2	1	6	9	5	8

Puzzle 688

1	6	7	8	9	3	4	2	5
4	2	5	6	7	1	8	3	9
8	9	3	4	5	2	1	6	7
5	7	1	2	8	9	3	4	6
9	3	8	1	4	6	7	5	2
6	4	2	5	3	7	9	8	1
2	5	4	7	1	8	6	9	3
7	8	9	3	6	5	2	1	4
3	1	6	9	2	4	5	7	8

Puzzle 689

5	7	6	4	8	2	9	3	1
3	9	2	5	6	1	7	4	8
1	4	8	9	7	3	5	2	6
4	8	9	6	5	7	3	1	2
2	3	5	8	1	4	6	9	7
7	6	1	2	3	9	8	5	4
6	5	4	3	2	8	1	7	9
8	2	7	1	9	5	4	6	3
9	1	3	7	4	6	2	8	5

Puzzle 690

9	3	7	2	4	1	6	8	5
5	2	8	9	7	6	3	1	4
4	1	6	8	3	5	9	2	7
6	7	9	3	5	8	1	4	2
3	5	1	4	2	9	7	6	8
8	4	2	1	6	7	5	3	9
1	9	3	7	8	2	4	5	6
7	8	5	6	1	4	2	9	3
2	6	4	5	9	3	8	7	1

Puzzle 691

3	8	1	5	4	2	6	7	9
6	7	9	1	3	8	4	2	5
2	4	5	9	6	7	8	3	1
4	2	6	8	9	3	1	5	7
8	9	3	7	1	5	2	4	6
5	1	7	4	2	6	9	8	3
7	5	4	6	8	9	3	1	2
1	6	2	3	7	4	5	9	8
9	3	8	2	5	1	7	6	4

Puzzle 692

8	7	9	5	1	3	2	6	4
2	1	6	4	8	7	9	3	5
4	3	5	2	9	6	8	7	1
6	5	7	3	4	2	1	8	9
9	2	4	8	7	1	3	5	6
3	8	1	9	6	5	4	2	7
1	4	2	6	5	8	7	9	3
5	9	8	7	3	4	6	1	2
7	6	3	1	2	9	5	4	8

Puzzle 693

2	7	8	1	6	5	4	3	9
6	9	4	7	2	3	5	1	8
3	1	5	9	8	4	7	6	2
4	2	9	8	7	6	3	5	1
5	8	1	3	4	2	6	9	7
7	6	3	5	1	9	2	8	4
9	4	7	6	5	8	1	2	3
8	5	2	4	3	1	9	7	6
1	3	6	2	9	7	8	4	5

Puzzle 694

9	7	1	8	6	2	4	5	3
6	3	5	9	4	1	7	2	8
2	4	8	3	7	5	9	6	1
3	6	7	4	5	9	8	1	2
5	8	9	1	2	7	6	3	4
4	1	2	6	3	8	5	9	7
7	2	3	5	9	4	1	8	6
8	9	4	2	1	6	3	7	5
1	5	6	7	8	3	2	4	9

Puzzle 695

1	6	2	5	3	4	8	7	9
8	7	4	9	1	2	5	6	3
5	3	9	8	7	6	4	1	2
3	8	7	1	4	9	2	5	6
2	1	6	7	5	8	9	3	4
4	9	5	2	6	3	1	8	7
9	2	3	6	8	1	7	4	5
6	5	1	4	2	7	3	9	8
7	4	8	3	9	5	6	2	1

Puzzle 696

9	5	2	4	1	8	6	3	7
8	6	1	9	7	3	2	4	5
7	3	4	5	2	6	1	9	8
4	7	3	1	5	2	9	8	6
2	9	8	6	4	7	3	5	1
6	1	5	8	3	9	7	2	4
5	4	7	2	9	1	8	6	3
3	2	6	7	8	4	5	1	9
1	8	9	3	6	5	4	7	2

Puzzle 697

7	8	6	4	9	5	1	2	3
1	2	9	3	7	8	6	4	5
3	4	5	1	6	2	9	7	8
2	9	7	8	3	4	5	1	6
5	1	4	6	2	9	8	3	7
6	3	8	5	1	7	4	9	2
4	5	1	2	8	3	7	6	9
9	6	2	7	5	1	3	8	4
8	7	3	9	4	6	2	5	1

Puzzle 698

8	6	9	2	7	1	4	3	5
1	7	4	9	3	5	6	2	8
5	2	3	4	6	8	9	7	1
7	8	1	5	2	6	3	4	9
6	9	5	3	4	7	1	8	2
3	4	2	8	1	9	7	5	6
4	1	8	6	5	3	2	9	7
2	5	7	1	9	4	8	6	3
9	3	6	7	8	2	5	1	4

Puzzle 699

7	2	5	9	4	3	1	8	6
4	3	6	7	1	8	2	9	5
1	9	8	6	2	5	3	4	7
6	8	2	4	9	1	7	5	3
3	5	1	2	8	7	4	6	9
9	7	4	3	5	6	8	1	2
5	1	3	8	7	9	6	2	4
2	6	9	1	3	4	5	7	8
8	4	7	5	6	2	9	3	1

Puzzle 700

4	6	3	2	9	5	7	8	1
2	8	1	7	3	6	5	9	4
5	9	7	8	1	4	2	3	6
3	1	8	5	4	2	9	6	7
7	4	2	3	6	9	1	5	8
6	5	9	1	7	8	4	2	3
1	7	6	9	2	3	8	4	5
8	2	4	6	5	7	3	1	9
9	3	5	4	8	1	6	7	2

Puzzle 701

2	9	4	6	3	1	5	7	8
8	3	5	4	7	2	9	6	1
6	7	1	8	5	9	2	4	3
5	6	9	2	1	8	4	3	7
4	8	7	3	6	5	1	2	9
3	1	2	7	9	4	8	5	6
9	4	3	5	8	6	7	1	2
7	2	8	1	4	3	6	9	5
1	5	6	9	2	7	3	8	4

Puzzle 702

5	2	1	7	3	8	6	4	9
3	4	7	6	2	9	8	1	5
9	6	8	5	1	4	2	3	7
8	9	6	3	5	1	4	7	2
2	5	3	9	4	7	1	6	8
1	7	4	8	6	2	9	5	3
4	8	5	1	9	3	7	2	6
6	1	9	2	7	5	3	8	4
7	3	2	4	8	6	5	9	1

Puzzle 703

9	7	1	3	4	5	6	8	2
6	5	3	8	7	2	1	4	9
2	4	8	6	1	9	5	7	3
1	9	6	2	3	7	4	5	8
8	2	4	9	5	1	7	3	6
7	3	5	4	6	8	2	9	1
5	1	2	7	9	3	8	6	4
3	6	7	1	8	4	9	2	5
4	8	9	5	2	6	3	1	7

Puzzle 704

8	7	3	9	1	4	6	2	5
9	5	6	2	8	7	4	3	1
1	2	4	5	6	3	7	8	9
4	9	8	1	3	5	2	6	7
2	1	7	4	9	6	8	5	3
3	6	5	7	2	8	1	9	4
5	4	2	8	7	9	3	1	6
6	8	9	3	4	1	5	7	2
7	3	1	6	5	2	9	4	8

Puzzle 705

6	2	9	4	5	8	1	7	3
5	8	7	6	3	1	2	4	9
3	1	4	9	7	2	8	6	5
9	6	8	5	2	3	7	1	4
1	7	5	8	6	4	3	9	2
2	4	3	7	1	9	5	8	6
4	3	6	1	8	5	9	2	7
8	9	2	3	4	7	6	5	1
7	5	1	2	9	6	4	3	8

Puzzle 706

6	9	2	3	8	5	7	4	1
5	4	7	2	9	1	8	3	6
8	3	1	6	7	4	2	5	9
7	5	3	8	4	6	1	9	2
4	8	9	1	3	2	6	7	5
2	1	6	9	5	7	4	8	3
9	6	4	5	1	8	3	2	7
3	2	8	7	6	9	5	1	4
1	7	5	4	2	3	9	6	8

Puzzle 707

5	7	9	6	1	2	8	3	4
1	6	3	8	5	4	2	7	9
4	8	2	9	3	7	5	6	1
3	4	1	2	9	6	7	5	8
6	2	5	7	8	1	9	4	3
7	9	8	3	4	5	1	2	6
8	5	4	1	7	3	6	9	2
9	3	6	5	2	8	4	1	7
2	1	7	4	6	9	3	8	5

Puzzle 708

9	2	7	5	1	3	6	4	8
8	6	4	2	9	7	5	1	3
5	1	3	6	4	8	2	7	9
7	4	5	1	6	9	3	8	2
1	8	9	3	2	4	7	5	6
6	3	2	8	7	5	4	9	1
3	9	1	7	5	6	8	2	4
4	5	6	9	8	2	1	3	7
2	7	8	4	3	1	9	6	5

Puzzle 709

6	5	9	7	3	4	1	2	8
3	7	8	1	2	5	9	6	4
2	4	1	9	8	6	7	3	5
7	3	2	5	4	8	6	9	1
9	1	5	6	7	2	4	8	3
4	8	6	3	9	1	5	7	2
5	2	7	4	6	3	8	1	9
8	9	4	2	1	7	3	5	6
1	6	3	8	5	9	2	4	7

Puzzle 710

5	4	3	7	8	9	2	1	6
1	2	9	6	5	4	3	8	7
8	6	7	3	1	2	4	5	9
9	1	4	8	6	7	5	2	3
7	3	6	5	2	1	8	9	4
2	8	5	9	4	3	7	6	1
4	5	8	1	3	6	9	7	2
6	7	2	4	9	8	1	3	5
3	9	1	2	7	5	6	4	8

Puzzle 711

7	2	5	9	3	1	6	8	4
3	4	9	6	5	8	1	2	7
6	8	1	7	4	2	9	5	3
5	9	7	3	8	6	2	4	1
2	6	4	5	1	7	3	9	8
8	1	3	2	9	4	5	7	6
4	7	2	1	6	5	8	3	9
9	5	6	8	7	3	4	1	2
1	3	8	4	2	9	7	6	5

Puzzle 712

6	5	1	8	3	7	9	4	2
4	3	8	6	9	2	5	1	7
9	7	2	4	1	5	6	8	3
8	6	3	7	4	9	2	5	1
2	1	5	3	6	8	4	7	9
7	4	9	2	5	1	8	3	6
1	8	7	9	2	4	3	6	5
5	9	6	1	8	3	7	2	4
3	2	4	5	7	6	1	9	8

Puzzle 713

4	3	7	5	9	8	2	1	6
9	2	1	3	4	6	7	8	5
6	8	5	1	7	2	9	3	4
2	4	6	7	1	5	3	9	8
7	9	3	6	8	4	5	2	1
1	5	8	9	2	3	6	4	7
8	7	4	2	5	9	1	6	3
5	6	9	4	3	1	8	7	2
3	1	2	8	6	7	4	5	9

Puzzle 714

7	1	9	2	8	3	4	5	6
6	3	5	4	7	9	2	1	8
8	4	2	5	1	6	7	9	3
9	8	1	7	5	4	6	3	2
4	5	6	1	3	2	9	8	7
3	2	7	6	9	8	1	4	5
2	6	3	8	4	1	5	7	9
5	9	4	3	2	7	8	6	1
1	7	8	9	6	5	3	2	4

Puzzle 715

9	7	2	6	3	4	8	5	1
8	4	1	2	9	5	6	3	7
3	6	5	7	1	8	2	9	4
2	1	7	5	4	6	3	8	9
4	5	8	3	2	9	7	1	6
6	3	9	8	7	1	5	4	2
5	8	4	9	6	2	1	7	3
1	2	3	4	5	7	9	6	8
7	9	6	1	8	3	4	2	5

Puzzle 716

2	3	6	8	4	7	1	9	5
9	8	1	3	2	5	7	6	4
5	4	7	6	9	1	8	3	2
7	6	2	1	5	3	9	4	8
1	9	4	2	7	8	6	5	3
3	5	8	9	6	4	2	1	7
6	1	3	4	8	2	5	7	9
8	7	9	5	3	6	4	2	1
4	2	5	7	1	9	3	8	6

Puzzle 717

5	8	7	1	3	6	4	2	9
3	6	9	5	2	4	8	1	7
4	1	2	9	7	8	6	3	5
6	9	4	2	5	3	1	7	8
2	7	5	4	8	1	3	9	6
1	3	8	7	6	9	5	4	2
7	5	1	6	4	2	9	8	3
9	2	3	8	1	5	7	6	4
8	4	6	3	9	7	2	5	1

Puzzle 718

1	5	4	9	8	7	3	2	6
3	8	7	1	2	6	5	9	4
2	9	6	4	5	3	7	8	1
4	6	5	3	9	2	8	1	7
7	2	3	6	1	8	4	5	9
9	1	8	5	7	4	6	3	2
8	4	9	2	6	5	1	7	3
6	7	2	8	3	1	9	4	5
5	3	1	7	4	9	2	6	8

Puzzle 719

2	8	3	7	4	1	5	9	6
1	7	5	2	6	9	3	8	4
6	4	9	5	3	8	2	1	7
8	2	7	6	5	4	1	3	9
4	5	6	9	1	3	8	7	2
3	9	1	8	7	2	6	4	5
5	1	2	3	9	7	4	6	8
9	3	8	4	2	6	7	5	1
7	6	4	1	8	5	9	2	3

Puzzle 720

8	3	1	7	9	2	4	6	5
2	5	7	6	3	4	9	1	8
6	4	9	8	5	1	3	2	7
5	8	4	9	6	7	1	3	2
1	6	3	5	2	8	7	4	9
7	9	2	1	4	3	5	8	6
9	7	8	4	1	6	2	5	3
3	1	6	2	7	5	8	9	4
4	2	5	3	8	9	6	7	1

Puzzle 721

1	4	5	8	6	9	2	3	7
3	2	8	4	7	5	9	6	1
7	9	6	2	3	1	8	5	4
9	3	7	1	8	2	6	4	5
2	5	4	7	9	6	3	1	8
8	6	1	3	5	4	7	2	9
5	1	9	6	2	7	4	8	3
6	7	3	5	4	8	1	9	2
4	8	2	9	1	3	5	7	6

Puzzle 722

6	4	8	3	5	9	2	1	7
3	5	1	8	7	2	4	6	9
2	9	7	6	1	4	8	3	5
9	2	4	7	3	1	6	5	8
8	7	3	2	6	5	1	9	4
1	6	5	9	4	8	7	2	3
4	1	2	5	8	3	9	7	6
5	8	6	1	9	7	3	4	2
7	3	9	4	2	6	5	8	1

Puzzle 723

2	9	7	1	4	5	3	8	6
6	1	8	9	3	2	4	7	5
4	5	3	7	6	8	2	9	1
5	8	4	6	1	7	9	3	2
1	6	9	2	8	3	5	4	7
7	3	2	5	9	4	1	6	8
8	4	6	3	2	1	7	5	9
3	7	1	8	5	9	6	2	4
9	2	5	4	7	6	8	1	3

Puzzle 724

9	4	7	1	2	5	8	6	3
1	6	3	9	8	4	5	2	7
2	5	8	7	3	6	9	4	1
6	8	2	3	9	1	7	5	4
4	7	5	2	6	8	3	1	9
3	1	9	4	5	7	2	8	6
5	9	4	8	1	3	6	7	2
8	2	1	6	7	9	4	3	5
7	3	6	5	4	2	1	9	8

Puzzle 725

2	1	6	4	5	8	9	7	3
5	8	3	2	7	9	1	6	4
9	4	7	6	3	1	2	8	5
8	5	4	9	1	2	6	3	7
3	9	1	7	8	6	4	5	2
6	7	2	5	4	3	8	1	9
1	3	5	8	2	4	7	9	6
7	2	9	1	6	5	3	4	8
4	6	8	3	9	7	5	2	1

Puzzle 726

3	5	8	7	2	6	4	1	9
9	6	1	4	8	5	3	7	2
4	7	2	3	9	1	5	6	8
1	3	4	9	7	8	6	2	5
7	8	6	5	1	2	9	4	3
2	9	5	6	3	4	7	8	1
5	1	9	8	6	7	2	3	4
6	2	3	1	4	9	8	5	7
8	4	7	2	5	3	1	9	6

Puzzle 727

5	9	4	8	2	6	1	3	7
3	6	2	4	7	1	5	9	8
1	7	8	3	5	9	2	6	4
2	4	9	6	1	8	3	7	5
7	3	6	5	9	2	4	8	1
8	1	5	7	3	4	9	2	6
6	5	7	9	4	3	8	1	2
4	2	3	1	8	7	6	5	9
9	8	1	2	6	5	7	4	3

Puzzle 728

4	9	7	3	5	6	2	1	8
3	5	6	8	1	2	9	7	4
8	2	1	4	9	7	5	3	6
9	1	3	7	8	4	6	2	5
7	6	2	9	3	5	4	8	1
5	8	4	2	6	1	7	9	3
2	7	5	1	4	8	3	6	9
6	3	8	5	7	9	1	4	2
1	4	9	6	2	3	8	5	7

Puzzle 729

1	2	6	5	4	7	3	8	9
5	8	4	9	2	3	6	7	1
7	9	3	8	6	1	4	2	5
2	7	8	3	9	4	5	1	6
6	1	9	7	5	2	8	3	4
3	4	5	6	1	8	2	9	7
8	3	1	4	7	5	9	6	2
4	6	7	2	8	9	1	5	3
9	5	2	1	3	6	7	4	8

Puzzle 730

6	9	1	5	7	3	2	4	8
7	2	8	9	1	4	5	6	3
4	5	3	8	6	2	1	7	9
1	6	2	7	5	9	3	8	4
3	8	9	2	4	6	7	1	5
5	4	7	1	3	8	6	9	2
2	1	5	4	9	7	8	3	6
9	7	6	3	8	5	4	2	1
8	3	4	6	2	1	9	5	7

Puzzle 731

8	5	1	6	2	3	9	7	4
6	3	2	4	9	7	8	5	1
7	4	9	5	8	1	6	2	3
1	7	6	3	5	2	4	8	9
9	8	5	1	4	6	7	3	2
4	2	3	9	7	8	1	6	5
3	1	8	2	6	9	5	4	7
5	9	7	8	3	4	2	1	6
2	6	4	7	1	5	3	9	8

Puzzle 732

9	7	3	4	6	1	5	2	8
1	6	2	5	9	8	4	3	7
8	5	4	3	7	2	9	6	1
6	8	7	1	5	3	2	4	9
2	1	9	7	4	6	8	5	3
3	4	5	8	2	9	1	7	6
4	9	8	2	3	7	6	1	5
7	2	1	6	8	5	3	9	4
5	3	6	9	1	4	7	8	2

Puzzle 733

3	2	4	9	1	5	6	7	8
1	7	8	2	6	3	4	9	5
6	5	9	7	8	4	1	2	3
7	6	5	8	9	2	3	1	4
2	4	1	3	7	6	5	8	9
8	9	3	4	5	1	7	6	2
5	3	7	1	2	9	8	4	6
9	8	6	5	4	7	2	3	1
4	1	2	6	3	8	9	5	7

Puzzle 734

6	1	7	9	2	3	8	5	4
5	4	3	8	1	7	6	9	2
8	2	9	6	5	4	1	3	7
3	7	2	1	9	8	4	6	5
4	5	8	3	6	2	9	7	1
9	6	1	4	7	5	3	2	8
2	9	6	5	8	1	7	4	3
7	8	4	2	3	6	5	1	9
1	3	5	7	4	9	2	8	6

Puzzle 735

7	8	6	4	1	5	9	3	2
5	3	2	6	8	9	7	1	4
1	9	4	2	7	3	5	8	6
8	4	9	5	2	1	6	7	3
2	6	5	8	3	7	1	4	9
3	1	7	9	6	4	2	5	8
9	7	8	1	4	2	3	6	5
6	2	3	7	5	8	4	9	1
4	5	1	3	9	6	8	2	7

Puzzle 736

5	2	1	9	3	4	6	8	7
9	6	4	7	1	8	3	2	5
7	3	8	2	6	5	4	9	1
8	5	2	4	7	1	9	3	6
3	4	7	8	9	6	5	1	2
1	9	6	5	2	3	7	4	8
6	1	5	3	8	9	2	7	4
4	7	9	1	5	2	8	6	3
2	8	3	6	4	7	1	5	9

Puzzle 737

2	4	5	1	3	7	8	9	6
9	7	3	2	8	6	5	4	1
6	1	8	5	9	4	7	3	2
3	8	4	7	2	5	1	6	9
7	5	2	9	6	1	4	8	3
1	9	6	8	4	3	2	7	5
5	6	1	4	7	9	3	2	8
8	3	7	6	1	2	9	5	4
4	2	9	3	5	8	6	1	7

Puzzle 738

9	2	5	3	8	6	4	7	1
8	1	7	9	5	4	3	2	6
3	6	4	1	2	7	5	8	9
2	8	3	7	9	5	1	6	4
4	5	9	2	6	1	8	3	7
6	7	1	4	3	8	2	9	5
1	9	2	6	4	3	7	5	8
5	4	6	8	7	2	9	1	3
7	3	8	5	1	9	6	4	2

Puzzle 739

1	5	7	9	6	3	4	2	8
3	4	6	2	5	8	7	9	1
9	2	8	4	1	7	6	3	5
5	6	2	3	7	4	8	1	9
7	9	1	8	2	5	3	4	6
8	3	4	6	9	1	2	5	7
2	1	5	7	3	6	9	8	4
6	8	9	5	4	2	1	7	3
4	7	3	1	8	9	5	6	2

Puzzle 740

4	5	6	1	7	9	2	8	3
2	9	8	5	4	3	6	1	7
7	1	3	2	8	6	9	4	5
8	7	2	9	3	1	4	5	6
6	4	5	8	2	7	1	3	9
9	3	1	4	6	5	8	7	2
3	2	9	7	1	4	5	6	8
5	6	4	3	9	8	7	2	1
1	8	7	6	5	2	3	9	4

Puzzle 741

6	9	1	8	2	4	3	5	7
8	4	7	3	9	5	2	6	1
2	5	3	1	6	7	4	9	8
5	8	6	2	3	1	7	4	9
9	3	4	5	7	6	1	8	2
1	7	2	4	8	9	5	3	6
3	6	8	7	4	2	9	1	5
7	1	9	6	5	3	8	2	4
4	2	5	9	1	8	6	7	3

Puzzle 742

7	9	8	6	4	2	5	1	3
5	4	2	7	3	1	8	6	9
3	6	1	5	8	9	2	4	7
8	7	6	4	9	5	1	3	2
4	5	3	2	1	7	9	8	6
2	1	9	3	6	8	7	5	4
6	8	7	9	5	3	4	2	1
1	2	4	8	7	6	3	9	5
9	3	5	1	2	4	6	7	8

Puzzle 743

9	7	3	1	8	2	5	4	6
6	5	1	4	9	3	8	2	7
8	2	4	5	7	6	9	3	1
3	8	9	6	4	7	2	1	5
7	4	5	3	2	1	6	9	8
2	1	6	8	5	9	4	7	3
5	9	2	7	3	8	1	6	4
4	6	7	2	1	5	3	8	9
1	3	8	9	6	4	7	5	2

Puzzle 744

5	3	4	7	2	6	8	1	9
1	8	9	5	3	4	2	7	6
7	2	6	9	8	1	4	5	3
3	9	8	1	7	2	6	4	5
2	4	1	3	6	5	7	9	8
6	5	7	8	4	9	3	2	1
9	6	2	4	5	8	1	3	7
8	7	5	2	1	3	9	6	4
4	1	3	6	9	7	5	8	2

Puzzle 745

4	8	9	6	1	3	7	2	5
3	5	2	9	8	7	4	6	1
7	6	1	5	2	4	3	8	9
9	1	4	2	5	8	6	3	7
6	7	3	4	9	1	8	5	2
5	2	8	3	7	6	9	1	4
2	4	5	8	6	9	1	7	3
1	3	6	7	4	2	5	9	8
8	9	7	1	3	5	2	4	6

Puzzle 746

6	3	7	8	4	9	1	2	5
8	5	4	6	1	2	3	9	7
9	2	1	7	3	5	8	6	4
5	1	9	4	8	7	6	3	2
3	8	6	2	5	1	7	4	9
7	4	2	3	9	6	5	8	1
2	9	5	1	6	3	4	7	8
4	7	3	5	2	8	9	1	6
1	6	8	9	7	4	2	5	3

Puzzle 747

1	7	5	4	2	3	8	6	9
3	8	4	6	5	9	1	2	7
2	9	6	1	8	7	4	5	3
9	3	8	5	1	6	2	7	4
6	4	1	2	7	8	3	9	5
7	5	2	9	3	4	6	8	1
8	1	9	3	6	5	7	4	2
5	2	7	8	4	1	9	3	6
4	6	3	7	9	2	5	1	8

Puzzle 748

3	7	4	6	5	1	8	2	9
6	2	8	4	3	9	7	1	5
9	1	5	8	7	2	4	3	6
1	8	7	5	6	3	2	9	4
4	9	6	2	8	7	3	5	1
5	3	2	1	9	4	6	8	7
8	6	3	9	4	5	1	7	2
7	5	1	3	2	6	9	4	8
2	4	9	7	1	8	5	6	3

Puzzle 749

1	7	6	8	9	2	5	4	3
2	9	4	1	3	5	7	8	6
5	3	8	4	6	7	9	2	1
8	6	9	7	5	4	1	3	2
4	1	2	3	8	9	6	7	5
3	5	7	2	1	6	8	9	4
7	2	5	9	4	1	3	6	8
6	4	3	5	7	8	2	1	9
9	8	1	6	2	3	4	5	7

Puzzle 750

6	8	4	3	1	7	5	9	2
2	5	7	9	6	8	3	1	4
1	9	3	4	5	2	7	6	8
8	1	5	2	3	6	9	4	7
4	7	6	1	8	9	2	5	3
3	2	9	5	7	4	6	8	1
5	6	1	8	2	3	4	7	9
7	4	2	6	9	1	8	3	5
9	3	8	7	4	5	1	2	6

Puzzle 751

3	8	1	6	5	2	4	7	9
6	7	2	4	1	9	8	5	3
9	4	5	3	7	8	1	2	6
8	9	3	1	6	7	2	4	5
5	2	4	9	8	3	6	1	7
1	6	7	5	2	4	3	9	8
7	1	6	8	4	5	9	3	2
4	5	9	2	3	6	7	8	1
2	3	8	7	9	1	5	6	4

Puzzle 752

9	5	4	7	6	8	1	2	3
8	2	7	1	3	5	9	6	4
1	3	6	9	4	2	5	7	8
3	6	9	5	1	7	8	4	2
4	7	2	6	8	9	3	5	1
5	8	1	3	2	4	6	9	7
2	9	5	8	7	3	4	1	6
7	1	3	4	5	6	2	8	9
6	4	8	2	9	1	7	3	5

Puzzle 753

5	6	2	7	8	3	1	9	4
9	4	3	1	2	5	8	7	6
8	1	7	9	6	4	2	5	3
4	2	5	3	7	6	9	1	8
3	7	6	8	1	9	5	4	2
1	9	8	4	5	2	6	3	7
2	8	1	5	3	7	4	6	9
7	5	4	6	9	8	3	2	1
6	3	9	2	4	1	7	8	5

Puzzle 754

9	3	1	4	5	7	6	8	2
8	5	4	9	2	6	7	1	3
2	7	6	3	1	8	4	5	9
7	9	5	6	4	1	3	2	8
3	1	2	8	7	5	9	6	4
4	6	8	2	3	9	1	7	5
5	8	9	7	6	4	2	3	1
1	2	7	5	9	3	8	4	6
6	4	3	1	8	2	5	9	7

Puzzle 755

3	1	5	7	4	6	2	8	9
4	2	8	1	9	5	7	6	3
7	9	6	8	2	3	5	1	4
2	4	9	6	7	1	8	3	5
1	5	3	9	8	2	4	7	6
8	6	7	3	5	4	1	9	2
9	3	4	5	1	7	6	2	8
5	8	1	2	6	9	3	4	7
6	7	2	4	3	8	9	5	1

Puzzle 756

1	9	8	6	3	4	7	2	5
4	2	7	9	8	5	3	1	6
5	6	3	2	7	1	9	8	4
8	5	9	3	4	2	6	7	1
3	1	2	8	6	7	5	4	9
6	7	4	5	1	9	2	3	8
2	4	6	1	5	3	8	9	7
7	3	5	4	9	8	1	6	2
9	8	1	7	2	6	4	5	3

Puzzle 757

6	4	8	3	9	5	2	1	7
1	3	9	7	2	8	5	6	4
7	5	2	6	1	4	3	9	8
4	8	7	1	3	9	6	5	2
5	9	6	8	7	2	1	4	3
3	2	1	4	5	6	7	8	9
8	1	3	9	6	7	4	2	5
2	7	4	5	8	1	9	3	6
9	6	5	2	4	3	8	7	1

Puzzle 758

6	3	2	8	5	9	1	4	7
5	4	9	1	6	7	3	8	2
8	7	1	3	2	4	5	9	6
2	6	3	5	7	8	4	1	9
9	8	4	6	3	1	2	7	5
7	1	5	4	9	2	8	6	3
4	5	7	2	8	6	9	3	1
1	2	6	9	4	3	7	5	8
3	9	8	7	1	5	6	2	4

Puzzle 759

4	8	6	7	1	9	5	2	3
5	7	3	2	8	4	1	6	9
1	9	2	3	5	6	7	4	8
8	3	1	5	4	7	6	9	2
2	4	5	9	6	1	3	8	7
9	6	7	8	3	2	4	1	5
6	5	9	1	2	3	8	7	4
7	1	8	4	9	5	2	3	6
3	2	4	6	7	8	9	5	1

Puzzle 760

2	4	7	9	3	1	5	6	8
6	9	5	2	7	8	1	4	3
8	3	1	6	4	5	7	9	2
3	6	9	7	5	4	2	8	1
1	8	2	3	9	6	4	7	5
7	5	4	8	1	2	6	3	9
4	1	6	5	8	3	9	2	7
5	7	8	4	2	9	3	1	6
9	2	3	1	6	7	8	5	4

Puzzle 761

1	8	6	3	2	7	9	4	5
9	3	2	4	8	5	7	1	6
7	5	4	6	9	1	2	3	8
5	7	1	9	4	2	6	8	3
2	4	3	1	6	8	5	9	7
8	6	9	7	5	3	1	2	4
3	1	5	8	7	9	4	6	2
4	9	7	2	3	6	8	5	1
6	2	8	5	1	4	3	7	9

Puzzle 762

8	4	1	5	2	3	7	9	6
9	7	5	1	8	6	4	2	3
6	3	2	7	9	4	5	1	8
7	8	3	9	5	2	1	6	4
2	6	4	3	1	8	9	7	5
1	5	9	4	6	7	8	3	2
4	2	8	6	7	1	3	5	9
3	9	7	2	4	5	6	8	1
5	1	6	8	3	9	2	4	7

Puzzle 763

2	8	1	3	4	9	6	7	5
5	4	7	2	6	1	9	8	3
3	6	9	7	5	8	1	4	2
7	2	4	9	8	6	3	5	1
6	5	8	1	2	3	4	9	7
9	1	3	5	7	4	8	2	6
8	3	2	6	9	5	7	1	4
1	9	5	4	3	7	2	6	8
4	7	6	8	1	2	5	3	9

Puzzle 764

2	8	7	5	6	1	4	9	3
4	6	5	2	9	3	7	1	8
9	3	1	7	4	8	5	6	2
6	5	2	8	1	4	3	7	9
3	9	4	6	5	7	2	8	1
1	7	8	9	3	2	6	5	4
8	2	6	3	7	9	1	4	5
5	1	9	4	2	6	8	3	7
7	4	3	1	8	5	9	2	6

Puzzle 765

9	2	4	8	7	1	5	6	3
8	6	5	9	2	3	7	1	4
7	3	1	5	4	6	2	9	8
1	7	3	2	8	4	9	5	6
6	4	8	1	9	5	3	2	7
2	5	9	6	3	7	8	4	1
5	9	7	4	6	8	1	3	2
4	8	2	3	1	9	6	7	5
3	1	6	7	5	2	4	8	9

Puzzle 766

3	9	4	1	2	8	5	6	7
7	1	6	3	4	5	2	8	9
8	2	5	6	9	7	3	4	1
4	3	2	7	5	9	6	1	8
6	8	7	4	1	3	9	2	5
1	5	9	2	8	6	7	3	4
5	4	1	9	3	2	8	7	6
9	7	3	8	6	4	1	5	2
2	6	8	5	7	1	4	9	3

Puzzle 767

4	3	7	5	6	2	1	9	8
5	1	6	9	8	4	3	7	2
9	2	8	7	3	1	5	4	6
7	9	5	6	4	3	2	8	1
6	8	2	1	7	9	4	5	3
3	4	1	8	2	5	7	6	9
8	6	3	2	5	7	9	1	4
2	5	9	4	1	6	8	3	7
1	7	4	3	9	8	6	2	5

Puzzle 768

3	1	9	5	6	7	8	4	2
4	8	2	1	9	3	6	7	5
5	6	7	4	8	2	3	1	9
2	5	1	7	4	6	9	3	8
8	9	4	2	3	5	1	6	7
6	7	3	8	1	9	5	2	4
7	3	6	9	5	4	2	8	1
9	4	8	3	2	1	7	5	6
1	2	5	6	7	8	4	9	3

Puzzle 769

5	8	2	6	1	3	7	9	4
7	4	9	2	8	5	3	1	6
6	3	1	4	7	9	8	2	5
2	5	3	1	4	7	6	8	9
8	1	6	9	5	2	4	3	7
4	9	7	3	6	8	2	5	1
1	2	8	7	9	4	5	6	3
3	6	4	5	2	1	9	7	8
9	7	5	8	3	6	1	4	2

Puzzle 770

5	1	9	4	3	8	2	6	7
7	2	3	5	6	1	9	8	4
8	4	6	7	9	2	1	3	5
2	8	4	9	7	3	5	1	6
1	6	7	2	8	5	4	9	3
3	9	5	6	1	4	7	2	8
9	3	8	1	4	7	6	5	2
6	7	2	8	5	9	3	4	1
4	5	1	3	2	6	8	7	9

Puzzle 771

6	9	2	7	1	3	4	5	8
3	8	5	4	2	9	7	1	6
7	4	1	6	5	8	3	9	2
2	1	3	8	4	5	9	6	7
4	7	8	9	3	6	5	2	1
5	6	9	2	7	1	8	3	4
9	3	7	1	8	2	6	4	5
1	5	4	3	6	7	2	8	9
8	2	6	5	9	4	1	7	3

Puzzle 772

7	1	4	6	3	5	2	9	8
3	8	9	4	1	2	7	6	5
6	5	2	7	9	8	3	4	1
2	9	5	1	7	4	6	8	3
8	3	1	5	2	6	4	7	9
4	6	7	3	8	9	5	1	2
5	7	3	9	6	1	8	2	4
9	4	8	2	5	7	1	3	6
1	2	6	8	4	3	9	5	7

Puzzle 773

6	5	8	1	7	2	3	4	9
3	7	1	8	9	4	2	5	6
9	4	2	5	6	3	1	7	8
8	2	5	9	3	6	7	1	4
1	3	6	7	4	8	9	2	5
7	9	4	2	1	5	8	6	3
2	6	7	4	8	9	5	3	1
4	1	9	3	5	7	6	8	2
5	8	3	6	2	1	4	9	7

Puzzle 774

8	2	5	6	3	4	7	9	1
9	3	7	2	1	5	6	4	8
6	1	4	9	7	8	3	5	2
2	5	8	4	6	9	1	3	7
3	4	1	8	2	7	5	6	9
7	9	6	3	5	1	2	8	4
4	7	3	1	9	6	8	2	5
1	6	9	5	8	2	4	7	3
5	8	2	7	4	3	9	1	6

Puzzle 775

8	9	6	4	3	1	5	7	2
1	3	2	9	7	5	8	6	4
4	7	5	6	8	2	1	3	9
2	8	4	3	9	6	7	1	5
9	6	1	8	5	7	4	2	3
3	5	7	1	2	4	9	8	6
6	1	3	7	4	9	2	5	8
7	2	9	5	6	8	3	4	1
5	4	8	2	1	3	6	9	7

Puzzle 776

2	4	7	5	9	3	6	8	1
1	3	9	7	8	6	5	2	4
5	8	6	1	4	2	7	9	3
9	1	2	4	6	7	3	5	8
8	6	5	2	3	1	4	7	9
3	7	4	9	5	8	2	1	6
4	2	8	6	1	5	9	3	7
6	5	3	8	7	9	1	4	2
7	9	1	3	2	4	8	6	5

Puzzle 777

9	8	5	7	1	4	2	3	6
1	4	7	2	3	6	9	5	8
2	6	3	9	5	8	7	4	1
6	7	1	5	4	3	8	2	9
4	3	2	1	8	9	6	7	5
5	9	8	6	7	2	4	1	3
7	1	6	4	9	5	3	8	2
8	2	4	3	6	1	5	9	7
3	5	9	8	2	7	1	6	4

Puzzle 778

4	5	3	7	6	8	1	9	2
1	8	9	5	3	2	4	6	7
2	7	6	1	9	4	3	8	5
9	6	7	4	5	1	2	3	8
5	2	1	9	8	3	7	4	6
8	3	4	2	7	6	9	5	1
3	9	5	8	1	7	6	2	4
6	1	2	3	4	5	8	7	9
7	4	8	6	2	9	5	1	3

Puzzle 779

7	3	4	9	2	8	6	1	5
2	1	8	5	6	3	4	7	9
5	6	9	7	4	1	8	3	2
1	2	3	8	5	4	9	6	7
6	9	7	3	1	2	5	8	4
4	8	5	6	7	9	3	2	1
3	7	6	1	9	5	2	4	8
9	4	1	2	8	6	7	5	3
8	5	2	4	3	7	1	9	6

Puzzle 780

8	5	4	1	7	6	9	2	3
7	9	1	8	3	2	4	6	5
6	2	3	4	9	5	7	8	1
9	7	8	3	5	4	2	1	6
1	4	5	6	2	7	8	3	9
2	3	6	9	1	8	5	7	4
3	1	2	5	8	9	6	4	7
4	8	9	7	6	3	1	5	2
5	6	7	2	4	1	3	9	8

Puzzle 781

2	5	9	4	8	7	3	1	6
3	4	6	2	1	9	5	7	8
1	8	7	3	6	5	2	9	4
5	7	1	6	2	8	4	3	9
9	6	3	7	4	1	8	2	5
8	2	4	5	9	3	1	6	7
7	3	8	9	5	2	6	4	1
4	1	2	8	7	6	9	5	3
6	9	5	1	3	4	7	8	2

Puzzle 782

6	1	7	9	5	2	4	3	8
3	2	8	7	1	4	9	5	6
9	5	4	8	6	3	1	2	7
1	4	2	5	9	6	7	8	3
8	7	3	4	2	1	6	9	5
5	6	9	3	8	7	2	1	4
4	8	5	2	7	9	3	6	1
2	3	1	6	4	5	8	7	9
7	9	6	1	3	8	5	4	2

Puzzle 783

1	5	2	7	9	4	3	8	6
8	6	9	5	3	1	7	2	4
3	4	7	2	6	8	5	9	1
7	1	4	8	5	6	9	3	2
6	3	5	4	2	9	1	7	8
9	2	8	1	7	3	6	4	5
5	7	6	3	8	2	4	1	9
2	9	1	6	4	7	8	5	3
4	8	3	9	1	5	2	6	7

Puzzle 784

1	6	3	8	2	5	4	7	9
9	7	8	6	4	1	2	5	3
2	5	4	9	3	7	8	1	6
4	2	7	5	1	9	6	3	8
5	1	6	2	8	3	7	9	4
8	3	9	4	7	6	1	2	5
7	9	1	3	6	8	5	4	2
3	8	2	7	5	4	9	6	1
6	4	5	1	9	2	3	8	7

Puzzle 785

6	7	4	8	5	1	2	9	3
2	9	1	6	4	3	8	7	5
8	3	5	7	2	9	4	1	6
1	4	8	5	9	7	6	3	2
9	6	2	4	3	8	7	5	1
7	5	3	2	1	6	9	8	4
3	2	6	9	7	5	1	4	8
5	8	9	1	6	4	3	2	7
4	1	7	3	8	2	5	6	9

Puzzle 786

6	8	4	9	1	2	5	3	7
7	9	2	8	5	3	4	1	6
5	1	3	4	6	7	8	2	9
2	6	7	1	4	8	9	5	3
3	4	9	6	2	5	1	7	8
8	5	1	3	7	9	6	4	2
1	3	6	7	8	4	2	9	5
9	2	8	5	3	1	7	6	4
4	7	5	2	9	6	3	8	1

Puzzle 787

4	5	7	3	1	2	9	8	6
1	6	3	9	7	8	2	4	5
8	9	2	5	4	6	3	1	7
5	3	6	7	2	1	8	9	4
2	7	4	8	5	9	1	6	3
9	1	8	6	3	4	7	5	2
7	2	1	4	9	5	6	3	8
3	8	5	1	6	7	4	2	9
6	4	9	2	8	3	5	7	1

Puzzle 788

5	2	6	8	3	9	4	1	7
9	3	7	5	1	4	2	8	6
4	1	8	2	7	6	3	9	5
6	8	3	4	9	5	1	7	2
7	9	2	1	6	3	5	4	8
1	4	5	7	8	2	6	3	9
2	5	1	9	4	8	7	6	3
8	6	4	3	2	7	9	5	1
3	7	9	6	5	1	8	2	4

Puzzle 789

1	4	8	9	5	3	6	2	7
9	5	3	6	7	2	1	8	4
2	7	6	8	1	4	5	3	9
6	2	7	3	9	5	8	4	1
5	9	4	2	8	1	3	7	6
8	3	1	7	4	6	2	9	5
7	1	2	5	3	9	4	6	8
3	8	5	4	6	7	9	1	2
4	6	9	1	2	8	7	5	3

Puzzle 790

6	3	9	5	2	7	1	8	4
8	2	7	1	9	4	6	3	5
4	5	1	8	3	6	7	9	2
3	4	2	6	7	8	9	5	1
5	1	8	2	4	9	3	7	6
9	7	6	3	5	1	2	4	8
1	9	4	7	8	2	5	6	3
2	8	3	9	6	5	4	1	7
7	6	5	4	1	3	8	2	9

Puzzle 791

5	8	9	7	1	3	2	4	6
2	7	6	5	4	9	1	3	8
3	4	1	2	8	6	7	9	5
6	5	4	8	2	7	3	1	9
7	9	8	6	3	1	4	5	2
1	3	2	9	5	4	6	8	7
4	2	7	1	9	5	8	6	3
9	6	3	4	7	8	5	2	1
8	1	5	3	6	2	9	7	4

Puzzle 792

6	4	8	7	9	1	5	2	3
3	1	9	6	2	5	7	4	8
2	7	5	8	4	3	9	1	6
9	6	2	5	8	4	3	7	1
1	8	4	3	7	2	6	5	9
7	5	3	9	1	6	4	8	2
5	2	7	1	6	9	8	3	4
8	9	1	4	3	7	2	6	5
4	3	6	2	5	8	1	9	7

Puzzle 793

6	4	7	2	1	3	9	8	5
8	1	5	9	4	7	2	6	3
2	3	9	5	8	6	7	1	4
4	6	8	7	2	5	3	9	1
1	9	2	4	3	8	5	7	6
5	7	3	1	6	9	4	2	8
3	5	6	8	7	2	1	4	9
7	8	4	3	9	1	6	5	2
9	2	1	6	5	4	8	3	7

Puzzle 794

5	4	6	2	1	7	8	3	9
7	1	2	3	8	9	5	6	4
3	9	8	5	6	4	1	7	2
9	3	4	7	5	2	6	1	8
2	5	1	8	3	6	4	9	7
8	6	7	4	9	1	3	2	5
4	7	3	1	2	5	9	8	6
1	2	9	6	4	8	7	5	3
6	8	5	9	7	3	2	4	1

Puzzle 795

6	8	4	3	1	7	2	9	5
1	9	2	6	5	4	8	7	3
5	7	3	8	2	9	6	1	4
9	5	1	7	6	2	4	3	8
8	3	6	9	4	5	7	2	1
4	2	7	1	8	3	9	5	6
2	6	8	5	7	1	3	4	9
7	1	9	4	3	8	5	6	2
3	4	5	2	9	6	1	8	7

Puzzle 796

5	9	4	3	6	8	1	7	2
1	7	8	9	5	2	3	6	4
6	3	2	1	4	7	8	5	9
9	6	5	8	3	1	2	4	7
2	1	7	6	9	4	5	3	8
8	4	3	2	7	5	6	9	1
7	8	6	4	2	3	9	1	5
3	5	1	7	8	9	4	2	6
4	2	9	5	1	6	7	8	3

Puzzle 797

4	3	5	8	2	9	1	6	7
1	8	6	4	7	5	9	2	3
9	7	2	3	6	1	5	4	8
5	2	4	7	1	8	6	3	9
3	9	8	6	5	2	7	1	4
7	6	1	9	4	3	8	5	2
2	4	9	5	8	6	3	7	1
8	5	7	1	3	4	2	9	6
6	1	3	2	9	7	4	8	5

Puzzle 798

3	7	2	5	6	8	1	4	9
1	5	6	3	9	4	2	8	7
9	8	4	2	7	1	3	5	6
6	9	3	4	2	5	8	7	1
8	2	1	7	3	9	5	6	4
7	4	5	1	8	6	9	3	2
5	1	9	6	4	3	7	2	8
2	6	8	9	5	7	4	1	3
4	3	7	8	1	2	6	9	5

Puzzle 799

2	5	3	4	9	8	1	6	7
8	6	7	5	2	1	9	3	4
4	9	1	3	6	7	8	5	2
9	8	4	1	5	6	2	7	3
7	3	5	8	4	2	6	1	9
6	1	2	7	3	9	5	4	8
5	7	6	2	8	4	3	9	1
3	4	8	9	1	5	7	2	6
1	2	9	6	7	3	4	8	5

Puzzle 800

6	5	8	9	4	7	2	1	3
4	1	9	3	5	2	7	6	8
7	2	3	1	6	8	4	9	5
3	8	4	7	1	5	9	2	6
2	7	5	6	9	3	8	4	1
9	6	1	2	8	4	3	5	7
1	9	7	2		6	5	3	4
8	4	2	5	3	1	6	7	9
5	3	6	4	7	9	1	8	2

Puzzle 801

9	3	4	7	1	5	8	2	6
7	8	5	6	2	9	3	4	1
6	1	2	4	3	8	5	9	7
3	7	9	8	5	6	4	1	2
2	5	1	9	4	3	7	6	8
4	6	8	1	7	2	9	3	5
1	2	3	5	9	7	6	8	4
8	9	7	2	6	4	1	5	3
5	4	6	3	8	1	2	7	9

Puzzle 802

3	9	1	5	2	4	7	8	6
7	2	6	9	1	8	3	5	4
8	4	5	6	3	7	2	9	1
1	5	3	2	6	9	4	7	8
6	7	2	8	4	1	5	3	9
9	8	4	7	5	3	6	1	2
5	1	9	4	7	6	8	2	3
2	6	8	3	9	5	1	4	7
4	3	7	1	8	2	9	6	5

Puzzle 803

8	9	4	2	6	7	3	1	5
1	2	7	3	4	5	6	8	9
3	6	5	1	8	9	2	7	4
4	8	9	7	2	1	5	3	6
5	7	3	4	9	6	8	2	1
6	1	2	8	5	3	9	4	7
9	3	1	6	7	8	4	5	2
7	4	6	5	3	2	1	9	8
2	5	8	9	1	4	7	6	3

Puzzle 804

4	7	9	8	2	1	3	5	6
2	6	1	9	3	5	8	7	4
5	8	3	4	6	7	1	9	2
9	5	8	1	7	6	2	4	3
7	2	4	5	8	3	6	1	9
3	1	6	2	9	4	7	8	5
1	3	5	7	4	2	9	6	8
8	4	2	6	1	9	5	3	7
6	9	7	3	5	8	4	2	1

Puzzle 805

2	7	9	5	8	6	3	4	1
6	8	4	2	1	3	5	9	7
5	3	1	7	4	9	2	6	8
9	2	5	3	6	1	8	7	4
3	1	8	4	7	5	9	2	6
7	4	6	9	2	8	1	3	5
1	6	2	8	9	7	4	5	3
8	9	3	6	5	4	7	1	2
4	5	7	1	3	2	6	8	9

Puzzle 806

3	8	7	1	6	9	5	4	2
5	2	9	7	3	4	8	1	6
1	6	4	8	5	2	9	3	7
6	3	2	4	8	5	7	9	1
8	9	1	6	2	7	4	5	3
7	4	5	3	9	1	6	2	8
2	5	8	9	1	6	3	7	4
9	7	3	2	4	8	1	6	5
4	1	6	5	7	3	2	8	9

Puzzle 807

1	3	8	9	5	7	4	2	6
2	7	6	3	4	1	5	9	8
5	9	4	8	2	6	1	3	7
3	8	7	6	9	4	2	5	1
9	4	1	2	8	5	6	7	3
6	2	5	7	1	3	8	4	9
4	5	3	1	7	8	9	6	2
8	6	9	5	3	2	7	1	4
7	1	2	4	6	9	3	8	5

Puzzle 808

8	2	6	7	5	3	4	9	1
5	7	3	1	4	9	6	2	8
9	4	1	6	8	2	5	7	3
1	5	8	2	9	6	7	3	4
7	3	9	4	1	8	2	5	6
2	6	4	5	3	7	8	1	9
3	9	2	8	7	4	1	6	5
4	1	7	3	6	5	9	8	2
6	8	5	9	2	1	3	4	7

Puzzle 809

7	3	1	8	6	5	4	9	2
4	8	9	1	3	2	7	5	6
5	6	2	9	7	4	1	3	8
6	7	3	2	1	9	5	8	4
2	9	8	4	5	3	6	1	7
1	5	4	6	8	7	9	2	3
3	2	7	5	9	6	8	4	1
9	1	6	3	4	8	2	7	5
8	4	5	7	2	1	3	6	9

Puzzle 810

8	9	1	5	6	2	7	4	3
5	3	4	8	9	7	2	6	1
7	6	2	3	1	4	5	8	9
2	5	3	6	8	1	9	7	4
6	7	9	4	3	5	8	1	2
4	1	8	7	2	9	6	3	5
9	4	6	2	7	3	1	5	8
1	8	5	9	4	6	3	2	7
3	2	7	1	5	8	4	9	6

Puzzle 811

7	4	6	5	2	1	8	9	3
8	3	1	9	7	6	2	4	5
5	9	2	8	4	3	1	7	6
1	8	5	7	6	4	9	3	2
2	6	3	1	8	9	4	5	7
9	7	4	2	3	5	6	8	1
6	5	7	4	1	8	3	2	9
3	2	8	6	9	7	5	1	4
4	1	9	3	5	2	7	6	8

Puzzle 812

8	6	7	9	3	1	5	4	2
2	4	3	5	6	8	7	1	9
1	9	5	4	2	7	3	8	6
9	2	1	3	4	5	8	6	7
7	8	4	6	1	9	2	5	3
5	3	6	8	7	2	1	9	4
3	7	8	1	9	4	6	2	5
6	5	9	2	8	3	4	7	1
4	1	2	7	5	6	9	3	8

Puzzle 813

2	1	3	8	5	7	9	6	4
9	6	7	3	4	2	5	1	8
5	8	4	1	6	9	3	7	2
8	4	6	5	9	1	2	3	7
7	3	2	4	8	6	1	5	9
1	5	9	2	7	3	8	4	6
6	2	8	7	1	5	4	9	3
3	9	5	6	2	4	7	8	1
4	7	1	9	3	8	6	2	5

Puzzle 814

8	9	5	1	2	7	6	3	4
3	7	1	6	5	4	9	8	2
6	2	4	8	3	9	1	5	7
5	3	6	2	8	1	7	4	9
7	8	9	3	4	6	5	2	1
4	1	2	9	7	5	3	6	8
9	4	8	7	6	3	2	1	5
1	5	3	4	9	2	8	7	6
2	6	7	5	1	8	4	9	3

Puzzle 815

6	4	9	2	1	5	7	8	3
3	2	7	8	9	6	5	1	4
8	5	1	4	7	3	9	2	6
9	3	2	6	5	4	8	7	1
5	7	8	1	3	9	4	6	2
4	1	6	7	2	8	3	9	5
1	8	5	9	4	2	6	3	7
7	9	4	3	6	1	2	5	8
2	6	3	5	8	7	1	4	9

Puzzle 816

1	9	5	6	8	3	7	2	4
2	7	6	4	5	9	3	1	8
4	3	8	1	2	7	5	9	6
6	8	9	7	3	4	2	5	1
3	1	2	8	9	5	4	6	7
7	5	4	2	1	6	8	3	9
9	2	7	3	6	8	1	4	5
5	4	3	9	7	1	6	8	2
8	6	1	5	4	2	9	7	3

Puzzle 817

9	8	7	2	3	4	6	5	1
2	6	1	5	7	8	4	3	9
4	3	5	1	6	9	7	2	8
6	1	9	8	5	2	3	4	7
5	4	8	3	9	7	2	1	6
7	2	3	6	4	1	8	9	5
1	7	6	9	2	3	5	8	4
3	9	4	7	8	5	1	6	2
8	5	2	4	1	6	9	7	3

Puzzle 818

4	3	1	2	6	8	5	9	7
5	2	6	7	9	4	3	8	1
7	9	8	1	3	5	4	6	2
1	5	7	8	2	9	6	4	3
9	8	4	3	1	6	7	2	5
3	6	2	5	4	7	9	1	8
2	1	9	4	5	3	8	7	6
8	4	5	6	7	1	2	3	9
6	7	3	9	8	2	1	5	4

Puzzle 819

9	5	2	1	3	7	8	6	4
1	3	6	2	8	4	5	9	7
4	8	7	6	9	5	1	2	3
8	2	3	4	7	1	6	5	9
6	7	1	9	5	2	4	3	8
5	4	9	3	6	8	7	1	2
3	1	4	8	2	6	9	7	5
7	9	8	5	1	3	2	4	6
2	6	5	7	4	9	3	8	1

Puzzle 820

3	5	4	8	1	6	7	2	9
1	8	2	5	7	9	4	3	6
9	7	6	3	2	4	8	5	1
6	3	7	9	4	2	5	1	8
4	9	1	7	8	5	2	6	3
8	2	5	1	6	3	9	4	7
5	1	9	4	3	7	6	8	2
2	4	3	6	9	8	1	7	5
7	6	8	2	5	1	3	9	4

Puzzle 821

1	2	4	9	3	5	7	8	6
5	6	9	2	7	8	3	4	1
7	3	8	1	4	6	9	5	2
6	1	7	5	8	4	2	3	9
2	4	3	7	9	1	5	6	8
9	8	5	3	6	2	1	7	4
3	9	6	4	2	7	8	1	5
4	5	2	8	1	3	6	9	7
8	7	1	6	5	9	4	2	3

Puzzle 822

7	3	5	4	2	6	1	8	9
9	6	2	7	8	1	4	5	3
4	1	8	3	5	9	2	6	7
2	5	6	1	9	7	3	4	8
3	9	1	5	4	8	6	7	2
8	4	7	2	6	3	9	1	5
5	7	9	6	1	2	8	3	4
6	8	3	9	7	4	5	2	1
1	2	4	8	3	5	7	9	6

Puzzle 823

8	1	9	7	2	3	4	6	5
3	7	5	1	6	4	9	8	2
6	4	2	9	5	8	7	1	3
4	9	8	6	7	5	3	2	1
7	2	6	4	3	1	5	9	8
1	5	3	2	8	9	6	4	7
2	8	7	5	9	6	1	3	4
9	3	1	8	4	7	2	5	6
5	6	4	3	1	2	8	7	9

Puzzle 824

3	1	4	9	2	6	5	7	8
9	8	7	1	3	5	2	6	4
2	5	6	4	7	8	1	9	3
6	4	2	7	9	1	8	3	5
1	3	9	5	8	4	6	2	7
5	7	8	3	6	2	4	1	9
7	2	5	8	1	9	3	4	6
8	9	1	6	4	3	7	5	2
4	6	3	2	5	7	9	8	1

Puzzle 825

4	1	2	6	8	7	9	3	5
9	5	6	2	4	3	7	8	1
8	3	7	1	5	9	4	6	2
2	9	5	3	7	1	8	4	6
3	8	4	9	6	5	1	2	7
7	6	1	8	2	4	3	5	9
5	2	9	7	3	8	6	1	4
6	7	3	4	1	2	5	9	8
1	4	8	5	9	6	2	7	3

Puzzle 826

3	8	6	7	5	1	2	4	9
5	1	7	2	4	9	8	3	6
9	2	4	3	8	6	7	5	1
6	5	9	4	3	7	1	2	8
8	7	2	1	6	5	4	9	3
4	3	1	9	2	8	6	7	5
1	9	5	6	7	4	3	8	2
2	4	8	5	1	3	9	6	7
7	6	3	8	9	2	5	1	4

Puzzle 827

2	5	6	3	7	8	1	4	9
7	3	9	2	4	1	5	8	6
8	1	4	9	5	6	7	3	2
3	2	8	7	1	9	6	5	4
6	4	7	8	2	5	9	1	3
1	9	5	6	3	4	8	2	7
9	7	3	1	8	2	4	6	5
5	8	2	4	6	7	3	9	1
4	6	1	5	9	3	2	7	8

Puzzle 828

3	1	6	2	4	7	9	8	5
8	2	9	1	3	5	7	6	4
4	7	5	9	8	6	1	2	3
5	3	4	7	9	8	6	1	2
9	8	2	3	6	1	4	5	7
7	6	1	5	2	4	3	9	8
2	4	3	6	5	9	8	7	1
6	5	7	8	1	3	2	4	9
1	9	8	4	7	2	5	3	6

Puzzle 829

3	1	6	9	8	5	7	4	2
9	7	5	1	2	4	6	3	8
8	2	4	3	7	6	9	1	5
4	9	2	8	6	3	5	7	1
6	8	3	5	1	7	4	2	9
7	5	1	2	4	9	8	6	3
5	4	9	6	3	2	1	8	7
1	3	7	4	9	8	2	5	6
2	6	8	7	5	1	3	9	4

Puzzle 830

4	1	2	3	7	5	9	6	8
9	8	7	2	6	1	5	4	3
5	3	6	4	9	8	7	2	1
2	6	9	1	3	4	8	7	5
3	7	4	5	8	6	2	1	9
8	5	1	9	2	7	4	3	6
7	4	8	6	1	9	3	5	2
1	9	3	7	5	2	6	8	4
6	2	5	8	4	3	1	9	7

Puzzle 831

4	7	5	1	9	3	6	2	8
3	2	9	7	6	8	5	4	1
6	8	1	4	2	5	3	9	7
2	6	3	9	8	7	4	1	5
7	9	4	2	5	1	8	6	3
1	5	8	6	3	4	2	7	9
5	1	6	3	7	2	9	8	4
8	4	2	5	1	9	7	3	6
9	3	7	8	4	6	1	5	2

Puzzle 832

2	3	7	8	4	5	9	1	6
5	9	6	3	1	7	2	8	4
1	8	4	2	6	9	3	5	7
4	1	9	7	2	8	5	6	3
8	5	3	4	9	6	1	7	2
7	6	2	5	3	1	8	4	9
6	2	5	1	7	3	4	9	8
3	7	1	9	8	4	6	2	5
9	4	8	6	5	2	7	3	1

Puzzle 833

8	6	2	7	5	3	1	4	9
3	5	9	1	8	4	7	2	6
1	7	4	6	2	9	3	8	5
7	1	8	4	6	2	5	9	3
9	2	6	8	3	5	4	1	7
5	4	3	9	1	7	8	6	2
4	3	1	2	7	6	9	5	8
2	8	7	5	9	1	6	3	4
6	9	5	3	4	8	2	7	1

Puzzle 834

9	5	4	7	2	6	3	1	8
3	8	2	9	1	5	4	7	6
7	6	1	8	3	4	9	2	5
5	7	6	2	9	8	1	3	4
4	1	3	5	6	7	8	9	2
2	9	8	3	4	1	5	6	7
6	2	5	1	8	3	7	4	9
1	4	7	6	5	9	2	8	3
8	3	9	4	7	2	6	5	1

Puzzle 835

1	9	6	4	3	7	8	2	5
3	8	5	6	1	2	4	7	9
4	2	7	5	9	8	1	6	3
2	6	8	7	4	3	9	5	1
5	1	9	8	2	6	7	3	4
7	4	3	1	5	9	2	8	6
8	3	4	9	7	5	6	1	2
6	5	1	2	8	4	3	9	7
9	7	2	3	6	1	5	4	8

Puzzle 836

1	2	6	8	3	4	7	5	9
3	4	7	2	5	9	1	6	8
9	8	5	1	7	6	4	3	2
4	9	1	3	6	7	8	2	5
5	3	2	4	8	1	9	7	6
6	7	8	5	9	2	3	1	4
2	1	3	6	4	8	5	9	7
8	6	9	7	1	5	2	4	3
7	5	4	9	2	3	6	8	1

Puzzle 837

5	1	6	3	4	8	2	7	9
2	3	8	7	1	9	4	5	6
7	9	4	6	2	5	8	3	1
1	2	9	5	7	4	6	8	3
3	6	5	2	8	1	7	9	4
8	4	7	9	3	6	1	2	5
6	5	2	1	9	7	3	4	8
9	8	3	4	6	2	5	1	7
4	7	1	8	5	3	9	6	2

Puzzle 838

7	4	5	8	9	3	2	1	6
8	6	1	5	7	2	3	4	9
3	9	2	1	4	6	5	8	7
5	1	7	9	8	4	6	3	2
6	2	8	7	3	5	1	9	4
4	3	9	6	2	1	8	7	5
1	8	6	4	5	9	7	2	3
2	7	4	3	6	8	9	5	1
9	5	3	2	1	7	4	6	8

Puzzle 839

4	7	5	6	3	8	2	1	9
8	2	1	7	9	5	4	6	3
9	3	6	4	2	1	7	5	8
1	4	9	8	7	6	5	3	2
2	6	8	9	5	3	1	4	7
3	5	7	2	1	4	8	9	6
6	8	3	5	4	2	9	7	1
5	9	2	1	6	7	3	8	4
7	1	4	3	8	9	6	2	5

Puzzle 840

5	9	3	2	8	4	1	6	7
4	7	6	1	3	5	8	2	9
8	1	2	7	6	9	5	3	4
9	4	1	8	7	2	3	5	6
2	8	5	6	4	3	9	7	1
3	6	7	9	5	1	2	4	8
6	2	8	3	9	7	4	1	5
7	3	4	5	1	8	6	9	2
1	5	9	4	2	6	7	8	3

Puzzle 841

3	6	5	2	8	7	9	1	4
4	1	7	6	9	5	8	3	2
9	8	2	4	3	1	7	6	5
7	5	9	1	4	8	3	2	6
8	4	1	3	6	2	5	7	9
2	3	6	5	7	9	1	4	8
5	9	3	7	2	6	4	8	1
1	2	4	8	5	3	6	9	7
6	7	8	9	1	4	2	5	3

Puzzle 842

6	7	1	2	4	8	5	9	3
4	8	5	6	3	9	7	1	2
3	2	9	5	7	1	8	6	4
5	3	2	8	9	4	1	7	6
7	6	8	1	5	3	2	4	9
9	1	4	7	6	2	3	5	8
1	4	3	9	2	5	6	8	7
2	5	7	4	8	6	9	3	1
8	9	6	3	1	7	4	2	5

Puzzle 843

7	2	4	5	1	8	3	9	6
6	5	3	7	2	9	1	8	4
8	1	9	4	3	6	7	2	5
3	8	5	2	4	7	9	6	1
9	4	2	8	6	1	5	3	7
1	6	7	3	9	5	2	4	8
4	3	1	6	5	2	8	7	9
2	9	8	1	7	4	6	5	3
5	7	6	9	8	3	4	1	2

Puzzle 844

2	3	4	6	7	1	9	8	5
7	9	8	4	5	3	6	2	1
1	6	5	2	8	9	7	3	4
8	5	3	9	2	7	1	4	6
6	2	1	3	4	5	8	9	7
4	7	9	1	6	8	2	5	3
3	8	6	7	9	4	5	1	2
9	4	2	5	1	6	3	7	8
5	1	7	8	3	2	4	6	9

Puzzle 845

7	1	5	3	9	8	2	4	6
2	8	6	1	4	5	7	3	9
4	9	3	6	2	7	8	5	1
6	4	2	8	1	9	3	7	5
3	7	8	4	5	6	1	9	2
9	5	1	7	3	2	6	8	4
8	3	9	2	6	4	5	1	7
1	2	4	5	7	3	9	6	8
5	6	7	9	8	1	4	2	3

Puzzle 846

1	8	9	3	5	6	4	7	2
7	4	5	9	2	1	8	3	6
3	6	2	8	4	7	1	5	9
4	3	1	5	7	2	6	9	8
8	9	7	4	6	3	2	1	5
5	2	6	1	9	8	3	4	7
9	7	3	6	8	4	5	2	1
2	1	8	7	3	5	9	6	4
6	5	4	2	1	9	7	8	3

Puzzle 847

9	3	2	6	7	5	1	8	4
8	6	5	2	4	1	7	9	3
1	4	7	8	9	3	2	5	6
3	9	4	7	8	2	6	1	5
7	2	1	9	5	6	3	4	8
5	8	6	1	3	4	9	7	2
4	1	8	3	2	7	5	6	9
2	7	9	5	6	8	4	3	1
6	5	3	4	1	9	8	2	7

Puzzle 848

3	8	2	9	5	1	4	7	6
5	4	9	7	6	8	1	2	3
1	6	7	3	4	2	8	5	9
4	1	5	2	9	3	7	6	8
8	2	3	1	7	6	9	4	5
7	9	6	5	8	4	3	1	2
2	3	8	4	1	5	6	9	7
9	5	1	6	3	7	2	8	4
6	7	4	8	2	9	5	3	1

Puzzle 849

1	9	5	6	2	4	8	7	3
4	2	8	9	7	3	1	5	6
7	6	3	8	5	1	4	2	9
9	1	6	7	3	5	2	4	8
3	4	7	2	6	8	5	9	1
8	5	2	1	4	9	3	6	7
5	3	9	4	1	6	7	8	2
6	7	1	5	8	2	9	3	4
2	8	4	3	9	7	6	1	5

Puzzle 850

3	4	5	7	6	2	1	8	9
1	9	8	3	4	5	7	2	6
2	6	7	1	9	8	3	5	4
8	2	4	9	7	3	6	1	5
7	1	6	2	5	4	8	9	3
5	3	9	6	8	1	4	7	2
4	7	3	8	2	9	5	6	1
6	5	2	4	1	7	9	3	8
9	8	1	5	3	6	2	4	7

Puzzle 851

9	5	3	4	8	2	6	7	1
1	2	4	5	6	7	9	8	3
6	8	7	9	1	3	5	2	4
7	6	9	1	3	8	4	5	2
3	1	8	2	5	4	7	9	6
5	4	2	7	9	6	1	3	8
2	3	5	6	7	1	8	4	9
4	9	6	8	2	5	3	1	7
8	7	1	3	4	9	2	6	5

Puzzle 852

2	4	7	8	6	1	5	3	9
8	9	6	2	5	3	1	4	7
3	1	5	4	7	9	8	6	2
4	7	9	1	2	5	3	8	6
6	5	2	9	3	8	7	1	4
1	8	3	6	4	7	2	9	5
5	2	4	3	1	6	9	7	8
9	6	1	7	8	2	4	5	3
7	3	8	5	9	4	6	2	1

Puzzle 853

7	2	8	1	4	6	9	3	5
4	1	5	8	9	3	2	6	7
6	3	9	2	7	5	1	8	4
3	5	4	7	1	8	6	2	9
2	6	1	3	5	9	7	4	8
9	8	7	4	6	2	3	5	1
1	7	3	6	8	4	5	9	2
5	4	6	9	2	7	8	1	3
8	9	2	5	3	1	4	7	6

Puzzle 854

7	5	3	6	8	1	2	9	4
1	9	6	5	2	4	7	8	3
8	4	2	9	3	7	6	5	1
4	7	5	2	1	3	8	6	9
9	6	8	4	7	5	3	1	2
3	2	1	8	9	6	5	4	7
6	8	7	1	4	2	9	3	5
2	1	9	3	5	8	4	7	6
5	3	4	7	6	9	1	2	8

Puzzle 855

3	5	8	6	2	7	1	9	4
4	9	7	3	1	8	6	2	5
2	1	6	4	5	9	7	8	3
6	7	5	1	9	3	2	4	8
9	2	3	7	8	4	5	6	1
1	8	4	2	6	5	9	3	7
8	6	1	5	3	2	4	7	9
7	3	2	9	4	1	8	5	6
5	4	9	8	7	6	3	1	2

Puzzle 856

2	6	7	9	5	3	8	4	1
1	3	4	2	8	7	6	9	5
9	8	5	6	1	4	7	3	2
5	7	2	4	6	1	9	8	3
3	4	1	8	2	9	5	7	6
6	9	8	7	3	5	1	2	4
4	5	9	1	7	2	3	6	8
8	2	3	5	9	6	4	1	7
7	1	6	3	4	8	2	5	9

Puzzle 857

7	1	2	8	3	9	5	4	6
5	4	3	2	7	6	8	9	1
6	9	8	4	5	1	2	3	7
4	7	6	1	9	8	3	2	5
9	8	1	5	2	3	7	6	4
2	3	5	6	4	7	9	1	8
8	5	4	9	1	2	6	7	3
1	2	7	3	6	5	4	8	9
3	6	9	7	8	4	1	5	2

Puzzle 858

1	4	6	7	3	8	2	5	9
2	9	3	6	5	1	8	4	7
7	5	8	4	2	9	3	6	1
9	3	5	8	7	4	6	1	2
4	8	7	1	6	2	5	9	3
6	1	2	5	9	3	4	7	8
3	2	1	9	4	5	7	8	6
5	6	9	3	8	7	1	2	4
8	7	4	2	1	6	9	3	5

Puzzle 859

3	6	8	9	7	5	1	2	4
2	4	9	8	6	1	3	5	7
5	7	1	3	4	2	6	8	9
1	2	3	4	9	6	5	7	8
7	5	4	2	1	8	9	6	3
9	8	6	5	3	7	4	1	2
8	3	7	6	5	4	2	9	1
4	1	5	7	2	9	8	3	6
6	9	2	1	8	3	7	4	5

Puzzle 860

7	2	3	1	4	9	5	6	8
4	6	9	2	5	8	3	7	1
5	1	8	7	6	3	4	9	2
1	7	5	9	3	2	6	8	4
2	8	6	4	7	1	9	3	5
3	9	4	5	8	6	1	2	7
8	5	2	3	9	4	7	1	6
9	4	1	6	2	7	8	5	3
6	3	7	8	1	5	2	4	9

Puzzle 861

3	7	4	2	5	9	8	6	1
6	8	2	4	7	1	3	9	5
1	5	9	8	3	6	7	2	4
5	6	1	9	4	3	2	7	8
4	3	8	7	2	5	9	1	6
9	2	7	1	6	8	4	5	3
8	1	3	5	9	2	6	4	7
2	4	6	3	1	7	5	8	9
7	9	5	6	8	4	1	3	2

Puzzle 862

6	9	4	5	7	1	3	8	2
1	5	8	9	3	2	6	7	4
7	3	2	6	8	4	5	1	9
2	8	7	1	4	3	9	6	5
5	4	1	7	9	6	8	2	3
9	6	3	2	5	8	1	4	7
8	7	5	4	1	9	2	3	6
3	2	9	8	6	7	4	5	1
4	1	6	3	2	5	7	9	8

Puzzle 863

3	8	9	2	1	5	7	4	6
5	1	6	7	9	4	3	8	2
2	7	4	3	6	8	5	9	1
9	3	2	1	4	7	8	6	5
8	5	1	9	2	6	4	7	3
6	4	7	5	8	3	2	1	9
4	9	3	8	5	1	6	2	7
7	2	8	6	3	9	1	5	4
1	6	5	4	7	2	9	3	8

Puzzle 864

8	9	4	1	7	5	6	2	3
2	1	6	8	4	3	7	9	5
3	7	5	2	6	9	8	4	1
9	2	3	4	5	7	1	8	6
6	8	7	9	1	2	3	5	4
5	4	1	6	3	8	2	7	9
7	6	2	3	9	4	5	1	8
4	3	8	5	2	1	9	6	7
1	5	9	7	8	6	4	3	2

Puzzle 865

9	8	1	3	4	2	7	5	6
2	7	4	8	6	5	1	9	3
3	5	6	7	9	1	4	8	2
1	6	7	2	8	9	5	3	4
8	9	2	4	5	3	6	1	7
5	4	3	1	7	6	9	2	8
4	2	5	9	3	7	8	6	1
6	1	8	5	2	4	3	7	9
7	3	9	6	1	8	2	4	5

Puzzle 866

3	1	7	4	8	6	2	9	5
4	8	6	2	9	5	1	7	3
5	9	2	7	1	3	6	8	4
8	7	9	5	3	1	4	2	6
6	3	1	9	4	2	7	5	8
2	5	4	6	7	8	9	3	1
1	6	8	3	2	7	5	4	9
9	2	5	8	6	4	3	1	7
7	4	3	1	5	9	8	6	2

Puzzle 867

8	4	3	9	5	7	6	2	1
1	2	5	8	4	6	7	9	3
6	7	9	2	3	1	8	5	4
7	5	6	1	2	3	4	8	9
4	3	1	7	9	8	5	6	2
2	9	8	5	6	4	1	3	7
9	8	7	3	1	5	2	4	6
5	6	2	4	7	9	3	1	8
3	1	4	6	8	2	9	7	5

Puzzle 868

3	8	6	5	2	7	1	4	9
7	5	2	1	4	9	8	6	3
1	9	4	6	3	8	7	5	2
6	4	8	3	5	1	2	9	7
2	1	9	8	7	4	5	3	6
5	3	7	2	9	6	4	8	1
4	6	5	7	1	3	9	2	8
9	7	3	4	8	2	6	1	5
8	2	1	9	6	5	3	7	4

Puzzle 869

7	8	2	6	9	4	5	1	3
1	6	3	2	5	7	9	4	8
4	5	9	1	8	3	7	6	2
5	1	8	7	3	2	4	9	6
2	7	4	8	6	9	1	3	5
3	9	6	5	4	1	2	8	7
6	4	1	3	2	5	8	7	9
8	2	7	9	1	6	3	5	4
9	3	5	4	7	8	6	2	1

Puzzle 870

5	6	8	2	9	3	1	4	7
7	4	2	8	1	6	9	5	3
1	3	9	7	5	4	8	2	6
3	7	4	6	8	1	5	9	2
6	9	1	3	2	5	4	7	8
2	8	5	4	7	9	6	3	1
9	2	7	5	6	8	3	1	4
4	5	6	1	3	2	7	8	9
8	1	3	9	4	7	2	6	5

Puzzle 871

3	9	6	8	1	4	5	2	7
8	5	7	6	3	2	4	9	1
2	1	4	7	5	9	8	3	6
4	6	5	3	7	8	9	1	2
1	3	9	5	2	6	7	8	4
7	8	2	4	9	1	3	6	5
5	7	8	2	6	3	1	4	9
6	4	1	9	8	5	2	7	3
9	2	3	1	4	7	6	5	8

Puzzle 872

8	3	2	5	7	9	4	1	6
6	4	7	8	2	1	9	3	5
1	9	5	4	6	3	2	7	8
3	7	4	2	1	6	5	8	9
2	8	9	3	5	4	7	6	1
5	1	6	7	9	8	3	4	2
9	2	8	6	3	7	1	5	4
4	5	3	1	8	2	6	9	7
7	6	1	9	4	5	8	2	3

Puzzle 873

1	9	2	6	3	4	5	7	8
4	6	7	1	5	8	2	3	9
5	3	8	7	2	9	1	6	4
2	7	5	9	4	6	8	1	3
3	4	9	8	1	2	7	5	6
8	1	6	5	7	3	9	4	2
9	2	1	3	6	7	4	8	5
7	8	3	4	9	5	6	2	1
6	5	4	2	8	1	3	9	7

Puzzle 874

8	7	4	5	1	9	3	2	6
1	3	2	8	4	6	9	7	5
5	9	6	3	7	2	8	4	1
2	5	1	7	9	8	4	6	3
3	6	8	1	2	4	7	5	9
7	4	9	6	3	5	2	1	8
6	8	7	4	5	3	1	9	2
9	1	3	2	6	7	5	8	4
4	2	5	9	8	1	6	3	7

Puzzle 875

9	3	2	6	1	8	5	4	7
4	5	6	9	3	7	8	2	1
1	7	8	5	2	4	6	3	9
2	4	3	7	9	6	1	5	8
8	9	5	3	4	1	2	7	6
6	1	7	8	5	2	4	9	3
5	2	9	1	8	3	7	6	4
7	8	4	2	6	9	3	1	5
3	6	1	4	7	5	9	8	2

Puzzle 876

8	1	4	7	3	2	9	5	6
2	9	3	4	6	5	1	7	8
7	6	5	8	1	9	2	3	4
4	2	6	5	7	8	3	1	9
1	8	7	6	9	3	5	4	2
5	3	9	1	2	4	6	8	7
3	4	2	9	8	1	7	6	5
9	7	8	3	5	6	4	2	1
6	5	1	2	4	7	8	9	3

Puzzle 877

2	9	8	4	3	5	6	7	1
4	5	1	7	6	2	3	9	8
6	7	3	8	1	9	4	5	2
3	1	4	5	7	6	8	2	9
9	8	2	1	4	3	5	6	7
7	6	5	2	9	8	1	4	3
8	4	9	3	5	7	2	1	6
1	2	7	6	8	4	9	3	5
5	3	6	9	2	1	7	8	4

Puzzle 878

4	5	7	9	2	1	6	3	8
9	3	1	7	8	6	4	2	5
2	6	8	4	3	5	7	1	9
3	9	2	6	7	8	1	5	4
1	8	6	5	4	2	3	9	7
7	4	5	3	1	9	8	6	2
8	2	4	1	9	3	5	7	6
6	1	9	8	5	7	2	4	3
5	7	3	2	6	4	9	8	1

Puzzle 879

3	5	4	7	1	9	2	6	8
9	2	1	8	3	6	7	5	4
8	7	6	2	4	5	1	3	9
2	8	3	1	9	4	5	7	6
7	4	5	3	6	8	9	2	1
6	1	9	5	2	7	4	8	3
1	9	8	6	7	2	3	4	5
4	6	7	9	5	3	8	1	2
5	3	2	4	8	1	6	9	7

Puzzle 880

3	5	7	8	4	9	6	2	1
9	1	8	6	5	2	7	3	4
2	6	4	3	1	7	9	8	5
4	2	1	9	3	6	8	5	7
7	8	5	1	2	4	3	6	9
6	9	3	5	7	8	1	4	2
1	7	2	4	8	3	5	9	6
8	4	9	7	6	5	2	1	3
5	3	6	2	9	1	4	7	8

Puzzle 881

7	2	5	9	6	1	8	3	4
8	6	9	7	4	3	1	5	2
4	3	1	2	5	8	7	6	9
9	7	3	5	8	6	2	4	1
1	8	2	3	9	4	5	7	6
6	5	4	1	7	2	3	9	8
3	4	7	8	2	9	6	1	5
5	9	8	6	1	7	4	2	3
2	1	6	4	3	5	9	8	7

Puzzle 882

9	6	3	7	8	5	4	1	2
5	7	2	9	1	4	3	6	8
4	1	8	3	2	6	9	7	5
3	2	1	8	7	9	5	4	6
7	8	5	4	6	1	2	3	9
6	4	9	5	3	2	7	8	1
1	5	4	6	9	7	8	2	3
2	3	7	1	5	8	6	9	4
8	9	6	2	4	3	1	5	7

Puzzle 883

6	7	4	2	5	3	9	8	1
2	5	1	8	7	9	4	3	6
9	3	8	1	6	4	7	2	5
3	1	7	9	2	5	6	4	8
5	4	6	3	8	7	1	9	2
8	2	9	6	4	1	3	5	7
1	8	3	5	9	6	2	7	4
7	6	2	4	3	8	5	1	9
4	9	5	7	1	2	8	6	3

Puzzle 884

4	2	3	9	7	5	6	1	8
5	1	8	2	6	3	9	7	4
9	6	7	4	1	8	3	5	2
6	7	4	8	9	1	2	3	5
8	3	1	6	5	2	4	9	7
2	9	5	7	3	4	1	8	6
7	4	9	3	8	6	5	2	1
3	5	6	1	2	7	8	4	9
1	8	2	5	4	9	7	6	3

Puzzle 885

9	1	5	7	3	4	6	8	2
4	8	6	5	1	2	9	7	3
7	2	3	9	8	6	1	5	4
2	7	1	3	6	8	5	4	9
3	5	4	2	9	1	7	6	8
8	6	9	4	7	5	2	3	1
6	4	7	8	2	9	3	1	5
1	9	8	6	5	3	4	2	7
5	3	2	1	4	7	8	9	6

Puzzle 886

4	6	5	7	1	3	8	9	2
8	1	2	4	5	9	3	7	6
9	7	3	6	8	2	4	1	5
5	2	7	8	9	1	6	3	4
1	9	4	2	3	6	5	8	7
6	3	8	5	7	4	1	2	9
2	4	9	3	6	8	7	5	1
3	5	1	9	4	7	2	6	8
7	8	6	1	2	5	9	4	3

Puzzle 887

7	3	8	2	4	9	6	5	1
6	4	1	3	7	5	2	9	8
5	9	2	6	8	1	3	4	7
1	2	6	5	3	7	9	8	4
9	7	3	8	2	4	1	6	5
8	5	4	9	1	6	7	2	3
4	8	9	7	6	3	5	1	2
3	1	5	4	9	2	8	7	6
2	6	7	1	5	8	4	3	9

Puzzle 888

7	8	3	4	1	9	6	5	2
5	6	2	7	3	8	1	9	4
9	4	1	2	5	6	8	7	3
1	7	4	9	2	3	5	6	8
3	9	8	6	4	5	2	1	7
2	5	6	1	8	7	3	4	9
4	3	5	8	7	1	9	2	6
8	2	9	5	6	4	7	3	1
6	1	7	3	9	2	4	8	5

Puzzle 889

5	2	4	3	1	8	7	6	9
6	9	7	2	5	4	3	1	8
3	8	1	7	9	6	5	2	4
4	3	5	6	2	1	8	9	7
7	1	2	8	3	9	4	5	6
9	6	8	4	7	5	1	3	2
2	5	3	9	8	7	6	4	1
1	7	6	5	4	2	9	8	3
8	4	9	1	6	3	2	7	5

Puzzle 890

1	5	9	7	3	4	6	2	8
4	7	3	8	6	2	9	1	5
8	6	2	9	1	5	4	3	7
7	8	6	5	9	1	2	4	3
9	1	5	4	2	3	8	7	6
2	3	4	6	7	8	1	5	9
6	4	8	2	5	7	3	9	1
3	2	7	1	8	9	5	6	4
5	9	1	3	4	6	7	8	2

Puzzle 891

8	5	7	3	2	4	1	9	6
3	9	1	8	5	6	4	7	2
2	4	6	1	9	7	5	8	3
5	2	4	7	1	9	6	3	8
1	6	9	2	3	8	7	5	4
7	8	3	4	6	5	9	2	1
6	7	8	9	4	3	2	1	5
4	3	2	5	7	1	8	6	9
9	1	5	6	8	2	3	4	7

Puzzle 892

6	4	7	8	5	9	1	2	3
3	1	5	4	6	2	9	8	7
2	9	8	1	3	7	6	4	5
1	8	4	2	7	5	3	6	9
9	5	3	6	4	1	8	7	2
7	6	2	3	9	8	4	5	1
5	7	6	9	1	4	2	3	8
8	3	1	5	2	6	7	9	4
4	2	9	7	8	3	5	1	6

Puzzle 893

5	9	7	3	6	8	4	1	2
2	1	6	4	7	9	8	3	5
4	3	8	2	1	5	6	9	7
7	5	1	9	2	6	3	8	4
3	8	9	7	4	1	2	5	6
6	2	4	5	8	3	9	7	1
8	7	2	1	3	4	5	6	9
9	4	3	6	5	7	1	2	8
1	6	5	8	9	2	7	4	3

Puzzle 894

6	9	8	3	1	4	5	2	7
5	1	2	9	8	7	4	3	6
3	7	4	2	6	5	9	8	1
1	4	6	8	9	2	7	5	3
8	2	3	7	5	1	6	9	4
7	5	9	6	4	3	2	1	8
4	3	5	1	2	6	8	7	9
2	8	7	4	3	9	1	6	5
9	6	1	5	7	8	3	4	2

Puzzle 895

1	3	5	8	4	6	7	9	2
6	7	8	3	9	2	5	1	4
2	9	4	1	7	5	3	6	8
8	2	3	9	6	4	1	7	5
5	1	7	2	8	3	6	4	9
4	6	9	7	5	1	8	2	3
3	8	2	6	1	9	4	5	7
7	4	1	5	2	8	9	3	6
9	5	6	4	3	7	2	8	1

Puzzle 896

9	6	7	4	1	5	3	2	8
1	2	5	6	3	8	7	9	4
3	8	4	2	7	9	1	6	5
2	5	9	7	6	3	8	4	1
4	1	6	8	5	2	9	7	3
8	7	3	9	4	1	2	5	6
7	4	8	3	9	6	5	1	2
5	9	2	1	8	4	6	3	7
6	3	1	5	2	7	4	8	9

Puzzle 897

1	6	9	4	7	5	2	3	8
8	3	5	6	9	2	4	7	1
4	7	2	8	1	3	6	5	9
2	9	7	3	5	8	1	4	6
5	8	3	1	6	4	9	2	7
6	4	1	7	2	9	5	8	3
3	1	8	2	4	6	7	9	5
9	2	6	5	3	7	8	1	4
7	5	4	9	8	1	3	6	2

Puzzle 898

9	5	3	4	6	8	7	1	2
2	1	6	7	3	9	5	8	4
4	7	8	5	1	2	9	6	3
6	2	1	9	4	7	3	5	8
7	3	9	8	5	6	2	4	1
8	4	5	1	2	3	6	7	9
3	8	2	6	7	1	4	9	5
1	6	4	3	9	5	8	2	7
5	9	7	2	8	4	1	3	6

Puzzle 899

1	4	2	8	5	9	6	3	7
6	3	8	7	4	2	9	5	1
5	7	9	1	3	6	2	8	4
8	5	1	2	7	4	3	9	6
3	6	7	9	1	8	5	4	2
2	9	4	5	6	3	1	7	8
9	8	3	6	2	7	4	1	5
7	2	5	4	9	1	8	6	3
4	1	6	3	8	5	7	2	9

Puzzle 900

7	2	9	4	6	3	8	1	5
8	6	3	2	1	5	7	4	9
1	4	5	7	9	8	6	2	3
9	3	7	1	5	4	2	6	8
6	8	2	3	7	9	1	5	4
5	1	4	8	2	6	3	9	7
2	9	6	5	8	7	4	3	1
3	7	1	9	4	2	5	8	6
4	5	8	6	3	1	9	7	2

Puzzle 901

3	6	4	2	5	9	8	7	1
5	2	1	6	7	8	4	9	3
8	7	9	1	3	4	5	2	6
9	1	7	3	8	6	2	4	5
4	3	2	9	1	5	7	6	8
6	8	5	4	2	7	3	1	9
7	5	6	8	4	1	9	3	2
1	4	3	5	9	2	6	8	7
2	9	8	7	6	3	1	5	4

Puzzle 902

7	1	3	5	2	6	8	4	9
2	9	4	1	8	7	3	5	6
6	5	8	9	3	4	7	1	2
5	8	9	2	6	1	4	3	7
1	4	7	8	9	3	2	6	5
3	6	2	4	7	5	9	8	1
4	7	5	3	1	9	6	2	8
9	2	1	6	4	8	5	7	3
8	3	6	7	5	2	1	9	4

Puzzle 903

5	6	1	2	3	8	9	7	4
8	9	7	4	1	6	3	5	2
3	2	4	9	5	7	8	1	6
6	7	5	3	4	1	2	9	8
9	4	2	6	8	5	1	3	7
1	3	8	7	2	9	4	6	5
7	5	3	8	9	4	6	2	1
4	1	9	5	6	2	7	8	3
2	8	6	1	7	3	5	4	9

Puzzle 904

3	6	5	7	4	8	2	1	9
8	7	4	2	1	9	3	6	5
1	2	9	3	6	5	7	4	8
9	3	2	5	8	1	6	7	4
4	8	7	6	3	2	9	5	1
6	5	1	4	9	7	8	3	2
5	4	8	9	7	3	1	2	6
7	1	6	8	2	4	5	9	3
2	9	3	1	5	6	4	8	7

Puzzle 905

4	6	7	9	5	8	1	2	3
3	2	9	4	1	6	5	8	7
5	1	8	7	3	2	9	6	4
2	5	6	3	9	7	8	4	1
9	8	3	1	2	4	6	7	5
7	4	1	6	8	5	2	3	9
1	9	4	8	6	3	7	5	2
6	7	5	2	4	9	3	1	8
8	3	2	5	7	1	4	9	6

Puzzle 906

8	6	5	2	7	1	4	9	3
7	4	2	8	3	9	6	5	1
9	3	1	4	6	5	7	2	8
4	7	6	9	2	8	3	1	5
3	5	8	7	1	4	9	6	2
2	1	9	6	5	3	8	7	4
6	2	3	5	8	7	1	4	9
1	9	7	3	4	2	5	8	6
5	8	4	1	9	6	2	3	7

Puzzle 907

2	7	8	6	9	4	1	5	3
5	6	4	3	2	1	9	8	7
3	9	1	7	8	5	6	2	4
4	5	3	2	6	9	7	1	8
8	2	7	1	5	3	4	9	6
6	1	9	4	7	8	5	3	2
7	4	5	9	3	2	8	6	1
9	3	6	8	1	7	2	4	5
1	8	2	5	4	6	3	7	9

Puzzle 908

2	7	4	1	8	3	6	5	9
1	5	6	9	2	4	3	7	8
9	3	8	7	6	5	2	4	1
8	9	5	2	3	6	7	1	4
4	6	7	8	9	1	5	3	2
3	2	1	5	4	7	8	9	6
5	4	9	6	7	8	1	2	3
6	1	2	3	5	9	4	8	7
7	8	3	4	1	2	9	6	5

Puzzle 909

9	8	2	1	3	7	5	4	6
7	5	6	8	9	4	2	1	3
1	4	3	5	2	6	7	8	9
6	2	9	3	1	5	4	7	8
5	3	4	6	7	8	1	9	2
8	1	7	9	4	2	6	3	5
4	9	5	7	6	3	8	2	1
2	6	1	4	8	9	3	5	7
3	7	8	2	5	1	9	6	4

Puzzle 910

5	4	6	1	2	9	3	8	7
8	2	1	5	3	7	6	4	9
9	3	7	8	4	6	5	1	2
6	5	4	2	9	8	1	7	3
1	9	2	6	7	3	4	5	8
3	7	8	4	5	1	9	2	6
4	6	9	7	1	2	8	3	5
2	8	5	3	6	4	7	9	1
7	1	3	9	8	5	2	6	4

Puzzle 911

1	9	3	7	2	6	4	8	5
4	7	6	3	8	5	9	1	2
2	8	5	9	4	1	6	3	7
5	1	2	4	9	3	8	7	6
6	4	7	1	5	8	2	9	3
8	3	9	6	7	2	5	4	1
9	6	8	2	3	7	1	5	4
7	5	1	8	6	4	3	2	9
3	2	4	5	1	9	7	6	8

Puzzle 912

4	1	8	7	5	6	2	9	3
9	6	2	3	1	8	4	7	5
5	7	3	4	9	2	1	8	6
2	5	7	1	4	9	3	6	8
6	4	1	8	3	7	9	5	2
8	3	9	6	2	5	7	4	1
7	2	5	9	6	1	8	3	4
3	8	6	2	7	4	5	1	9
1	9	4	5	8	3	6	2	7

Puzzle 913

4	2	3	1	9	8	6	5	7
7	1	8	5	4	6	2	9	3
6	5	9	3	7	2	4	8	1
8	3	4	7	2	9	1	6	5
5	6	7	8	3	1	9	4	2
1	9	2	4	6	5	7	3	8
9	7	6	2	8	3	5	1	4
2	8	1	9	5	4	3	7	6
3	4	5	6	1	7	8	2	9

Puzzle 914

7	3	2	5	6	4	1	8	9
8	4	9	2	3	1	7	5	6
5	6	1	9	8	7	2	3	4
1	7	6	8	9	3	5	4	2
2	5	3	1	4	6	8	9	7
9	8	4	7	5	2	6	1	3
6	9	7	3	1	8	4	2	5
3	2	8	4	7	5	9	6	1
4	1	5	6	2	9	3	7	8

Puzzle 915

4	9	2	5	8	6	1	7	3
1	6	5	4	3	7	2	8	9
8	7	3	9	2	1	6	4	5
6	5	7	1	4	2	9	3	8
2	4	1	8	9	3	5	6	7
9	3	8	6	7	5	4	1	2
5	8	4	7	6	9	3	2	1
7	2	9	3	1	4	8	5	6
3	1	6	2	5	8	7	9	4

Puzzle 916

2	3	7	8	9	1	5	6	4
8	9	6	2	4	5	7	3	1
1	4	5	7	6	3	9	2	8
7	2	3	6	8	4	1	9	5
6	5	1	3	7	9	8	4	2
9	8	4	5	1	2	6	7	3
5	1	2	9	3	7	4	8	6
3	6	9	4	5	8	2	1	7
4	7	8	1	2	6	3	5	9

Puzzle 917

8	5	9	3	4	1	2	6	7
7	4	3	8	2	6	1	5	9
2	6	1	9	5	7	8	4	3
9	8	4	2	3	5	6	7	1
5	1	7	4	6	8	3	9	2
6	3	2	7	1	9	5	8	4
1	7	8	6	9	2	4	3	5
3	9	5	1	8	4	7	2	6
4	2	6	5	7	3	9	1	8

Puzzle 918

3	4	9	1	8	2	6	7	5
2	6	1	5	9	7	8	3	4
5	8	7	6	3	4	9	1	2
7	9	8	3	5	6	2	4	1
1	3	6	4	2	8	5	9	7
4	5	2	9	7	1	3	8	6
9	7	3	2	1	5	4	6	8
8	2	4	7	6	9	1	5	3
6	1	5	8	4	3	7	2	9

Puzzle 919

6	1	8	3	9	5	7	4	2
4	2	5	7	1	6	9	3	8
9	7	3	4	8	2	6	5	1
5	4	7	8	2	1	3	9	6
3	6	1	9	5	7	2	8	4
2	8	9	6	4	3	5	1	7
1	3	2	5	6	8	4	7	9
8	5	4	2	7	9	1	6	3
7	9	6	1	3	4	8	2	5

Puzzle 920

2	7	5	1	3	8	9	6	4
4	8	3	5	9	6	1	7	2
1	6	9	2	4	7	5	8	3
5	9	7	4	2	1	8	3	6
6	1	4	7	8	3	2	5	9
8	3	2	9	6	5	4	1	7
3	2	1	8	7	4	6	9	5
7	4	8	6	5	9	3	2	1
9	5	6	3	1	2	7	4	8

Puzzle 921

1	7	5	9	8	2	6	3	4
6	8	3	4	5	1	9	7	2
2	4	9	3	6	7	1	5	8
4	2	6	1	7	5	8	9	3
8	3	1	6	9	4	5	2	7
5	9	7	2	3	8	4	1	6
9	5	2	8	4	3	7	6	1
3	6	8	7	1	9	2	4	5
7	1	4	5	2	6	3	8	9

Puzzle 922

2	9	4	3	7	6	8	5	1
5	7	8	4	9	1	3	2	6
6	1	3	5	8	2	7	4	9
4	6	5	9	2	3	1	8	7
3	8	1	6	5	7	2	9	4
9	2	7	1	4	8	5	6	3
7	3	2	8	6	9	4	1	5
8	5	6	7	1	4	9	3	2
1	4	9	2	3	5	6	7	8

Puzzle 923

6	7	4	3	9	2	5	1	8
5	3	2	6	8	1	7	4	9
9	8	1	4	5	7	6	3	2
7	6	3	9	1	4	2	8	5
8	2	5	7	6	3	4	9	1
4	1	9	8	2	5	3	6	7
2	9	8	5	3	6	1	7	4
3	5	7	1	4	8	9	2	6
1	4	6	2	7	9	8	5	3

Puzzle 924

7	3	6	8	2	9	4	1	5
1	4	2	6	3	5	7	8	9
5	8	9	4	1	7	6	2	3
8	9	5	1	6	3	2	4	7
6	1	4	7	9	2	3	5	8
2	7	3	5	4	8	1	9	6
4	5	7	2	8	6	9	3	1
3	2	8	9	7	1	5	6	4
9	6	1	3	5	4	8	7	2

Puzzle 925

5	7	2	1	4	8	9	3	6
3	9	4	7	6	2	1	8	5
6	8	1	3	9	5	4	2	7
4	2	3	6	7	1	8	5	9
9	6	5	2	8	3	7	4	1
7	1	8	9	5	4	3	6	2
2	4	7	5	3	9	6	1	8
8	5	6	4	1	7	2	9	3
1	3	9	8	2	6	5	7	4

Puzzle 926

5	1	2	3	8	7	6	9	4
8	9	4	6	2	1	3	5	7
3	7	6	5	4	9	1	2	8
6	4	3	2	7	8	9	1	5
9	5	1	4	6	3	8	7	2
2	8	7	9	1	5	4	3	6
4	6	9	1	5	2	7	8	3
7	3	5	8	9	6	2	4	1
1	2	8	7	3	4	5	6	9

Puzzle 927

5	1	4	9	7	3	6	2	8
2	3	8	4	1	6	7	9	5
6	9	7	2	5	8	1	4	3
1	4	3	7	8	9	2	5	6
7	8	2	5	6	4	9	3	1
9	6	5	3	2	1	4	8	7
4	5	9	1	3	7	8	6	2
3	7	6	8	9	2	5	1	4
8	2	1	6	4	5	3	7	9

Puzzle 928

7	3	5	2	1	8	6	9	4
8	4	1	7	9	6	5	2	3
9	6	2	4	5	3	7	8	1
4	1	3	9	7	2	8	6	5
5	8	7	1	6	4	9	3	2
6	2	9	8	3	5	1	4	7
2	5	6	3	8	1	4	7	9
3	9	8	5	4	7	2	1	6
1	7	4	6	2	9	3	5	8

Puzzle 929

9	3	4	7	1	6	2	5	8
1	8	7	5	9	2	6	3	4
6	5	2	3	4	8	1	9	7
5	4	6	2	7	9	8	1	3
3	7	1	8	5	4	9	2	6
2	9	8	6	3	1	7	4	5
4	1	5	9	6	7	3	8	2
8	6	3	1	2	5	4	7	9
7	2	9	4	8	3	5	6	1

Puzzle 930

4	5	1	2	6	3	8	7	9
3	7	6	4	9	8	5	2	1
9	2	8	5	1	7	6	3	4
1	6	2	8	5	9	7	4	3
8	4	9	3	7	2	1	5	6
7	3	5	6	4	1	2	9	8
5	9	3	1	2	6	4	8	7
6	8	4	7	3	5	9	1	2
2	1	7	9	8	4	3	6	5

Puzzle 931

6	4	2	8	1	9	5	3	7
9	7	1	6	3	5	2	4	8
8	5	3	4	7	2	1	6	9
1	9	5	7	6	4	8	2	3
3	2	8	9	5	1	4	7	6
7	6	4	2	8	3	9	1	5
4	8	9	3	2	6	7	5	1
5	3	7	1	4	8	6	9	2
2	1	6	5	9	7	3	8	4

Puzzle 932

9	2	3	5	4	6	7	1	8
6	1	8	7	2	9	3	4	5
4	5	7	1	8	3	2	6	9
1	7	4	3	6	8	9	5	2
5	8	2	4	9	1	6	7	3
3	9	6	2	7	5	4	8	1
2	3	5	6	1	7	8	9	4
7	4	9	8	5	2	1	3	6
8	6	1	9	3	4	5	2	7

Puzzle 933

6	7	3	2	4	1	8	9	5
4	9	2	5	8	6	1	3	7
8	5	1	9	7	3	6	4	2
5	1	6	7	9	8	4	2	3
7	4	9	1	3	2	5	8	6
3	2	8	4	6	5	9	7	1
9	3	4	6	5	7	2	1	8
1	6	7	8	2	9	3	5	4
2	8	5	3	1	4	7	6	9

Puzzle 934

8	7	5	1	9	6	2	3	4
3	2	1	8	4	5	6	7	9
6	4	9	3	2	7	1	8	5
5	3	2	6	1	8	9	4	7
4	9	6	5	7	2	8	1	3
1	8	7	9	3	4	5	6	2
7	6	4	2	8	9	3	5	1
9	5	3	4	6	1	7	2	8
2	1	8	7	5	3	4	9	6

Puzzle 935

8	9	1	5	4	6	7	2	3
7	3	2	9	1	8	5	4	6
6	5	4	7	2	3	1	9	8
1	4	9	2	8	7	3	6	5
2	6	5	4	3	1	8	7	9
3	8	7	6	9	5	2	1	4
4	2	8	3	7	9	6	5	1
9	1	6	8	5	2	4	3	7
5	7	3	1	6	4	9	8	2

Puzzle 936

9	6	4	2	7	1	8	3	5
7	8	2	9	5	3	1	4	6
5	3	1	8	6	4	2	7	9
6	1	8	4	9	2	7	5	3
2	7	5	1	3	6	4	9	8
4	9	3	7	8	5	6	2	1
3	5	7	6	4	8	9	1	2
8	2	9	5	1	7	3	6	4
1	4	6	3	2	9	5	8	7

Puzzle 937

1	2	7	4	8	5	9	6	3
3	6	5	9	1	7	2	8	4
9	4	8	6	3	2	7	1	5
2	3	6	7	5	4	8	9	1
7	9	4	1	2	8	3	5	6
5	8	1	3	6	9	4	2	7
6	7	3	2	9	1	5	4	8
4	5	9	8	7	6	1	3	2
8	1	2	5	4	3	6	7	9

Puzzle 938

1	6	4	9	5	3	7	2	8
2	8	7	6	1	4	9	3	5
3	9	5	7	8	2	6	4	1
4	5	2	1	3	7	8	6	9
9	7	3	8	4	6	1	5	2
6	1	8	5	2	9	4	7	3
8	4	6	3	9	5	2	1	7
7	3	9	2	6	1	5	8	4
5	2	1	4	7	8	3	9	6

Puzzle 939

2	4	1	7	8	3	5	9	6
8	5	6	9	2	1	3	7	4
3	9	7	4	5	6	1	8	2
5	3	8	6	9	2	4	1	7
9	7	2	3	1	4	6	5	8
6	1	4	5	7	8	9	2	3
7	2	3	1	4	5	8	6	9
1	6	9	8	3	7	2	4	5
4	8	5	2	6	9	7	3	1

Puzzle 940

3	5	1	2	4	8	6	9	7
6	2	4	5	7	9	3	1	8
9	8	7	6	3	1	5	2	4
8	4	9	1	6	7	2	3	5
2	7	5	8	9	3	1	4	6
1	6	3	4	5	2	8	7	9
4	1	8	7	2	6	9	5	3
5	3	6	9	1	4	7	8	2
7	9	2	3	8	5	4	6	1

Puzzle 941

7	6	3	2	9	1	4	8	5
8	1	4	5	6	7	3	2	9
9	2	5	3	8	4	1	6	7
3	9	1	8	4	5	6	7	2
2	7	8	6	1	9	5	4	3
4	5	6	7	3	2	8	9	1
1	4	2	9	5	6	7	3	8
5	3	9	4	7	8	2	1	6
6	8	7	1	2	3	9	5	4

Puzzle 942

6	9	1	8	5	2	3	4	7
2	4	3	9	6	7	1	8	5
8	5	7	4	1	3	6	9	2
9	3	4	7	2	6	8	5	1
7	8	2	5	9	1	4	6	3
5	1	6	3	8	4	2	7	9
4	2	9	6	3	5	7	1	8
1	7	8	2	4	9	5	3	6
3	6	5	1	7	8	9	2	4

Puzzle 943

7	8	5	3	4	6	2	9	1
2	1	6	7	9	8	4	3	5
4	3	9	1	5	2	6	8	7
3	6	1	4	8	5	9	7	2
8	4	7	2	3	9	5	1	6
5	9	2	6	1	7	8	4	3
1	2	3	9	6	4	7	5	8
9	7	8	5	2	3	1	6	4
6	5	4	8	7	1	3	2	9

Puzzle 944

8	5	4	2	6	7	9	1	3
9	2	1	4	5	3	8	7	6
3	7	6	9	1	8	2	5	4
5	8	7	1	3	6	4	2	9
2	1	3	8	9	4	5	6	7
4	6	9	7	2	5	1	3	8
6	3	2	5	4	9	7	8	1
7	4	5	3	8	1	6	9	2
1	9	8	6	7	2	3	4	5

Puzzle 945

6	2	3	1	4	7	8	5	9
4	9	5	3	6	8	7	1	2
8	7	1	2	5	9	3	4	6
9	1	4	6	8	2	5	7	3
7	8	2	5	9	3	4	6	1
5	3	6	4	7	1	2	9	8
3	4	7	8	1	6	9	2	5
1	5	8	9	2	4	6	3	7
2	6	9	7	3	5	1	8	4

Puzzle 946

6	3	5	4	8	1	2	7	9
8	1	2	5	7	9	6	4	3
7	9	4	6	2	3	8	5	1
9	4	1	7	6	8	5	3	2
3	2	6	9	4	5	1	8	7
5	8	7	3	1	2	4	9	6
1	7	8	2	3	4	9	6	5
4	5	3	1	9	6	7	2	8
2	6	9	8	5	7	3	1	4

Puzzle 947

8	3	5	2	6	4	7	9	1
2	9	1	3	7	8	4	6	5
4	7	6	5	9	1	2	8	3
5	6	8	9	1	7	3	2	4
3	4	7	6	2	5	9	1	8
1	2	9	4	8	3	5	7	6
7	8	3	1	4	2	6	5	9
6	1	4	7	5	9	8	3	2
9	5	2	8	3	6	1	4	7

Puzzle 948

6	4	2	3	9	5	7	8	1
5	9	8	1	4	7	2	3	6
1	7	3	2	6	8	4	5	9
8	5	9	4	1	6	3	2	7
7	6	1	5	3	2	9	4	8
3	2	4	7	8	9	1	6	5
2	8	7	9	5	4	6	1	3
4	3	6	8	7	1	5	9	2
9	1	5	6	2	3	8	7	4

Puzzle 949

1	3	2	7	4	6	8	9	5
9	6	5	8	2	1	4	7	3
8	4	7	9	3	5	1	2	6
7	9	4	6	8	2	3	5	1
2	8	6	1	5	3	9	4	7
5	1	3	4	7	9	6	8	2
6	2	8	5	1	4	7	3	9
4	5	1	3	9	7	2	6	8
3	7	9	2	6	8	5	1	4

Puzzle 950

5	6	2	4	3	9	1	7	8
1	9	8	5	6	7	2	3	4
7	4	3	1	8	2	5	6	9
9	5	4	2	7	6	3	8	1
6	3	7	8	1	4	9	5	2
2	8	1	9	5	3	6	4	7
8	1	9	3	4	5	7	2	6
4	7	5	6	2	1	8	9	3
3	2	6	7	9	8	4	1	5

Puzzle 951

9	4	7	3	6	5	2	8	1
2	6	1	8	4	9	3	7	5
8	5	3	7	1	2	4	6	9
6	1	4	2	3	7	9	5	8
7	9	2	6	5	8	1	3	4
3	8	5	1	9	4	7	2	6
1	7	8	4	2	6	5	9	3
5	3	6	9	7	1	8	4	2
4	2	9	5	8	3	6	1	7

Puzzle 952

5	3	4	2	8	6	1	9	7
2	6	7	3	9	1	5	4	8
1	8	9	7	4	5	3	6	2
3	7	1	6	2	4	9	8	5
9	5	2	8	7	3	6	1	4
6	4	8	1	5	9	7	2	3
4	9	6	5	3	8	2	7	1
8	2	5	9	1	7	4	3	6
7	1	3	4	6	2	8	5	9

Puzzle 953

6	5	1	3	2	7	9	4	8
2	9	4	8	1	6	3	5	7
3	8	7	4	5	9	6	1	2
1	3	6	5	4	8	2	7	9
5	4	9	2	7	3	1	8	6
7	2	8	9	6	1	5	3	4
4	7	2	1	9	5	8	6	3
9	1	3	6	8	4	7	2	5
8	6	5	7	3	2	4	9	1

Puzzle 954

8	4	7	6	1	5	3	9	2
9	2	3	4	7	8	1	5	6
1	5	6	3	9	2	8	4	7
7	1	2	5	8	9	6	3	4
3	9	5	2	6	4	7	8	1
4	6	8	7	3	1	9	2	5
2	8	9	1	4	6	5	7	3
5	7	1	9	2	3	4	6	8
6	3	4	8	5	7	2	1	9

Puzzle 955

7	4	9	3	5	1	6	8	2
5	8	2	9	7	6	1	4	3
6	1	3	8	4	2	5	7	9
3	2	4	7	8	5	9	6	1
8	9	7	6	1	3	2	5	4
1	6	5	2	9	4	8	3	7
9	5	8	4	2	7	3	1	6
2	7	6	1	3	8	4	9	5
4	3	1	5	6	9	7	2	8

Puzzle 956

5	3	8	2	6	4	7	9	1
1	4	7	5	9	8	3	6	2
2	9	6	1	7	3	5	4	8
8	5	1	4	2	7	9	3	6
7	2	4	6	3	9	1	8	5
9	6	3	8	1	5	2	7	4
6	8	9	7	5	1	4	2	3
4	7	5	3	8	2	6	1	9
3	1	2	9	4	6	8	5	7

Puzzle 957

5	7	4	8	3	9	2	6	1
3	1	2	4	5	6	7	9	8
6	9	8	1	2	7	4	3	5
7	6	1	2	8	4	9	5	3
8	2	9	3	6	5	1	7	4
4	5	3	9	7	1	8	2	6
9	3	6	7	1	8	5	4	2
1	4	5	6	9	2	3	8	7
2	8	7	5	4	3	6	1	9

Puzzle 958

8	4	9	3	2	7	6	1	5
7	5	3	6	4	1	8	9	2
2	6	1	9	5	8	7	4	3
6	7	5	1	3	4	9	2	8
1	9	8	2	7	6	3	5	4
4	3	2	8	9	5	1	6	7
5	1	6	7	8	2	4	3	9
3	8	4	5	1	9	2	7	6
9	2	7	4	6	3	5	8	1

Puzzle 959

7	9	3	2	8	5	6	4	1
6	8	4	7	9	1	3	2	5
1	5	2	3	6	4	9	7	8
4	2	6	8	1	7	5	3	9
5	3	8	4	2	9	1	6	7
9	1	7	6	5	3	4	8	2
8	4	5	1	3	2	7	9	6
3	6	1	9	7	8	2	5	4
2	7	9	5	4	6	8	1	3

Puzzle 960

8	5	4	3	6	9	1	2	7
6	7	2	8	5	1	4	9	3
9	1	3	7	4	2	8	5	6
7	8	5	4	2	6	3	1	9
3	4	9	1	7	8	5	6	2
2	6	1	9	3	5	7	4	8
4	2	8	6	1	3	9	7	5
5	3	7	2	9	4	6	8	1
1	9	6	5	8	7	2	3	4

Puzzle 961

3	1	5	9	7	2	6	4	8
8	7	6	1	3	4	2	5	9
4	9	2	5	6	8	1	3	7
6	8	3	4	1	5	9	7	2
9	4	1	7	2	6	3	8	5
5	2	7	8	9	3	4	6	1
1	6	9	3	5	7	8	2	4
2	5	4	6	8	1	7	9	3
7	3	8	2	4	9	5	1	6

Puzzle 962

3	6	5	2	1	4	7	9	8
7	2	9	6	5	8	4	3	1
8	1	4	3	9	7	5	6	2
9	7	3	5	8	6	2	1	4
4	8	1	7	2	9	6	5	3
2	5	6	4	3	1	8	7	9
5	3	7	9	4	2	1	8	6
1	9	2	8	6	5	3	4	7
6	4	8	1	7	3	9	2	5

Puzzle 963

4	7	9	8	3	5	1	6	2
5	2	6	1	7	4	9	3	8
3	1	8	2	6	9	7	4	5
2	3	4	5	9	8	6	7	1
6	8	5	4	1	7	3	2	9
7	9	1	6	2	3	5	8	4
8	5	7	3	4	1	2	9	6
1	6	3	9	8	2	4	5	7
9	4	2	7	5	6	8	1	3

Puzzle 964

3	4	1	6	8	7	5	9	2
9	8	5	2	3	1	6	7	4
7	2	6	4	5	9	1	8	3
2	3	9	5	6	4	8	1	7
8	1	7	9	2	3	4	6	5
5	6	4	1	7	8	2	3	9
1	7	2	8	9	5	3	4	6
6	9	8	3	4	2	7	5	1
4	5	3	7	1	6	9	2	8

Puzzle 965

7	3	5	8	9	6	1	4	2
6	9	8	2	4	1	7	5	3
4	1	2	7	5	3	9	8	6
1	2	9	3	8	5	6	7	4
3	7	6	1	2	4	8	9	5
8	5	4	6	7	9	2	3	1
9	6	7	4	3	2	5	1	8
5	4	1	9	6	8	3	2	7
2	8	3	5	1	7	4	6	9

Puzzle 966

2	8	5	6	3	4	7	9	1
1	3	9	2	8	7	6	4	5
4	7	6	5	9	1	2	3	8
7	6	3	4	2	8	5	1	9
8	5	4	1	6	9	3	2	7
9	1	2	7	5	3	8	6	4
5	9	7	3	4	2	1	8	6
3	4	1	8	7	6	9	5	2
6	2	8	9	1	5	4	7	3

Puzzle 967

2	9	6	7	8	1	4	3	5
4	7	1	3	9	5	8	2	6
3	8	5	2	4	6	1	9	7
7	3	8	1	6	2	9	5	4
5	4	9	8	7	3	6	1	2
1	6	2	9	5	4	7	8	3
8	5	3	6	1	7	2	4	9
6	1	4	5	2	9	3	7	8
9	2	7	4	3	8	5	6	1

Puzzle 968

2	9	7	8	4	1	3	5	6
6	1	5	2	9	3	7	4	8
4	8	3	7	5	6	1	2	9
8	6	4	3	2	5	9	1	7
1	5	9	6	8	7	4	3	2
7	3	2	4	1	9	8	6	5
3	4	8	9	6	2	5	7	1
9	2	1	5	7	4	6	8	3
5	7	6	1	3	8	2	9	4

Puzzle 969

5	1	3	8	7	2	9	4	6
9	7	6	3	1	4	8	2	5
8	4	2	5	6	9	3	1	7
7	3	1	6	8	5	2	9	4
4	6	5	9	2	1	7	3	8
2	8	9	4	3	7	6	5	1
6	5	7	1	9	3	4	8	2
1	9	8	2	4	6	5	7	3
3	2	4	7	5	8	1	6	9

Puzzle 970

3	4	2	5	1	6	8	9	7
5	9	6	4	7	8	2	1	3
8	7	1	9	3	2	5	4	6
9	2	7	3	8	4	6	5	1
6	5	8	2	9	1	7	3	4
1	3	4	7	6	5	9	8	2
4	6	9	8	2	3	1	7	5
2	8	3	1	5	7	4	6	9
7	1	5	6	4	9	3	2	8

Puzzle 971

8	6	3	4	2	1	9	7	5
9	1	2	8	7	5	3	4	6
7	4	5	3	6	9	8	1	2
1	7	9	2	3	6	4	5	8
2	5	6	1	4	8	7	3	9
4	3	8	5	9	7	6	2	1
3	8	1	6	5	4	2	9	7
5	2	7	9	8	3	1	6	4
6	9	4	7	1	2	5	8	3

Puzzle 972

8	9	3	7	2	6	1	5	4
6	4	7	5	9	1	3	8	2
1	2	5	4	8	3	6	9	7
2	6	8	3	7	9	5	4	1
4	3	1	6	5	8	7	2	9
5	7	9	2	1	4	8	3	6
9	1	4	8	6	5	2	7	3
7	5	6	9	3	2	4	1	8
3	8	2	1	4	7	9	6	5

Puzzle 973

5	6	8	4	2	7	1	9	3
7	2	4	3	9	1	5	8	6
3	9	1	8	5	6	4	7	2
6	7	2	1	8	3	9	4	5
8	4	5	7	6	9	3	2	1
1	3	9	5	4	2	8	6	7
4	8	7	6	1	5	2	3	9
9	5	6	2	3	4	7	1	8
2	1	3	9	7	8	6	5	4

Puzzle 974

5	3	8	9	6	4	1	2	7
2	4	7	3	8	1	6	5	9
6	9	1	2	5	7	8	3	4
9	6	2	4	3	5	7	8	1
7	1	5	6	9	8	3	4	2
4	8	3	7	1	2	9	6	5
3	5	9	1	4	6	2	7	8
1	2	4	8	7	3	5	9	6
8	7	6	5	2	9	4	1	3

Puzzle 975

8	6	3	5	7	9	1	4	2
5	1	2	3	8	4	7	9	6
4	9	7	2	1	6	8	3	5
1	8	4	6	3	7	2	5	9
7	3	6	9	5	2	4	1	8
2	5	9	1	4	8	6	7	3
6	7	8	4	9	5	3	2	1
9	2	1	7	6	3	5	8	4
3	4	5	8	2	1	9	6	7

Puzzle 976

2	1	3	9	8	6	7	5	4
4	7	9	1	5	2	3	6	8
5	6	8	4	3	7	2	1	9
7	8	2	6	1	4	9	3	5
1	9	4	5	2	3	6	8	7
6	3	5	8	7	9	1	4	2
9	5	6	7	4	1	8	2	3
3	4	1	2	9	8	5	7	6
8	2	7	3	6	5	4	9	1

Puzzle 977

2	1	9	6	5	4	8	7	3
6	7	5	9	3	8	4	2	1
4	3	8	7	1	2	6	5	9
1	5	3	4	7	9	2	6	8
7	4	2	8	6	1	3	9	5
9	8	6	3	2	5	1	4	7
3	2	1	5	9	6	7	8	4
5	6	4	1	8	7	9	3	2
8	9	7	2	4	3	5	1	6

Puzzle 978

7	4	3	2	9	6	8	1	5
9	8	1	3	4	5	2	7	6
5	2	6	8	1	7	9	3	4
6	9	4	7	2	8	1	5	3
3	7	8	6	5	1	4	2	9
2	1	5	9	3	4	6	8	7
1	5	7	4	8	9	3	6	2
8	3	9	5	6	2	7	4	1
4	6	2	1	7	3	5	9	8

Puzzle 979

8	7	4	9	5	3	2	6	1
3	6	1	4	7	2	8	5	9
5	9	2	1	8	6	3	4	7
4	3	7	2	1	9	6	8	5
2	8	6	7	4	5	1	9	3
9	1	5	3	6	8	4	7	2
6	2	3	8	9	7	5	1	4
1	5	9	6	2	4	7	3	8
7	4	8	5	3	1	9	2	6

Puzzle 980

8	9	5	3	7	2	4	6	1
2	3	6	9	1	4	8	7	5
7	1	4	5	6	8	3	9	2
4	7	3	8	9	1	5	2	6
6	5	9	7	2	3	1	4	8
1	8	2	4	5	6	7	3	9
9	2	8	1	4	7	6	5	3
3	6	7	2	8	5	9	1	4
5	4	1	6	3	9	2	8	7

Puzzle 981

1	6	5	4	9	3	8	7	2
2	7	8	5	6	1	9	3	4
9	4	3	2	8	7	6	5	1
8	9	1	3	5	2	4	6	7
7	5	6	9	4	8	2	1	3
3	2	4	7	1	6	5	8	9
4	8	9	1	3	5	7	2	6
6	1	7	8	2	4	3	9	5
5	3	2	6	7	9	1	4	8

Puzzle 982

1	8	2	9	4	3	7	5	6
4	9	6	8	5	7	2	3	1
5	7	3	2	1	6	8	4	9
2	4	7	6	3	5	9	1	8
9	5	1	4	2	8	3	6	7
6	3	8	1	7	9	4	2	5
3	6	9	5	8	2	1	7	4
8	2	4	7	6	1	5	9	3
7	1	5	3	9	4	6	8	2

Puzzle 983

8	4	9	7	2	3	6	1	5
5	1	3	9	6	8	2	4	7
2	7	6	4	1	5	8	3	9
4	9	1	3	8	2	7	5	6
3	6	5	1	7	9	4	8	2
7	8	2	5	4	6	1	9	3
9	2	4	6	5	1	3	7	8
1	3	8	2	9	7	5	6	4
6	5	7	8	3	4	9	2	1

Puzzle 984

9	1	2	5	8	3	4	7	6
5	7	6	4	9	1	8	3	2
4	3	8	2	7	6	1	9	5
7	5	3	6	4	2	9	1	8
1	8	4	3	5	9	6	2	7
6	2	9	7	1	8	5	4	3
2	4	5	1	6	7	3	8	9
8	6	7	9	3	4	2	5	1
3	9	1	8	2	5	7	6	4

Puzzle 985

9	1	4	5	8	3	7	2	6
7	5	8	6	2	4	3	9	1
3	2	6	9	1	7	8	4	5
1	4	5	3	6	2	9	7	8
8	9	7	1	4	5	2	6	3
6	3	2	7	9	8	1	5	4
5	6	1	8	7	9	4	3	2
4	8	9	2	3	6	5	1	7
2	7	3	4	5	1	6	8	9

Puzzle 986

5	9	7	4	3	2	6	1	8
8	2	6	5	7	1	4	3	9
1	4	3	9	8	6	7	5	2
9	5	1	8	4	3	2	7	6
2	7	4	6	5	9	3	8	1
6	3	8	1	2	7	5	9	4
4	1	5	3	6	8	9	2	7
3	8	2	7	9	4	1	6	5
7	6	9	2	1	5	8	4	3

Puzzle 987

1	9	2	8	5	6	3	7	4
5	7	8	4	3	9	1	6	2
3	6	4	7	2	1	8	9	5
6	8	5	3	9	4	7	2	1
7	4	9	2	1	8	5	3	6
2	1	3	5	6	7	9	4	8
8	2	6	9	7	5	4	1	3
4	3	7	1	8	2	6	5	9
9	5	1	6	4	3	2	8	7

Puzzle 988

4	6	8	7	2	1	9	5	3
3	7	9	4	6	5	1	2	8
1	5	2	8	9	3	6	4	7
5	9	6	1	3	8	2	7	4
2	1	4	5	7	9	8	3	6
7	8	3	2	4	6	5	1	9
9	4	5	3	8	2	7	6	1
6	2	7	9	1	4	3	8	5
8	3	1	6	5	7	4	9	2

Puzzle 989

4	6	5	1	7	9	8	3	2
2	9	8	6	5	3	7	4	1
1	7	3	2	8	4	5	9	6
9	5	2	7	1	8	4	6	3
8	3	4	9	6	5	2	1	7
6	1	7	4	3	2	9	8	5
5	4	1	8	2	6	3	7	9
3	8	6	5	9	7	1	2	4
7	2	9	3	4	1	6	5	8

Puzzle 990

8	5	9	6	7	1	3	2	4
4	2	7	5	3	8	6	9	1
3	1	6	9	2	4	8	5	7
1	8	5	4	9	3	2	7	6
2	9	4	8	6	7	5	1	3
6	7	3	1	5	2	4	8	9
7	3	1	2	8	6	9	4	5
9	4	2	3	1	5	7	6	8
5	6	8	7	4	9	1	3	2

Puzzle 991

4	2	5	8	9	3	6	7	1
9	1	8	4	7	6	5	2	3
3	7	6	2	1	5	8	4	9
2	9	4	6	8	7	1	3	5
8	3	7	5	2	1	4	9	6
5	6	1	9	3	4	2	8	7
1	5	2	3	4	9	7	6	8
7	4	9	1	6	8	3	5	2
6	8	3	7	5	2	9	1	4

Puzzle 992

8	4	3	1	5	6	9	7	2
9	2	5	4	8	7	6	3	1
7	1	6	9	3	2	5	4	8
1	5	8	3	7	9	2	6	4
4	7	9	2	6	1	3	8	5
3	6	2	8	4	5	7	1	9
5	8	1	6	9	3	4	2	7
6	9	4	7	2	8	1	5	3
2	3	7	5	1	4	8	9	6

Puzzle 993

6	1	8	5	7	3	4	9	2
5	9	2	6	1	4	8	3	7
7	4	3	8	9	2	1	5	6
9	7	5	2	4	1	6	8	3
8	6	1	3	5	9	7	2	4
3	2	4	7	8	6	9	1	5
2	5	9	1	6	7	3	4	8
1	3	7	4	2	8	5	6	9
4	8	6	9	3	5	2	7	1

Puzzle 994

2	1	8	3	4	7	9	6	5
6	5	4	8	1	9	7	2	3
3	9	7	2	5	6	4	1	8
8	6	5	1	3	4	2	9	7
7	3	2	6	9	5	1	8	4
9	4	1	7	8	2	5	3	6
5	8	9	4	6	1	3	7	2
1	2	6	5	7	3	8	4	9
4	7	3	9	2	8	6	5	1

Puzzle 995

8	6	5	3	1	2	4	7	9
2	9	3	4	8	7	5	1	6
4	7	1	5	6	9	8	3	2
3	8	4	7	2	1	9	6	5
1	2	7	9	5	6	3	8	4
6	5	9	8	4	3	1	2	7
5	1	2	6	3	4	7	9	8
9	3	8	2	7	5	6	4	1
7	4	6	1	9	8	2	5	3

Puzzle 996

6	5	1	8	9	7	2	3	4
4	9	3	6	5	2	7	1	8
8	7	2	1	4	3	6	9	5
1	6	5	4	3	9	8	7	2
9	3	4	7	2	8	5	6	1
2	8	7	5	6	1	9	4	3
3	4	8	9	7	5	1	2	6
7	1	6	2	8	4	3	5	9
5	2	9	3	1	6	4	8	7

Puzzle 997

5	1	2	4	8	7	9	6	3
9	8	3	1	5	6	4	2	7
4	7	6	2	9	3	5	8	1
2	6	9	3	1	4	8	7	5
7	3	5	8	6	9	1	4	2
1	4	8	5	7	2	3	9	6
6	9	1	7	3	8	2	5	4
8	5	4	6	2	1	7	3	9
3	2	7	9	4	5	6	1	8

Puzzle 998

3	6	4	7	8	9	1	5	2
1	2	7	3	4	5	9	8	6
9	8	5	6	2	1	3	7	4
7	1	3	5	6	2	8	4	9
5	9	2	8	7	4	6	1	3
6	4	8	1	9	3	5	2	7
2	7	1	9	5	6	4	3	8
8	3	6	4	1	7	2	9	5
4	5	9	2	3	8	7	6	1

Puzzle 999

1	5	6	7	3	8	4	9	2
2	7	4	5	1	9	6	3	8
8	3	9	6	4	2	1	7	5
6	4	3	8	9	7	5	2	1
9	2	8	3	5	1	7	6	4
5	1	7	4	2	6	9	8	3
4	8	2	9	7	5	3	1	6
7	6	5	1	8	3	2	4	9
3	9	1	2	6	4	8	5	7

Puzzle 1000

8	9	2	5	7	1	4	3	6
4	3	7	2	8	6	5	1	9
6	1	5	9	4	3	8	2	7
3	7	9	1	2	4	6	8	5
5	6	4	8	3	9	1	7	2
2	8	1	7	6	5	9	4	3
1	5	8	3	9	2	7	6	4
9	2	6	4	1	7	3	5	8
7	4	3	6	5	8	2	9	1

Puzzle 1001

2	1	6	7	8	9	5	4	3
5	4	9	3	2	1	7	6	8
3	7	8	5	4	6	9	2	1
9	5	3	4	7	8	2	1	6
4	6	2	9	1	5	3	8	7
1	8	7	6	3	2	4	5	9
7	3	5	1	6	4	8	9	2
8	9	1	2	5	7	6	3	4
6	2	4	8	9	3	1	7	5

Puzzle 1002

3	9	1	8	4	6	5	2	7
4	2	7	5	9	3	8	1	6
8	5	6	2	7	1	3	4	9
1	6	9	3	8	7	4	5	2
5	8	4	9	1	2	7	6	3
7	3	2	6	5	4	1	9	8
2	7	8	4	6	5	9	3	1
6	1	5	7	3	9	2	8	4
9	4	3	1	2	8	6	7	5

Puzzle 1003

7	9	5	1	4	8	6	2	3
6	4	8	9	3	2	1	5	7
1	3	2	5	6	7	9	4	8
8	5	7	4	9	1	3	6	2
9	2	6	7	8	3	4	1	5
3	1	4	6	2	5	8	7	9
5	8	3	2	1	6	7	9	4
2	6	9	3	7	4	5	8	1
4	7	1	8	5	9	2	3	6

Puzzle 1004

4	8	2	9	5	6	7	3	1
5	1	7	3	8	4	9	2	6
3	9	6	1	2	7	8	5	4
6	5	3	2	9	1	4	7	8
1	4	8	5	7	3	2	6	9
7	2	9	4	6	8	5	1	3
8	7	5	6	3	9	1	4	2
2	3	4	8	1	5	6	9	7
9	6	1	7	4	2	3	8	5

Puzzle 1005

3	2	6	9	1	4	5	7	8
9	4	1	5	8	7	3	6	2
5	7	8	3	2	6	4	1	9
1	6	5	2	3	8	7	9	4
4	8	2	1	7	9	6	3	5
7	9	3	4	6	5	8	2	1
8	3	4	7	9	2	1	5	6
2	5	7	6	4	1	9	8	3
6	1	9	8	5	3	2	4	7

Puzzle 1006

5	6	9	1	2	3	8	4	7
8	3	1	4	9	7	6	5	2
7	4	2	6	5	8	3	9	1
1	9	5	2	7	6	4	3	8
2	8	3	5	1	4	9	7	6
4	7	6	8	3	9	1	2	5
9	1	8	7	4	5	2	6	3
3	2	7	9	6	1	5	8	4
6	5	4	3	8	2	7	1	9

Puzzle 1007

5	2	6	8	7	1	4	9	3
4	3	7	9	5	2	8	6	1
9	8	1	3	4	6	7	5	2
2	1	5	7	8	9	6	3	4
6	4	8	2	1	3	9	7	5
7	9	3	5	6	4	2	1	8
1	7	4	6	3	8	5	2	9
8	6	9	1	2	5	3	4	7
3	5	2	4	9	7	1	8	6

Puzzle 1008

9	3	2	8	1	7	5	4	6
4	5	6	9	2	3	8	7	1
8	7	1	6	5	4	2	9	3
3	8	4	7	9	1	6	5	2
7	2	5	4	6	8	3	1	9
1	6	9	5	3	2	4	8	7
2	1	7	3	4	5	9	6	8
6	4	3	1	8	9	7	2	5
5	9	8	2	7	6	1	3	4